Communications and Control Engineering Series

Editors: A. Fettweis · J. L. Massey · M. Thoma

Scientific Fundamentals of Robotics 2

M. Vukobratović
D. Stokić

Control of Manipulation Robots

Theory and Application

With 111 Figures

Springer-Verlag Berlin Heidelberg New York 1982

D. Sc., Ph. D. MIOMIR VUKOBRATOVIĆ, corr. member of
Serbian Academy of Sciences and Arts
Institute »Mihailo Pupin«, Beograd,
Volgina 15, POB 15, Yugoslavia

Ph. D. DRAGAN STOKIĆ
Institute »Mihailo Pupin«, Beograd,
Volgina 15, POB 15, Yugoslavia

ISBN 3-540-11629-X Springer-Verlag Berlin Heidelberg NewYork
ISBN 0-387-11629-X Springer-Verlag NewYork Heidelberg Berlin

Library of Congress Cataloging in Publication Data
Vukobratović, Miomir.
Control of manipulation robots.
(Scientific fundamentals of robotics ; 2)
(Communications and controls engineering series)
Bibliography: p.
Includes index.
1. Automata. 2. Manipulators (Mechanism)
I. Stokić, Dragan. II. Title. III. Series.
iV. Series: Communications and control engineering series.
TJ211.V837 629.8'92 82-5591 AACR2

Offsetprinting: fotokop wilhelm weihert KG, Darmstadt. – Bookbinding: Konrad Triltsch, Würzburg.
2061/3020-543210

About the Series: »Scientific Fundamentals of Robotics«

The age of robotics is the present age. The study of robotics requires different kinds of knowledge multidisciplinary in nature, which go together to make robotics a specific scientific discipline. In particular, manipulator and robot systems possess several specific qualities in both a mechanical and a control sense. In the mechanical sense, a feature specific to manipulation robots is that all the degrees of freedom are "active", i.e., powered by their own actuators, in contrast to conventional mechanisms in which motion is produced primarily by the so-called kinematic degrees of freedom. Another specific quality of such mechanisms is their variable structure, ranging from open to closed configurations, from one to some other kind of boundary conditions. A further feature specific of spatial mechanisms is redundancy reflected in an excess of the degrees of freedom for producing certain functional movements of robots and manipulators.

From a control viewpoint, robot and manipulator systems represent redundant, multivariable, essentially nonlinear automatic control systems. A manipulation robot is also an example of a dynamically coupled system, and the control task itself is a dynamic task.

The basic motivation for establishing the conception of this series has consisted in an intention to clearly define the role of dynamics and dynamic control of this class of system. The associates who have been engaged in the work on this series have primarily based their contributions on the development of mathematical models of dynamics of these mechanisms. They have thus created a solid background for systematic studies of robot and manipulator dynamics as well as for the synthesis of optimal characteristics of these mechanisms from the point of view of their dynamic performances. Having in mind the characteristics of robotic systems, the results concerning the problems of control of manipulation robots represent one of the central contributions of this series. In trying to bridge, or at least reduce, the gap existing between theoretical robotics and its practical application, considerable efforts have been made towards synthesizing such algorithms as would be suitable for implementation and, at the same time, base them on sufficiently accurate models of system dynamics.

The main idea underlying the conception of the series will be realized: to begin with books which should provide a broad education for engineers and "create" specialists in robotics and reach texts which open up various possibilities for the practical design of manipulation mechanisms and the synthesis of control algorithms based on dynamic models, by applying today's microelectronics and computer technologies.

Those who have initiated the publication of this series believe they will thus create a sound background for systematic work in the research and application of robotics in a wider sense.

Belgrade, Yugoslavia, February 1982 M.Vukobratović

Preface

This monograph represents the second book of the series entitled: "SCIENTIFIC FUNDAMENTALS OF ROBOTICS". While the first volume provides a study of the dynamics of spatial mechanisms and its application to the design of these mechanisms, the present one focuses on the synthesis of control based on the knowledge of dynamic models (presented in detail in the first volume). In this way a logical continuity is formed in which one may easily recognize a "dynamic" approach to the design of manipulation robots and the synthesis of control algorithms based on exact mathematical models of dynamics of open spatial mechanisms.

When writing the monograph, the authors had the following objective: to prove that a study of dynamic properties of manipulation mechanisms is justifiable, to use the dynamic properties in the synthesis of control algorithms, and to determine, from one case to another, a proper measure of dynamics depending on the type of manipulation task, the velocity at which it is carried out, and on the type of the manipulation mechanisms itself. The authors believe they have thus made the study of dynamics, aimed at synthesizing algorithms for dynamic control, free from unnecessary academicism and allowed the readers to apply all the results presented here to practical purposes of manipulator design in the broader sense of the word. At this point, the authors would like to present some concepts which were their guidelines in preparing this text.

The first concept involves a statement that practical applications require the existence of such methods for the synthesis and control of manipulators and robots in general as would be actually applicable and acceptable by the designers of such systems. First of all, we have started from a class of the so-called nonredundant manipulators, i.e., six-degree-of-freedom manipulators, because practically all tasks encountered in industrial practice can be performed by a manipulation mechanism with such a degree of complexity. In a certain way, this concept has implied the very philosophy of automatic control of manipulation systems, and even robots in general. Namely, the previously mentioned reason as well as the highly deterministic nature of manipulation tasks, mostly industrial, have resulted in a concept of two-level synthesis of control of these systems and, in a broader sense, of large-scale mechanical systems. The first level of the synthesis of this suboptimal procedure involves the formation of nominal motions or programmed trajectories, which are synthesized on the basis of a conditionally exact, centralized model of dynamics of the active mechanism under consideration. With the help of computer-oriented methods of forming the mathematical models of mechanism dynamics and the synthesis of unique positioning and orientation, control synthesized in this way takes into account the real dynamic coupling among subsystems. At the second level of synthesis decoupled control is accepted, on the basis of which local control is synthesized, while the problem of dynamic coupling at the level of perturbed dynamics is solved by global control. Such a simple control concept allows the synthesis to be performed on the basis of sufficiently accurate dynamic models, so that one may speak of a kind of a new approach to suboptimal control of large-scale dynamic systems in general, and in robotics in particular. It is our firm belief that such a control concept makes mathematical models very close to practical applications and engineering practice. This is very important because today´s practice in robotics, and industrial robotics in particular, suffers from the lack of modern design methods and methods for the synthesis of control systems.

It is for this reason that the authors of this monograph have made their greatest efforts to apply very strictly all theoretical results, based on the established control concept, to a wide variety of typical manipulation tasks and appropriate manipulation mechanisms.

The monograph consists of introductory considerations, three chapters and four appendices.

The introductory considerations present the requirements imposed in synthesizing the control of robots and manipulators and a brief survey of the results achieved up to date in the synthesis of control of manipulation robots.

Chapter 1 presents a computer-oriented procedure for the construction of mathematical models of dynamics of active spatial mechanisms based on the general theorems of mechanics, and a computer procedure for the linearization of dynamic models formed in this way. This chapter is included here because of the authors´ wish to make the monograph autonomous as regards the dynamic models of the class of systems under consideration.

Chapter 2 gives a general procedure for the synthesis of control of large-scale mechanical systems. The possibilities of optimal synthesis of decentralized control are analyzed and a procedure for suboptimal two-level synthesis is described. The stability and suboptimality of decentralized control is considered. The problem of the suboptimality of various ways of implementing global control is also treated. The decoupled control of manipulation robots using asymptotic regulator properties is also presented as well as the synthesis of control in a time-discrete domain. For comparison of the proposed decentralized control with centralized control, the synthesis of classical centralized optimal linear regulator for manipulation systems is briefly presented.

In Chapter 3 the presented synthesis procedure is illustrated by several six-degree-of-freedom manipulators. These examples have been used to present the procedure of synthesizing nominal control and trajectories by decoupling the system into functional subsystems and to analyze the characteristics of selected control. This chapter contains the results of applying force feedback to solving two characteristic manipulation tasks: the transfer of a vessel filled with liquid and the assembly task, a typical task in which the manipulator behaves as a variable-structure system, due to changes in boundary conditions, under the action of external forces. The problem of implementing the proposed control form of robots and manipulators by parallel processing is also considered. The results of the synthesis of a decentralized observer for manipulation systems are also presented.

The portions of text which are not essential for a continuous reading of the results presented, but which might be of interest to readers wishing to obtain detailed explanations of certain concepts, are given in separate appendices. Appendix 2.A. contains the derivation of the relation for suboptimality estimation of decentralized and global control. Appendix 2.B. presents an example of the analysis of distributing the system model into subsystems and coupling from the standpoint of control suboptimality. Appendix 2.C. gives a procedure for stability analysis of the system with decentralized regulator and observer. Appendix 3.A. presents the algorithm for assembly process simulation with a manipulation system.

This monograph is intended for all researchers in applied and theoretical robotics, robot designers, and for post graduate students of robotics. The background required for following the text includes a knowledge of the basic results from linear and nonlinear systems theory as well as a fundamental knowledge of mechanism dynamics. Although the procedures from modern theory of large-scale nonlinear systems are used and developed in the monograph, the book has been conceived so as to allow easy following of the text and application of results to readers who are not very familiar with this theory. Moreover, the authors have done their best to enable those who are not interested in detailed theoretical explanations of the proposed algorithms for control synthesis to use the results presented in the examples of the synthesis of control of manipulation systems contained in Chapter 3.

The authors believe that the results presented in the first volume of the series together with those given in the present monograph offer actual possibilities for the creation of such procedures for the design of manipulation systems and synthesis of control systems as would be an adequate expression of the need for practical work on the problems of industrial robotics.

The following books of the series will therefore be even more application-oriented as far as robotics in a broader sense is concerned.

The authors are grateful to Dr D.Hristić, Mrs R.Nikolić, M.Sc and Miss G.Aleksić for their help in preparing English version of this book.

Our thanks also go to Dr T.Flannagan for improving the translation. Finally, our special appreciation goes to Miss V.Ćosić for her careful and excellent typing of the whole text.

Belgrade, Yugoslavia, February 1982 The Authors

Contents

General Principles of Control Synthesis of Robots and Manipulators

Introduction - System Class Definition

Robots and manipulators belong to the class of nonlinear, multivariable mechanical systems with several inputs and outputs. If considered to have no environmental effects (or known, determined environmental effects), they represent a deterministic, dynamic system which often has a variable structure. As regards mechanical characteristics, robots and manipulators belong to the class of large-scale spatial mechanisms and to the type of open and closed active kinematic chains, since certain degrees of freedom are powered by appropriate actuators.

Robots and manipulators are used in various fields of human activities ranging from industry, agriculture, transportation, underwater and space research to medicine, where they are used either for rehabilitation purposes or reinforce certain human muscular abilities. Research in the field of robotics is therefore highly significant from both techno-economic and humanitarian standpoints. Taking into account the nature of these systems and various application possibilities, the approach to robot and manipulator research and design has to be multidisciplinary since the field of robotics incorporates the results of numerous scientific and engineering disciplines such as applied mathematics, technical mechanics, control and systems theory, electronics, informatics and computer techniques, large-scale-mechanism theory and mechanical engineering in general, medicine, economy, sociology, etc.

This book treats the problems of control synthesis of robots and manipulators. Attention will be mainly devoted to the control synthesis of industrial manipulators. Nevertheless, all the results presented are applicable, with no great difficulty, to other robot and manipulator types. In addition, this research is easily expandable to other large-scale mechanical systems, such as flying vehicles, hydraulic systems, etc., and even to other systems (economic, traffic, etc.).

A brief classification of robots and manipulators is provided so as to give an idea of the scope of robotics and, thus, of practical applicability of the results presented in the book. The main principles underlying the control systems of robots and manipulators are then considered as well as the requirements imposed on control synthesis. The results achieved up to now in solving the tasks of control synthesis of robots and manipulators are briefly surveyed. Finally, essential ideas applied to the control synthesis of robots and manipulators are presented.

Classification of Robots

The considerably different definitions of the notion of a robot found in the literature arise from the fact that robotics is a comparatively young discipline characterized by extemely rapid development and expansion. That is why each definition and classification of robots easily becomes incomplete or too narrow. In general, robots are a class of technical systems which imitate or substitute human locomotion and intellectual functions [1]. Such a broad definition of robots covers not only various types of manipulation and locomotion systems, but also various types of computing and logic machines performing certain intellectual functions, such as chess-playing machines, automatic machine translators etc.; even an autopilot might be regarded as a robot. However, this book will treat only those robot types which necessarily involve the execution of certain mechanical movements, i.e., which execute certain functional movements, as well as being capable of performing some intellectual functions.

These robots may be classified in a number of ways. The classification into manipulation and mobile (locomotion) robots and complex (mobile--manipulation) robots is the most general one. Robots in a wider sense also include the so-called information robots, as well as other variants of complex robots.

According to their function, mobile robots may be divided into legged--locomotion, "self-mobile" systems and exoskeleta. According to the control mode, walking mechanisms are classified into programmed (automatic) mechanisms, which move in accordance with a program specified in advance, semi-automatic or biotechnical mechanisms, with a human operator participating in the control, and intelligent mechanisms, which possess certain elements of artificial intelligence for automatic recognition of obstacles and making decisions about their motion under unknown or variable external conditions. "Self-mobile" systems include robots intended for motion in underwater environments and on other planets and they utilize different modes of motion (wheeled or legged-machines, caterpillars, etc.). They are classified as programmed, adaptive and intelligent robots. Exoskeleta are a special type of locomotion robot serving to substitute or amplify the locomotion functions of a man in the exoskeleton (they are most widely applied in rehabilitation, though they are also used for space and underwater research and the like). According to their configuration (the number of legs), locomotion mechanisms are divided into statically stable and statically unstable systems (biped and quadruped systems).

Manipulation robots, whose principal characteristic is the imitation of the function of a human arm, are classified into three types according to control principles: automatic, biotechnical and interactive.

Automatic robot-manipulators perform their function without direct human participation in the control, except for the system operation which is supervised, started and interrupted by a man. With a view to their adaptabillity to external operating conditions, automatic robots-manipulators are divided into three generations.

The first generation involves robots with "fixed programs", or the so--called programmed robots. They are capable of repeating a strictly specified set of operations under conditions completely determined in advance, with no external perturbations. They are not adaptable to environmental changes and are most often intended for industrial applications for stereotype repeatative operations. Such a robot has practically no information about the external environment. However, it should be borne in mind that even these robots are designed so as to allow easy changes of programs according to which they operate. Such robots should be reprogrammable and should allow easy switching to the execution of some other set of operations according to another "fixed program" (if this were not the case, these robots would not differ from various mechanical arms which are completely inflexible and are made only for a strictly specified set of operations).

The second generation involves "adaptive" robot-manipulators capable of operating under variable or partially unknown conditions and being able to adapt to the action of perturbations due to environmental changes. These robots incorporate sensors which provide information about changing external conditions. They also perform a set of operations defined in advance but are capable of accomplishing the same operations under changing operating conditions.

"Intelligent" robots, namely, robots possessing certain features of artificial intelligence, belong to the third generation. They are capable of defining instantaneous tasks, taking into account the available information regarding their operating environment, of solving particular problems and of changing their own action in accordance with variations in operating conditions. In a word, these robots may be completely autonomous and the degree of their intelligence varies according to the functions they are intended for.

The number of robots belonging to the third generation is relatively small and they require a relatively high-cost and complex control system. That is why these robots are often replaced by so-called hybrid systems in which a human operator is directly incorporated in the on--line control process. Robots of the first two generations also require the presence of a man but his presence, in the case of first generation robots, is limited to a comparatively short period of preparing (teaching) a robot to perform a particular set of operations, or, in the case of second generation robots, imposing a particular task that a robot is capable of accomplishing in accordance with instantaneously changing external conditions. Briefly, a man is not directly involved in the on-line control process of these robots. His presence is reqired to the degree to which external conditions vary (with first-generation robots only), namely, as often as the task a robot is to perform varies.

Since "intelligent" robots are still rare and expensive, biotechnical and interactive manipulation robots are often used in practice. It should be borne in mind that the presence of a man in the control process is sometimes required by the purpose of the robot (e.g., rehabilitation robots, and the like). The so-called biotechnical robots require permanent participation of a human operator in the control process. These robots include the so-called remote controlled robots, master--slave robots and semi-automatic robots. In contrast to these, interactive manipulation robots require only periodical participation of a human operator in the control process. This takes the form of various types of interaction between a human operator and the computer which controls the robot. These interactive manipulation robots include robots with supervisory control or robots with dialogue control. They are most widely used in rehabilitation, ocean and space research and complex operations in industry.

Requirements of Robot Control Synthesis

We now focus on the problem of control synthesis of robots and manipulators. Numerous other problems of robotics (robot design problems, development of sensors, sociological and economic aspects of robotics, medical aspects of robot usage, etc.) will not be dealt with here.

The basic principle underlying the control systems of robots and manipulators, as well as those of large-scale systems in general, is hierarchical structure. Hierarchical organization of robot and manipulator control systems is most often vertical so that each control level deals with wider aspects of overall system behaviour than a lower level. A higher control level always refers to slower system aspects. It controls those system parameters which vary more slowly. A higher level communicates with a lower level giving it instructions and receiving from it relevant information required for decision-making. After obtaining information from a lower level, each level makes decisions taking into account general decisions obtained from a higher level and forwards them to a lower level for execution.

With robots and manipulators, the highest hierarchical level is related to the operator who may communicate with the control system in various modes, by means of various interrupt levels. As we said in the preceding pharagraph, with biotechnical systems a man is constantly involved in the control system. It is the man who represents the highest control level, just as in the case of interactive robots where his participation for decision-making purposes is only periodical. With automatic robots, the human operator only periodically interrupts the automatic control. A fact to be kept in mind is that the operator´s decisions may override any decision made by the automatic control system (it is the human operator who prescribes new operating conditions or intervenes in various emergency situations).

A hierarchical control system may have different numbers of levels depending on robot type and the complexity of tasks for which the robot is intended. Four control levels are encountered most often [2, 3]: the highest (which recognizes the obstacles in the operating space and the conditions under which a task is being performed and makes decisions on how the task imposed is to be accomplished), the strategical level (which divides the imposed operation into elementary movements), the tactical level (which performs the distribution of an elementary movement to the motion of each degree of freedom of the robot), and the executive level (which executes the imposed motion of each degree of freedom). All robots and manipulators have the two lowest control levels. The tactical level generates the trajectories of each degree of freedom which perform the desired functional movement, while the executive level executes these trajectories by means of appropriate actuators incorporated in each degree of freedom. With first-generation manipulators, the trajectories of certain degrees of freedom are memorized (tactical level), and the lowest level realizes them. With second-and third-generation manipulators, the higher control levels prescribe the elementary movements in accordance with external conditions, which are to be realized by the lowest levels. The lowest control levels may be realized in various modes, and their capabilities for realizing sufficiently accurately the movements prescribed by higher control levels will determine the organization and complexity of higher control levels (and vice versa). It should be borne in mind that each of the levels stated may be organized in a number of separate hierarchical levels.

This book is concerned with the problems of synthesizing the two lowest control levels for robot-manipulators and primarily with the synthesis of the executive level. It is not particularly important whether the executive level refers to automatic first-generation robots or to second - and third-generation robot-manipulators. Moreover, the same problems arise in the synthesis of the lowest control levels of biotechnical and interactive manipulators as well as of other robot types (e.g., locomotion systems). In other words, the book deals with the problems of the execution of particular movements, which are prescribed either in advance or by higher control levels. Decision-making levels, which are related to the field of artificial intelligence (the problems of communication with the operator) are out of the scope of this book.

The task imposed on the lowest control levels of robots and manipulators consists of the realization of a particular functional movement. In general, this task may be defined as the task of practical system stability with prescribed settling time. The most frequent task is to transfer the system state (the system comprises the mechanical portion of the robot and actuators) from one bounded region of initial states into another bounded region in the state space within finite settling time. The system state should belong to the bounded region in the state space during the transfer. Constraints on the regions are imposed by the conditions of the task (particularly with respect to the region into which the system state is to be transferred), by the constraints on the system itself (kinematic constraints on mechanism capabilities, constraints on system inputs, constraints on actuator capabilities, etc.) and the constraints imposed by environment in which a particular task is imposed (various obstacles, etc.). The control task thus reduces to the task of practical stability under the assumption that the perturbations acting on the system may be reduced to the type of perturbation of the initial conditions which may bring the state of the system within the bounded region of the state space. A precise definition of practical system stability with prescribed settling time, which is a requirement imposed in control synthesis of robots and manipulators, is provided in paragraph 2.2. Special attention will be dedicated to such perturbations as may not be reduced to the type of perturbation of the initial conditions (e.g., perturbations due to changed system parameters and the like).

Systems theory provides various possibilities of robot control synthesis at the lowest control levels, namely, for solving the task imposed. Since robots belong to the class of large-scale mechanical systems, those solutions relating to the control of large-scale mechanical systems should be examined. Namely, mechanical non-linear systems have a great number of specific features which require these systems to be regarded as a separate system class in control synthesis. Moreover, robots as active mechanisms also have their specific characteristics which must be taken into account in control synthesis. These specific features arise from mechano-structural, dynamic and control considerations. One characteristic of robots is their variable structure: when not in contact with external obstacles, manipulators represent open kinematic chains but when they come into contact with the external environment (e.g., during an assembly process), they change their structure and become closed kinematic chains. With locomotion robots, the system constantly changes its structure during the gait from an open to a closed mechanical scheme [4]. One of the specific features of locomotion robots is the presence of uncontrollable degrees of freedom (d.o.f. of the foot with respect to the support). The main specific characteristic of robots-active mechanisms is the redundancy feature reflected in multifold possible forms for the execution of a given functional movement. That is, robots often have more degrees of freedom than is required for the implementation of a particular functional movement, which makes the mechanism more adaptable to various tasks and facilitates robot reprogrammability. However, it adds to the complexity of control synthesis.

All these and many other robot characteristics [5], require this system class to be considered separately in the control synthesis. All the specific requirements referring to the system class under observation have to be taken into account when applying general methods of control synthesis of large-scale non-linear systems.

The main problem arising in the solution of the control task mentioned is to what extent one should take into account real robot dynamics in control synthesis. The highest levels (decision-making levels) do not usually consider actual system dynamics but they prescribe particular actions to be implemented by the lowest levels. To be more precise, the highest levels must have global knowledge of dynamic system capabilities (an aggregate system model) but they need not necessarily employ precise, dynamic, overall system models. The highest levels should, knowing dynamic robot capabilities, impose on lower levels such tasks as may be realized by the actual system. However, the two lowest control levels should be synthesized (taking into account dynamic robot behaviour) so as to provide the implementation of functional movement prescribed by higher levels.

When deciding on the functional movements that are to be performed by the mechanism in accomplishing a particular task, the highest control levels minimize those criteria which do not usually take into account overall system dynamics. However, the lowest control levels should guarantee the desired practical system stability by minimizing such criteria as would take into account actual system dynamics, so as to provide the best possible system behaviour by using actuators with no surplus power reserve.

As regards the extent of the knowledge of their model, large-scale non-linear mechanical systems, and particularly robots intended for industrial applications, belong to the systems whose models may be set precisely enough, since their behaviour is described by the known laws of mechanics. It may therefore be considered that the solutions of the models of these systems are in sufficiently precise accordance with the responses of an actual system to identical input signals. That is why this knowledge of the system model should be used in the control synthesis as extensively as possible. It should be used to synthesize such control as would ensure the required practical stability conditions and as would not be too suboptimal in relation to a chosen criterion. However, a fact to be borne in mind is that the models of certain mechanical systems, and particularly of active mechanisms, are very complex and their implementation for the purpose of calculating real-time control requires powerful computers so as to achieve a sampling time compatible with system dynamics. In other words, the problem arising is how to use the information about system dynamics and synthesize as simple and as low-cost a control form as possible which is at the same time sufficiently reliable and robust.

To synthesize the simplest possible control, approximate system models are usually considered which allow the synthesis of simple and low-cost control. A linearized system model is frequently considered, on the basis of which control may easily be synthesized using the results of the extensively elaborated linear systems theory. Since a chosen approximate model is assumed to be a sufficiently accurate approximation to the actual system model, it is usually considered in the literature that the control synthesized on the basis of an approximate model will also ensure the imposed requirements for a precise system model. However, this need not always be true since the validity of approximate models is related to bounded regions in the system state space. It is therefore always necessary to test to what extent the control, synthesized using an approximate model, will prove satisfactory when applied to the actual system [6]. Obviously, the closer the model to the actual system model, the higher the probability that synthesized control will satisfy the actual system as well. On the other hand, the closer the system model to the actual model, the more complex it becomes. Accordingly, the control required so as to satisfy the conditions of the task set is also more complex. If a centralized linearized system model is under consideration, and if a control is being synthesized which is to minimize the standard criterion for the practical system stability conditions to be satisfied, the resulting control is a centralized linear regulator which may be of a very complex structure.

To achieve as simple control synthesis as possible and to synthesize simple, reliable and low-cost control, it is convenient to apply decentralized control. That is, the approximate system model is taken to be a set of decoupled subsystems and autonomous local control satisfying the imposed control task is synthesized for each isolated subsystem [7, 8]. As far as a large-scale mechanical system is concerned, it is usual to treat either single degrees of freedom or groups of degrees of freedom of the mechanism as subsystems. Such an approximate system model practically neglects the coupling among certain degrees of freedom, namely, the dynamics of the mechanism itself. On the other hand, this results in a simple control synthesis procedure and a simple control law. However, with certain robot and manipulator types the effect of the coupling among subsystems may be too strong, so, when only local control is applied, overall system behaviour may prove unsatisfactory [9, 10]. That is, if the control structure chosen is such that the system with local control is controllable, it is possible to synthesize such local controls as will provide global system stability [11]. However, if the effect of coupling is strong, local gains may be too great and the control may be "too suboptimal" with respect to the accepted criterion through the optimization of which the control task imposed is solved (e.g., power consumption criterion, standard quadratic criterion, etc.). In such a case, when local control is applied to the actual system, the behaviour of the actual system may prove unsatisfactory. It is for this reason that the coupling among certain degrees of freedom of the mechanism, namely robot dynamics, has to be taken into account, in order to synthesize such control as will not be "too suboptimal" with respect to a given criterion.

Moreover, with some systems, the decentralized structure chosen may be uncontrollable in such a way that the desired practical system stability cannot be provided by applying only local control. For instance, with systems having directly uncontrollable degrees of freedom, decentralized control synthesized so as to stabilize locally those degrees of freedom which are powered by their own actuators usually proves insufficient for providing practical, overall system stability. Similarly, with systems operating under variable conditions, or those coming into contact with external objects (e.g., manipulators during an assembling process), the local control usually does not suffice for ensuring the execution of a particular functional movement. In a great number of tasks imposed on robots and manipulators control synthesis is inseparable from mechanism dynamics, i.e., there are tasks which are essentially dynamic by themselves, so that practical system stability may be achieved only if the complete system dynamics is considered. For example, in gait synthesis for biped locomotion systems, system dynamics has to be taken into account in order to ensure dynamic system equilibrium during the gait [5]. In certain manipulation tasks, such as the drink-test, dynamic system behaviour has to be taken into account so as to synthesize such control as will satisfy the functional requirements imposed [12].

It is clear so far that the synthesis of the lowest control levels of robots and manipulators requires careful consideration of the effect of system dynamics in order to synthesize as simple and reliable a control as possible, which will satisfy the actual conditions of practical system stability.

This book will treat the problem of the synthesis of a centralized optimal regulator and of decentralized control of robots and manipulators as an extremely simple control structure suitable for implementation. As we said, the main problem arising is how to use the information on system dynamics and yet maintain control synthesis simplicity and ensure that the resulting control is not too complicated to implement. On the other hand, the control is required to provide practical system stability, i.e. it is required not to be "too suboptimal" with respect to the accepted criterion (if the minimization of a certain criterion is required in control synthesis).

Various approaches to the control synthesis of robots and manipulators which may be found in the literature will be briefly presented in the following paragraph.

Brief Survey of Results from the Domain of Robot Control Synthesis

Although it was only a few years ago that robotics was founded as a separate engineering and scientific discipline, a great number of papers treating various aspects of robotics have been published up to now. In addition, several textbooks and monographs dealing with both industrial and rehabilitation robotics have been published [1, 3, 5, 13 - 16] etc.

Since robotics is a multidisciplinary field, the problems of robotics have been approached from various standpoints: from the standpoint of mathematical modelling of robots, robot design, from the standpoint of control of robots and manipulators, from the standpoint of actuators and sensors, as well as from informatics, medicine, sociology, economy, etc. We will only deal here with the results concerning the control of robots, and particularly with those referring to the synthesis of tactical and executive control levels of robots and manipulators. Chapter 1 will present one computer oriented method of mathematical modelling of active mechanisms.

As far as the control of robots is concerned, what is characteristics is that different approaches to the control synthesis of different robot classes and types have been developed and that there have been relatively few attempts to apply a unique approach to the control of robots as a separate system class [4, 5, 10, 17 - 20]. As a consequence, approaches to the control of legged locomotion robots have developed independently of the results achieved in the control of manipulators. Moreover, separate approaches have been developed within single robot types depending on their purpose and mechanical configuration. For example, within locomotion robots, approaches to the control synthesis of statically unstable robots (mostly of biped robots intended for rehabilitation, of exoskeleton type) have developed independently of the approach to the control of statically stable robots (intended primarily for space exploration and motion over hostile enviroment). At first sight, these separate developments of the approaches to the control synthesis of biped and multilegged robots seem to be a natural consequence of the fact that in implementing biped gait it is necessary to provide dynamic system stability and to prevent the system from collapsing. This problem does not arise with multilegged machines since they are statically stable. However, this holds only if relatively slow gait is in question, while faster gait requires dynamic system stability to be provided for statically stable locomotion robots as well. As a result of the assumption that the gait of multilegged machines is slow enough to be satisfied by static system stability during the gait, the majority of papers have treated the problems of the synthesis of higher control levels, problems of the recognition of obstacles and adaptation to them [2], while only a few papers have dealt with the problems of the stabilization of multilegged machines [21] and minimization of the reaction forces at the system´s feet [22].

In the domain of control synthesis of a biped based on using the complete model of a biped system and dynamic equilibrium of gait and posture, distinguished papers are those by Vukobratović and associates [4, 5, 18, 23 - 32]. Similar, very interesting results may be found in the papers by Beletski and associates [33, 34].

As for the control synthesis of manipulators, there are different approaches depending on the purpose of the manipulator, although numerous problems of the synthesis of the lowest control levels of robots are essentially identical for all manipulator types. Leaving aside the numerous attempts to design and synthesize the control of various types of manipulation aids intended for the rehabilitation of certain classes of patients, where the main problems relate to the implementation of the interaction between the patient and aid, we will primarily consider the approaches to the control synthesis for industrial manipulators.

In the field of control synthesis of manipulators, a great number of papers has been devoted to the problems of synthesizing the tactical control level, i.e., to the problem of the distribution of motion to single degrees of freedom so as to realize a particular motion of the object being transferred by the manipulator. Most of the papers devotet to the control synthesis for manipulators have treated so-called redundant manipulators, i.e., those having a greater number of degrees of freedom than is required for the execution of the task imposed. With manipulators, the general task of practical stabilization reduces to the task of transferring an operating object along a particular trajectory and providing particular object orientation during the transfer (see ch. 3). In the general case, the execution of this task requires six degrees of freedom. Manipulators are often made with a greater number of degrees of freedom (so as to ensure their flexibility), or else, six-degree-of-freedom manipulators are used to solve only the task of transferring the object along a prescribed trajectory without requiring particular object orientation during the transfer. In both cases the redundancy problem arises, i.e., how to distribute the motion to single degrees of freedom of the manipulator so as to achieve the desired motion of the operating object. Various solutions to the redundancy problem have been proposed, but it should be said that most approaches are based on the kinematic model of manipulators.

A trivial solution of the redundancy problem would consist in using a non-redundant manipulator for the given task, but this results in reduced manipulator reprogrammability, namely limited possibilities for employing a single manipulator in solving tasks of different complexity degrees.

Another approach to the redundancy problem consists in introducing, in advance, constraints on certain degrees of freedom of the manipulator, or in introducing constriants on the linear combinations of internal coordinates of the manipulator (or velocities of internal coordinates). This reduces the number of degrees of freedom of the manipulator that are to perform a desired motion [35]. It is also possible to solve the redundancy problem by introducing certain criteria that are to be minimized in the distribution of the motion to the degrees of freedom. Such a criterion may involve only kinematic parameters of the manipulator (positions and velocities), so that such distribution of motion is made as will provide minimum velocities of certain degrees of freedom and the desired motion of the object [36]. The distribution of motion may also be performed by minimizing the criterion which takes into account dynamic system characteristics [37], or by minimizing the power consumed by implementing the prescribed motion [38]. It should be stated that the distribution of motion to single degrees of freedom is made either by directly calculating the internal mechanism coordinates which ensure the desired trajectories of external coordinates (coordinates of the c.o.g. of the object and of the angles formed by the accepted object axis and the axes of the absolute coordinate system), or by calculating the velocities of internal manipulator coordinates as a function of prescribed object velocities with respect to the absolute system.

The results mentioned refer to the problems of redundancy of coordination of motion in the case when object trajectories to be implemented are prescribed, i.e., when the higher control levels completely define trajectories in external coordinates. However, several papers deal with the problems of the synthesis of manipulator trajectories which are not uniquely defined; e.g., terminal end-points on the trajectory are prescribed, or a set of points in space through which the object is to pass (with allowable deviation), while the trajectories of internal manipulator coordinates are to be synthesized. These trajectories may be synthesized so as to minimize certain criteria. Evidently, the problem of synthesizing such trajectories as satisfy certain requirements and are not uniquely defined, arises with both redundant and non-redundant manipulators. Without trying to consider in detail the need for optimization when synthesizing the trajectories (this will be dealt with in ch. 2), let us only mention several attempts to synthesize "optimal" trajectories of various types of manipulation systems in which approximate system models have usually been applied.

So, in the approach [39] planning is done directly via internal coordinates, which is more convenient from the computational point of view (since the inversion of the Jacobian matrix which relates the velocities of internal and external system coordinates is avoided), but the trajectory of the manipulator tip cannot be controlled (except for prescribed points). This means there is the danger of collision between the manipulator and obstacles. In this procedure the power criterion (or the criterion of time taken by the transfer from one prescribed point to another) is minimized (without taking into account any dynamic parameters) by considering the kinematic model of a non-redundant manipulator only. When the task of the practical stability is imposed, the problem of synthesizing internal manipulator coordinates, by prescribing a set of points of environment through which the manipulator is to pass, was treated by Paul [40]. The trajectories were synthesized so as to minimize the time taken to execute the task imposed taking into account the constraints on the velocities and accelerations of internal manipulator coordinates but without considering manipulator dynamics. In a paper by Luh and Lin [41] a method of preplanning time schedule of velocities and accelerations along an adopted path to obtain a minimum travelling time is considered. The constraints of composite Cartesian limit on linear and angular velocities and acceleration of the hand are imposed and a linear performance index is involved. Various levels of "approximate programming" are considered in order to ensure the convergence of the solution to the optimum feasible point and to reduce computing time. Hovewer, only the kinematical model of the system is involved. Kahn and Roth [42] dealt with the problem of synthesizing the trajectory of the manipulator tip between two prescribed points in space with the minimization of the transfer time using a dynamic system model. However, an approximate manipulator model in which some dynamic phenomena are neglected (e.g., Coriolis´s forces, etc.) was used for the simplification of the problem of time criterion minimization.

Apart from these approaches to the problem of synthesizing manipulator trajectories, i.e., of synthesizing the tactical control level, a rather large number of papers have appeared which treat specific problems of motion distribution to single degrees of freedom of a manipulator for various types of biotechnical and interactive manipulators, depending on the mode in which tasks are prescribed by the operator (via object position or velocity, etc.). However, these approaches are outside the scope of this book.

The number of results achieved in synthesizing the level of the implementation of trajectories is considerably smaller. Authors usually accept the assumption that desired trajectories may be realized via local servosystems, without analyzing dynamic system behaviour. However, a number of papers have been published in which control synthesis problems are approached from the point of view of the dynamics of the overall manipulation system.

Paul [16] considered several control variants taking into account dynamic system parameters but without analyzing system behaviour with the proposed control forms. He investigated the "inverse problem" technique which was called the "computed torque" technique by Bejczy [43]. In the "inverse problem" technique, the input torques necessary to drive the manipulator hand along a preplanned path are on-line computed as a function of desired joint accelerations, velocities and positions and their actual counterparts. A similar approach is taken by Pavlov and Timofeyev [44] where the stabilization of the manipulator trajectory is performed by control which involves on-line calculation of the complete model of manipulator dynamics, but it does this without treating the problem of whether such a solution is justifiable and without considering the possibilities of implementation in the case of a manipulator with a greater number of degrees of freedom and a more complicated dynamics model. Moreover, a paper by Timofeyev and Ekalo [45] proposes an adaptive algorithm for the stabilization of a programmed trajectory when a manipulation system with variable or uncertain parameters is in question. It is assumed that the control is implemented by on-line calculation of the complete model of manipulator dynamics. The compensation of nonlinearities of the manipulation system by on-line calculation of the complete model of manipulator dynamics is also proposed in a paper by Saridis and Lee [46]. In this paper an approximative algorithm is presented for optimal choice of feedback gains for the remaining linear system. A similar approach to the realization of trajectories of supervisory controlled manipulators was proposed by Kulakov [47]. Thus, on the basis of the complete model of manipulator dynamics and by testing the system for Liapunov's stability, a non-linear and complex control law is synthesized which is to provide the desired manipulator motion, but its implementation in case of complex manipulator types requires a powerful computer system. In a paper by Takengaki and Arimoto [48], Liapunov´s theory is used to synthesize a linear feedback of generalized coordinates and their derivatives but no precise analysis of the system performance (stability regions, choice of feedback gains etc.) is presented.

It should also be noted that there were few attempts to achieve on-line computation of dynamics of robots and manipulators. As we said, the models of dynamics of manipulators with many d.o.f. might be very complex and their on-line computation involves implementation of high--speed computers. However, if only generalized forces are to be calculated, when angles, velocities and accelerations of the mechanisms are given, the computational time can be significantly reduced and on-line calculation can be achieved [49]. However, the opposite task (i.e. to calculate accelerations when driving torques are given), which is frequently encountered when different control schemes are to be implemented, requires a much longer computational time. Luh, Walker and Paul [50] used their algorithm for on-line computation of a dynamic model in order to implement the "inverse problem" concept of position control. Their approach differs from "resolved - motion - rate" control of Whitney [37] and the "inverse problem" of Paul [16] in that all the feedback control is done at the hand level i.e. it deals directly with the position and orientation of the hand (and its velocity and acceleration). With this approach, significant reduction of computational time is obtained but a relatively fast computer is still needed to obtain satisfactory sampling frequency.

The conventional procedure for synthesizing the control of large-scale non-linear systems on the basis of a linearized model has also been used in synthesizing the control of manipulators. By linearizing the complete dynamic model of the manipulator about the trajectory to be implemented, Popov and his associates [51] have synthesized a linear centralized regulator by minimizing the standard quadratic criterion. However, it should be borne in mind that the implementation of the linear regulator for manipulation systems involves numerous problems due to complex control structure (a large number of feedback loops), and its application is limited by the fact that it has been synthesized on the basis of an approximative linear model. The synthesis of a complete servo-loop for single d.o.f. of a manipulator has been considered where the dynamic nonlinear model of the actuator takes into account dry friction, clearances and elastic deformations [52]. However, all the approaches mentioned require a relatively complex implementation of the lowest control level, which is why they are used in practice less frequently than the conventional solution using local servosystems.

Medvedov, Leskov and Yuschenko [3] have analyzed overall system behaviour when the implementation of trajectories prescribed by the tactical level is entrusted to local servosystems only. Again a linearized overall system model in a centralized form was used. The synthesis of local servosystems involved a decoupled system model so that the authors analyzed the effect of coupling among subsystems (the degrees of freedom of the manipulation system) and tested the overall system stability by analysis in the frequency domain. As in the previous case, the validity of this analysis is limited by the validity domain of a linearized system model. In addition, there still remains the problem of the introduction of global feedback loops if decentralized (local) control cannot provide satisfactory system behaviour. It has been shown that the conventional assumption that a manipulation system may be stabilized by local control only need not apply to cases of extremely fast movements or those manipulator types in which the coupling among degrees of freedom is strong [53], (see ch. 3). To compensate the effect of coupling among subsystems, decoupled control of manipulators has also been synthesized by introducing, on the basis of a linearized manipulator model, cross-feedback loops among selected subsystems [54]. However, the proposed synthesis of the compensator for decoupling the system into subsystems is easy to implement only when simple manipulation systems are in question, and its justification is again limited by the validity of the linearized system model in a wider zone of the state space. Non-linear control law for complete decoupling of the dynamics of manipulators and arbitrary pole assignement was proposed by Freund [55]. However, only relatively simple manipulation configurations with 3 degrees of freedom, where moments of inertia are neglected, were investigated. The application of such nonlinear control to more complex configurations would lead to complex control laws. Finally, in a paper by Roessler [56] the manipulation system was decoupled by introducing a global feedback loop. This was implemented on the basis of real-time calculation of the complete non-linear model of the coupling among selected subsystems but did not analyze overall system behaviour when the chosen local control form is applied and did not consider actual requirements or possibilities for the introduction of a complex and complicated mode of calculating the global control (which is a serious problem in case of complex manipulation configurations). An attractive idea of how to achieve decoupling of a manipulation system by force-feedback (feedback with respect to loads measured at each joint of the manipulator), which has been recognized and elaborated by the authors of this book (and which will be presented in detail in chs. 2 and 3), has also been noticed recently by Hewit and Burdess [57]. However, they have also introduced accelerometers which provide information about accelerations of the manipulator joints. They calculate on-line the matrix of inertia of the system in order to obtain information on gravitational, Coriolis and friction forces, so they have unnecessarily complicated the control law. The problem of stability of the nonlinear system and suboptimality of the chosen control also has not been treated in their paper. It should be noticed that (contrary to the approach in this book), in all the above papers on decoupled control of manipulators, control should compensate the influence of coupling in all cases. However, the coupling among chosen subsystems might be stabilizing. This is why, in this book, the control law is proposed which compensates only the destabilizing influence of coupling (see chap. 2).

Mention should also be made of a few attempts to perform the stabilization of manipulation systems by non-linear control laws, but these results are hardly applicable since they result in unsatisfactory system performance (e.g., Young's paper [58] based on variable-structure systems theory).

In the last few years, several attempts have been made to synthesize adaptive control for manipulation systems. Besides the above paper by Tymofeyev and Ekalo [45], in the paper by Dubowsky and DesForges [59], a model-referenced adaptive control is applied to manipulators in order to maintain uniformly good performance over a wide range of motions and payloads. However, only the linear decoupled reference-model is considered. No analysis of the range of the manipulator dynamics for which the proposed control is valid is presented. In a paper by Dubowsky [60], a time-discrete version of this control is developed. It should be said that few efforts have been made to analyze the necessity of adaptive control for manipulation systems. It seems that most of the parameter variations in practice could be compensated by sufficiently robust control. However, the problem of suboptimality of such robust control in comparison to adaptive control is still open. In this book we shall not treat the problem of adaptive control, since we think that most of the manipulation tasks can be solved by non-adaptive control, and adaptive control might be too complex and expensive to implement.

Several papers devoted to various specific manipulation tasks are also worth mentioning. For example, a number of papers treating an assembly process with an industrial manipulator have been published [61, 62] etc.; (see para. 3.5.). Numerous results refer to actual applications (papers by Nevins et al., [63]; Bejczy, [64], and by Nishimoto et al. [65]). There are also a few papers dealing with the problem of flexibility of the manipulator [66]. Book, Maizza-Neto and Whitney [67] have considered a simple two beam, two joint manipulation with distributed flexibility. A more general case has been treated in a paper by Truckenbrodt [68], where a manipulator with rigid and flexible elements is properly modelled and, by means of a series expansion of the distributed deformation coordinates, the model is obtained in a form of ordinary differential equations. Linearization of the model is then performed and standard control methods are applied. Finally, several papers have considered the possible application of force feedback as a significant possibility for solving the problem of control synthesis of robots, since the information about forces may facilitate solving the problem of dynamic control and manipulator/environment interaction [3, 61, 69]. The application of force feedback will be referred to further in the text.

This brief survey of results in the field of the control of robots and manipulators provides a partial insight into different attempts to synthesize the control at tactical and executive control levels. However, as may be seen from the above, no complete and practically acceptable answer to the question of how to use the knowledge of robot dynamics so as to synthesize such control as would be simple enough to implement in practice and guarantee satisfactory system behaviour has so far been given. That is why the majority of these results have found no application, while in practice designers have usually accepted the control synthesized without taking into account system dynamics. Accordingly they consider only those systems and tasks in which such control suffices to provide the execution of tasks, without considering the extent to which the applied control is suboptimal. Moreover, numerous tasks of robotics which are essentially dynamic in nature (e.g., biped gait, transfer of liquid by a manipulator, etc.) have so far not been appropriately implemented in practice.

Principle of Robot Control Synthesis

One may conclude from this review of the control synthesis of robots and manipulators that the problem of the synthesis of dynamic control of robots has not been properly solved. Nor have the numerous possibilities offered by systems theory been considered. On the other hand, the solutions given by systems theory have not been so elaborated as to make them directly applicable to the control synthesis of robots.

The principles on which we base our approach to the control synthesis of robots and manipulators will be briefly presented in this paragraph.

Taking into account the fact that for the majority of robots, and particularly manipulators in industrial practice, operating conditions are known in advance and that the functional movement to be performed by the system repeat in cycles and are therefore foreseeable, a two--stage control synthesis has been proposed [4, 9, 10, 70]. The first control-synthesis stage consists of synthesizing nominal, programmed control, which implements the desired system motion for a particular, selected initial state and assumes that no perturbations are acting on the system. Such programmed control is synthesized using a centralized system model. Since mathematical models of mechanical systems may be set precisely enough, the response of the system to the programmed control, synthesized on the basis of a complete model, may be expected to accord with the desired nominal system motion, assuming that no perturbations act on the system and that the initial state ideally coincides with an assumed nominal initial state. The synthesis of nominal programmed control is performed off-line by first defining in advance the trajectory to be implemented by the robot in such a way as to satisfy functional requirements and second by calculating in advance the appropriate programmed control for the complete system model. This means that the tactical control level of robots reduces to a set of trajectories prepared and memorized in advance and appropriate programmed controls. Higher control levels define the trajectory being implemented on-line. This allows a computer which is not involved in on-line process control to be used for calculating the control at the tactical level, which means that the control calculation time is not of major importance. For real-time control only a programming device with memorized and calculated trajectories is used (it may be implemented in several modes but it is, in any case, considerably cheaper than using more powerful computers, which would be employed in implementing real-time calculation of the dynamics of large-scale systems).

This solution is satisfactory for industrial applications of robots where a powerful computer may be used for off-line calculation of nominal trajectories for a large number of manipulators, while the robots themselves are controlled by mini-or micro-computers. At the preparation stage of the process to be accomplished by manipulators, the so-called teach-in stage, the operator prescribes the functional requirements to be satisfied by a particular manipulator. The computer calculates the nominal trajectories and control which satisfy the imposed functional requirements and these trajectories are stored in control systems of certain robots.

However, this solution is unsatisfactory in many robot applications where the possibility of prescribing the functional requirements on--line has to be provided, which means that real-time calculation of trajectories at the tactical level must also be ensured (as is the case with robot applications in unknown environments, where the conditions under which a task is accomplished cannot be defined in advance, etc.). Such cases require computers to be used for on-line calculation of trajectories, while the synthesis of programmed control usually needs to be simplified so as to avoid the calculation of system dynamics, except for particular cases already mentioned, where the task is dynamic in nature.

The second control-synthesis stage consists of synthesizing the control for the tracking of nominal trajectories when the initial state deviates from the nominal initial state (but belongs to a bounded region of initial states) and when perturbations of the initial conditions are acting on the system. Two modes of control synthesis will be considered at this stage: the first by means of a centralized optimal regulator and the second by decentralized control. Since large-scale non-linear systems are in question, the synthesis of decentralized control by minimizing a certain accepted criterion involves numerous problems, even if a particular decentralized control structure and a concrete control law are accepted (see para. 2.2.). That is why a different approach mentioned above is applied. Thus, during the second stage of control synthesis an approximative system model in the form of a set of decoupled subsystems (corresponding to single (or groups of) degrees of freedom) is considered. The control is synthesized by minimizing the selected criterion (e.g. standard quadratic criterion) respecting the constraint imposed by the decoupled system model or such control is chosen as will stabilize only local subsystems. The behaviour of the overall, complete system model with applied decentralized control synthesized on the basis of the decoupled system model is then analyzed. It analyzes whether or not practical system stability is satisfied and to what extent the control synthesized is suboptimal. Since practical system stability in a bounded region in the state space is in question, the introduction of programmed control at the first stage of control synthesis reduces the effect of coupling among subsystems. That is why there are further possibilities of stabilizing the system by decentralized control in the second stage than in the case if decentralized control were applied without programmed (nominal) control. That is, the decoupled system model is evidently a better approximation to the complete system model when the model of deviation from the nominal dynamics is considered than it would be without introducing the nominal control.

However, in spite of the reduced effect of coupling among selected subsystems achieved by nominal control, the effect of coupling in certain states and for certain tasks may still be too strong, and the decentralized control, synthesized without taking into account actual system dynamics at the executive level, may prove inappropriate. In that case, the decentralized control structure chosen has to be rejected and additional global feedback loops have to be introduced. The problem is how to introduce such global feedback loops as will ensure the desired system behaviour without making the control much more complex. One specific feature of mechanical systems may be used for this purpose, namely, if the degrees of freedom of the mechanism are regarded as subsystems, then forces (moments) which are directly measurable represent the coupling among subsystems. The introduction of global control in the form of force feedback loops may minimize the destabilizing effect of the coupling on global system stability [9, 70], that is, the suboptimality of the decentralized control applied [53]. This does not greatly reduce the simplicity of control structure, since the number of feedback loops required is not much greater, but additional tehnical difficulties arise and the control system becomes more expensive because of the introduction of force transducers. Another possibility for introducing global feedback loops is on-line calculation of the forces acting in single degrees of freedom in accordance with the mathematical system model. However, this calculation might prove very complex if a complete system model were used, so the coupling may be calculated using approximative system models. The complexity of the approximative calculation of coupling, i.e., of forces, which might provide a satisfactory effect of global control on system behaviour should be studied for each actual system. In addition, one should also investigate whether each actual system and task may be satisfied by local control only or whether global control has to be introduced and, if so, in what form.

The executive level is thus implemented either by a centralized regulator or by local control synthesized at the level of single actuators and degrees of freedom and by global control synthesized by considering the behaviour of the complete system model. Such control synthesis for the executive level may be performed regardless of whether or not the programmed control introduced at the tactical level has been synthesized on the basis of a complete centralized system model. If it has not, the need for the introduction of global control is more evident, since, in the opposite case, the control would be synthesized without the dynamic system parameters having any effect.

This two-stage control synthesis is evidently suboptimal, since, instead of synthesizing such control as would minimize a particular criterion and satisfy practical stability requirements, first, nominal control is synthesized for an "arbitrary" selected initial nominal state and second, the control is synthesized which will provide the tracking of the nominal state on the basis of the model of deviation from the nominal state. Moreover, even the synthesis of nominal control is not "an optimal synthesis". The nominal dynamics is synthesized so as to provide a functional and reliable solution. It is for this reason that control synthesis at the stage of the deviation from the nominal is not aimed at synthesizing optimal control but primarily at obtaining a simple synthesis procedure which would incorporate engineering experience of system designers and provide a control form with satisfactory characteristics. The main reason for deciding not to synthesize "optimal control", apart from there being considerable numerical problems involved in such a synthesis, which make it unsuitable for practical applications, lies in the fact that the choice of a criterion is conditional and this criterion can hardly incorporate all the requirements imposed by control synthesis of robots, such as a simple and realible structure. "Optimal" solutions are therefore often unacceptable in practice [71]. On the other hand, as we have stressed, if control "optimality" is neglected, the gains synthesized may be higher than required. The proposed two-stage synthesis procedure and the synthesis of decentralized control by means of a decoupled system model, with subsequent checking of control suboptimality and its eventual reduction by introducing global control, represent a compromise between the requirements for "optimality" and those for simple synthesis and suitable and reliable control. When control "optimality" is unimportant, the synthesis of decentralized control is performed by stabilizing decoupled subsystems. Only global system stability is then considered. This simplifies the synthesis further and allows various conventional procedures to be applied to the control synthesis of local low-order subsystems.

A characteristic problem arising in synthesizing the nominal control at the tactical level is that of manipulator and robot redundancy. Since solution optimality has been accepted as not being of primary importance, no particular criterion is minimized in solving the redundancy problem. When synthesizing, the nominal system is decoupled into two or more functional subsystems and the tasks to be realized by the system are divided into these subsystems. As a result, one redundant problem reduces to two or more nonredundant problems. This procedure will be analyzed in detail in para. 2.3. and its application to the manipulation system in para. 3.3. The advantages of such a procedure for the synthesis of nominal control are particularly recognizable in those tasks and systems which require the consideration of total system dynamics, for in such cases the synthesis of optimal solutions usually involves insurmountable numerical difficulties. What may be achieved by dividing the system into two functional subsystems is that one provides the desired system motion in space, while the other satisfies the dynamic conditions imposed on the system. Such a suboptimal synthesis procedure considerably simplifies being able to obtain such nominal control as will satisfy all functional requirements and be sufficiently reliable.

However, considerable simplification of both control synthesis procedure and control law itself may be achieved by using force feedback loops. With robots and manipulators, forces not only give information about the coupling among single degrees of freedom but also provide information about the overall system dynamics. The incorporation of forces into the control may therefore ensure that the control can stabilize the system dynamically and implement particular tasks which are dynamic in nature. Such an application of force feedback loops, which is completely different from conventional control laws, allows certain complex control tasks to be solved using comparatively simple control algorithms. The application of force feedback loops also appears suitable for those tasks in robotics where the system has a variable structure due to the contact with environment, since the information on forces yields the information about environmental effects on the robot, which allows system adaptation to changing conditions by means of appropriate control.

References

[1] Popov E.P., Vereschagin A.F., Zenkevich S.L., Manipulation Robots: Dynamics and Algorithms, (in Russian), Series "Scientific Fundamentals of Robotics", Nauka, Moscow, 1978.

[2] Okhotsimskii D.E., et al., "Control of Integral Locomotion Robots", (in Russian), Proc. VI IFAC Symp. on Autom. Contr. in Space, Erevan, USSR, 1974.

[3] Medvedov B.S., Leskov A.G., Yuschenko A.S., Systems of Manipulation Robots Control, (in Russian), Series "Scientific Fundamentals of Robotics", edited by E.P.Popov, Nauka, Moscow, 1978.

[4] Vukobratović K.M., "How to Control the Artificial Anthropomorphic Systems", IEEE Trans. on Systems, Man and Cybernetics, SMC-3, 497-507, 1973.

[5] Vukobratović K.M., Legged Locomotion Robots and Anthropomorphic Mechanisms, Monograph, Institute "M.Pupin", (in English) Beograd, 1975, Also Published by Mir, Moscow, 1976 (in Russian).

[6] Vukobratović K.M., V.S.Cvetković, D.M.Stokić, Airplane Flight Dynamics and its Application, (in Serbian), Monograph "Mihailo Pupin" Institute, Beograd, 1980.

[7] Sandell N.R., P.Varaiya, M.Athans, M.G.Safonov, "Survey of Decentralized Control Methods for Large-Scale Systems", IEEE Trans. on Automatic Control AC-23, 108-128, 1978.

[8] Šiljak D.D., Large-Scale Dynamic Systems: Stability and Structure, North-Holland, New York, 1978.

[9] Vukobratović K.M., D.M.Stokić, "Simplified Control Procedure of Strongly Coupled Complex Nonlinear Mechanical Systems", (in Russian), Avtomatika and telemechanika, No 11, also in English, Autom. and Remote Control, Vol. 39, No 11, 1978.

[10] Vukobratović K.M., D.M.Stokić, "Contribution to the Decoupled Control of Large-Scale Mechanical Systems", Automatica, Jan. No 1, 1980.

[11] Davison E.J., "The Robust Decentralized Control of a General Servomechanism Problem", IEEE Tran. on Automatic Control, AC-21, 14-24, 1976.

[12] Vukobratović K.M., D.M.Stokić, D.S.Hristić, "Algorithmic Control of Anthropomorphic Manipulators", (in Russian), Izvestiya AN USSR, Teknicheskaya kibernetika, No 3, 1976.

[13] Kuleshev V.S., Lakhota N.A., Dynamics of Manipulator Control Systems, (in Russian), Energiya, Moscow, 1971.

[14] Artobolevskii I.I., Theory and Design of Manipulators, (in Russian), Proc. papers, edited by Artobolevskii, Nauka, Moscow, 1970.

[15] Popov E.P., Robots-Manipulators, (in Russian), Znanie, Moscow, 1974.

[16] Paul C.R., Modelling, Trajectory Calculation and Servoing of a Computer Controlled Arm. (in Russian), Nauka, Moscow, 1976.

[17] Vukobratović K.M., D.S.Hristić, D.M.Stokić, "Algorithmic Control of Anthropomorphic Manipulators", Proc. of V Intern. Symp. on Industrial Robots, Chicago, Illinois, Sept. 1975.

[18] Vukobratović K.M., et al., "New Method of Artificial Motion Synthesis and its Application to Locomotion Robots and Manipulators", Proc. of 7th IFAC Symp. on Automatic Control in Space, Munich, 1976.

[19] Vukobratović K.M., D.M.Stokić, N.V.Gluhajić, D.S.Hristić, "One Method of Control for Large-Scale Humanoid Systems", Mathematical Biosciences, Vol. 36, No 3/4, 1977.

[20] Vukobratović K.M., D.M.Stokić, "Significance of Force Feedback in Controlling Artificial Locomotion-Manipulation Systems", Transaction on Biomedical Engineering, December, 1980.

[21] Okhotskimskii D.E., et al., "Stabilization Algorithm of Automatic Locomotion System", (in Russian), Proc. VI IFAC Symp. on Automatic Control in Space, Erevan, USSR, 1974.

[22] McGhee B.R., C.S.Chao, V.C.Jaswa, D.E.Orin, "Real-Time Computer Control of a Hexapod Vehicle", Proc. of III CISM - IFToMM Symposium on Theory and Practice of Robots and Manipulators, Udine, 1978.

[23] Vukobratović K.M., A.Frank, D.Juričić, "On the Stability of Biped Locomotion", Trans. IEEE, Biomedical Engineering, January, 1970.

[24] Vukobratović K.M., Y.Stepanenko, "On the Stability of Anthropomorphic Systems", Mathematical Biosciences, Vol. 15, October, 1972.

[25] Vukobratović K.M., et al., "Analysis of Energy Demand Distribution within Anthropomorphic Systems", Trans. of the ASME, Journal of Dynamic Systems, Measurement and Control, Dec., 1973.

[26] Vukobratović K.M., "Dynamics and Control of Anthropomorphic Active Mechanisms", Proc. 1st IFToMM Symp. on Theory and Practice of Robots and Manipulators, Udine, Italy, 1973.

[27] Vukobratović K.M., D.M.Stokić, "Dynamic Stability of Unstable Legged Locomotion Systems", Mathematical Biosciences, Vol. 24, No 1/2, 1975.

[28] Vukobratović K.M., D.M.Stokić, "Postural Stability of Anthropomorphic Active Mechanisms", Mathematical Biosciences, Vol. 25., No 3/4, 1975.

[29] Vukobratović K.M., D.M.Stokić, "Dynamic Control of Unstable Locomotion Robots", (in Russian), Izvestiya AN USSR, Teknicheskaya kibernetika, No 5, 1975.

[30] Vukobratović K.M., et al., "Synergetic Control Principle of Anthropomorphic Movements", Proc. of Second IFToMM Symp. on Theory and Practice of Robots and Manipulators, Warsaw, 1976.

[31] Vukobratović K.M., D.M.Stokić, "Postural Stability of Anthropomorphic Systems", (in Russian), Izvestiya AN USSR Teknicheskaya kibernetika, No 5, 1978.

[32] Stokić M.D., M.K.Vukobratović, "Dynamic Control of Biped Posture", Mathematical Biosciences, Vol. 44, No 2, 1979.

[33] Beletski V.V., P.S.Tshudinov, "Nonlinear Models of Biped Gait", (in Russian), Repr. No. 19 of the USSR Academy of Sciences, 1975.

[34] Beletski V.V., Golubkov E.A., "Task of Modelling the Dynamics of Underwater Biped Gait", (in Russian), Reprint No. 42 of the USSR Academy of Sciences, 1979.

[35] Lawrence P.D., Cybernetic Coordination in The Control of a Powered Arm, Ph. Dissertation, Case Western Reserve University Cleveland, Ohio, 1970.

[36] Kobrinskii A.A., Kobrinskii A.E., "About Motion Design of Manipulation Systems", (in Russian), Izvestiya AN USSR, Teknicheskaya kibernetika No 2, 1978.

[37] Whitney D.E., "Resolved Motion Rate Control of Manipulators and Human Prosthesis", IEEE Trans. on M.M.S., Vol. 10, No 2, June, 1969.

[38] Renaud M., Contribution a L´etude de la Modelisation et de la Commande des Systemes Mechaniques Articules, Ph. Thesis, University "Paul Sabatier", Toulouse, 1975.

[39] Vereschagin A.F., V.L.Generozov, "Planning of the Actuator Trajectories of Manipulation Robot", (in Russian), Izvestiya AN USSR, Teknicheskaya kibernetika, No 2, 1978.

[40] Paul C.R., "Cartezian Coordinate Control of Robots in Joint Coordinates", Proc. III IFToMM Internat. Symp. on Theory and Practice of Robots and Manipulators, Udine, 1978.

[41] Luh S.J.Y., C.S.Lin, "Optimum Path Planning for Mechanical Manipulators", Journal of Dynamic Systems Measurement and Control, Trans. of the ASME, Vol. 103, No 2, pp. 142-151, 1981.

[42] Kahn M.E., B.Roth, "The Near Minimum Time Control of Open Loop Articulated Kinematic Chains", Trans. of the ASME, Journal of Dynamic Systems, Measurement and Control, September, 164-172, 1971.

[43] Bejczy A.K., Robot Arm Dynamic and Control, JET Propulsion Laboratory NASA Technical Memorandum 33-669 - February 15, 1974.

[44] Pavlov B.A., Timofeyev A.B., "Calculation and Stabilization of Programmed Motion of a Moving Robot-Manipulator", (in Russian), Teknicheskaya kibernetika, No. 6, 91-101, 1976.

[45] Timofeyev A.V., J.V.Ekalo, "Stability and Stabilization of Programmed Motion of Robots-Manipulators", (in Russian), Avtomatika and Telemechanika, No 10, 148-156, 1976.

[46] Saridis N.G., C.S.G.Lee, "An Approximation Theory of Optimal Control for Trainable Manipulators", IEEE Trans. on Systems, Man, and Cybernetics, Vol. SMC-9, No 3, pp. 152-159, 1979.

[47] Kulakov F.M., "Organization of Supervisory Control of Robots--Manipulators", (in Russian), Teknicheskaya kibernetika, Part I: No 5, 37-46, Part II: No 6, 78-90, 1976, Part III: No 1, 51-66, 1977.

[48] Takengaki M., S.Arimoto, "A New Feedback Method for Dynamic Control of Manipulators", Journal of Dynamic Systems, Measurement and Control, Trans. of the ASME, Vol. 103, No 2, pp. 119-125, 1981.

[49] Luh J.Y.S., Walker M.W., R.P.C.Paul, "On-line Computational Scheme for Mechanical Manipulators", Trans. of the ASME, Journal of Dynamic Systems, Measurement and Control, 1980.

[50] Luh J.Y.S., M.W.Walker and R.P.C.Paul, "Resolved-Acceleration Control of Mechanical Manipulators", IEEE Trans, on Automatic Control, Vol. AC-25, No 3, pp. 468-474, 1980.

[51] Popov E.P., Vereschagin A.F., Ivkin A.M., Leskov A.S., Medvedov V.S., "Synthesis of Control System of Robots Using Dynamic Models of Manipulation Mechanisms", (in Russian), Proc. VI IFAC Symp. on Autom. Contr. in Space, Erevan, USSR, 1974.

[52] Filaretov B.F., "Synthesis of Nonlinear Tracking Systems for Manipulator Control with Clearance, Dry Friction and Elastic Deformation", (in Russian), Teknicheskaya kibernetika, No 6, 58-67, 1976.

[53] Vukobratović K.M., D.M.Stokić, "Choice of Decoupled Control Law of Large Scale Mechanical Systems", Proc. of Second Symp. on Large-Scale Systems: Theory and Applications, Toulouse, 181-192, 1980, also in Large-Scale Systems-Theory and Application, June, 1981.

[54] Yuan J.S.C., "Dynamic Decoupling of a Remote Manipulator System", JACC, San Francisco, 1977.

[55] Freund E., "A Nonlinear Control Concept for Computer-Controlled Manipulators", Proc. of VII Intern. Symposium on Industrial Robots, Tokyo, 1977.

[56] Roessler J., "A Decentralized Hierarchical Control Concept for Large-Scale Systems", Proc. II Sympos. on Large-Scale Systems: Theory and Applications, Toulouse, 171-180, 1980.

[57] Hewit R.J., J.S.Burdess, "Fast Dynamic Decoupled Control for Robotics, Using Active Force Control", Mechanism and Machine Theory, Vol. 16, No 5, pp. 535-542, 1981.

[58] Young K.K.D., "Controller Design for a Manipulator Using Theory of Variable-Structure Systems", IEEE Trans. on Systems, Man and Cybernetics, Vol. SMC-8, 1978.

[59] Dubowsky S., D.T.DesForges, "The Application of Model-References Adaptive Control to Robotic Manipulators", Trans. of the ASME, Journal of Dynamic Systems, Measurement and Control, Vol. 101, No. 3, pp 193-200, 1979.

[60] Dubowsky S., "On the Dynamics of Computer Controlled Robotic Manipulators", Preprints of IV CISM - IFToMM Symp. on Theory and Practice of Robots and Manipulators, pp. 89-98, Warsaw, 1981.

[61] Nevins J., D.Whitney et al., Exploratory Research in Industrial Modular Assembly, Report R-996, The Charles Stark Draper Laboratory, Cambridge, Mass, 1976.

[62] Nevins J., D.Whitney, "Research Issues for Automatic Assembly", Proc. I IFAC/IFIP Symp. on Information Control Problems in Manufacturing Technology, Tokyo, 1977.

[63] Nevins J., T.B.Sheridan, D.F.Whitney, A.F.Woodin, "The Multi--Model Remote Manipulating System", Proc. I Conference on Remotely Manned Systems, Calif. Inst. of Technology, 1973.

[64] Bejczy A., "Manipulation of Large Objects", Proc. of III IFToMM Internat. Symp. on Theory and Practice of Robots and Manipulators, Udine, 1978.

[65] Nishimoto K., T.Uchiyama, K.Tamamushi, T.Akita, "Small-Sized Robot With High Positioning Accuracy", Preprints of IV CISM - IFToMM Symposium on Theory and Practice of Robots and Manipulators, pp. 170-179, Warsaw, 1981.

[66] Vukobratović M., Potkonjak V., "Elastic Properties in Dynamic Models of Industrial Manipulators", IFToMM Journal of Mechanism and Machine Theory, Vol. 16, No 2, 1982.

[67] Book J.W., O.Maizza-Meto, D.E.Whitney, "Feedback Control of Two Beam, Two Joint Systems With Distributed Flexibility", Journal of Dynamic Systems Measurement and Control, Trans. of the ASME, Vol. 97, pp. 424-431, December, 1975.

[68] Truckenbrodt A., "Modelling and Control of Flexible Manipulator Structure", Preprints of IV CISM-IFToMM Symposium on Theory and Practice of Robots and Manipulators, pp. 110-120, Warsaw, 1981.

[69] Raibert H.M., J.J.Craig, "Hybrid Position/Force Control of Manipulators", Journal of Dynamic Systems, Measurement and Control, Trans. of the ASME, Vol. 103, No 2, pp. 126-133, 1981.

[70] Vukobratović K.M., D.M.Stokić, "One Engineering Concept of Dynamic Control of Manipulators", Journal of Dynamic Systems, Measurement and Control, Trans. of the ASME, Vol. 103, No 2, pp. 108-118, 1981.

[71] Rosenbrock, N.H., P.D.McMorran, "Good, Bad, or Optimal", IEEE Trans. on Automatic Control, Vol. AC-16, December, 1971.

Chapter 1
Dynamics of Manipulators

The reason for writing this chapter on dynamics of manipulators lies in the author´s idea that this book should be autonomous in regard to the problems of setting the mathematical models of dynamics of manipulators and active mechanisms in general. Dynamics of manipulators and active spatial mechanisms has been presented in detail in [1].

In this chapter we present one of the existing computer-oriented methods for forming the mathematical models of manipulator dynamics. The chosen method is the one which was first formulated. It is based on the general theorems of mechanics. In para. 1.2. the complete model of dynamics of manipulation systems is presented, and in para. 1.3. we describe the computer-oriented method for linearization of dynamic models of open kinematic chains.

1.1. Method for the Construction of Dynamic Equation Based on the Basic Theorems of Mechanics

This is a fully automatic computer method, developed by Stepanenko [2] and Vukobratović, [3, 4]. The following facts should be noted:

1. The active spatial articulated mechanisms are mechanical multivariable systems of extremely complex and variable structure. The principle features are a large number of members and degrees of freedom of their relative motion and frequent changes of the kinematic chain configuration.

2. All previously developed methods are dedicated only for one narrow class of spatial or active spatial mechanisms [5 - 13].

3. Methods which enable one to explicitly obtain the motion equations demand lengthy analytic calculations on account of the complexity

of the models considered. The only exception is the method of E.P. Popov and associates [14], which has been adapted to digital computer programming under certain conditions.

The variable structure of active spatial mechanisms necessitates considering during the analysis of some locomotion or manipulation structure, two or more different kinematic chains. The many members and degrees of freedom even in the case of only one configuration of kinematic chain, requires tedious work during the analytic setting of motion equations. The need to consider various designs in the design phase of these mechanisms render this task practically pointless.

This means it is necessary to form some general calculating algorithm for mathematical modelling of the active spatial mechanisms on a digital computer.

According to the above, such an algorithm should fulfill the following requirements. It is only necessary to provide the information about the kinematic scheme of the mechanism, its parameters and the type of task to be solved. Based on this information, the algorithm should assemble the mechanism at the basis of the kinematic scheme, calculate the displacements, the velocities and the accelerations of the mechanism members, formulate the differential equations of motion and integrate them.

In the formulation of the algorithm, it is necessary to keep in mind the characteristics of the digital computers, particularly in regard to the extremely complicated numerical differentiation. The formulation of motion equations at the basis of the algorithm should not be connected with differentiation of any analytical expression. Hence, instead of the 2nd-order Lagrange equations [15], the kinetostatic procedure is used [16]. By applying this method, the problem of the dynamic analysis reduces to three phases.

The first phase is the kinematic analysis, in which the positions, linear velocities and accelerations of the centers of gravity, and the angular velocities and accelerations of all the mechanism members are calculated.

In the second phase, by means of particular values of the angular velocities and accelerations of the members, the main vector of the inertial forces and the main moment of the inertial forces are determined.

The third phase is the static analysis, in which the equilibrium problem of the mechanism is solved, given the action of the external and inertial forces. Here a specificity arises. In order to determine the linear and angular accelerations of the members, it is necessary to know the relative accelerations at each joint. However, when the formulation of the model is begun, these values are unknown. What is more, the determination of them is the first task. In this case, one can use the fact that the mechanical equations are linear with respect to their second derivatives. The vectors of the relative accelerations can then be calculated up to the constant and relative to these one can calculate the corresponding forces and moments of the inertia forces. According to D´Alembert´s principle, the mechanism must be in equilibrium under the influence of the external and inertial forces, so one can obtain the necessary and sufficient conditions for determining the unknown constants (multiplicators) and the real values of the relative accelerations. There are five phases of the algorithm. The first four phases serve for the formulation of differential motion equations and the fifth for their integration in one step.

1. For the given time instant, based on known relative displacements of the members at each joint (these displacements are the changes of the generalized coordinates q^i), the positions of all members and joint axes in the absolute coordinate system are determined. In this phase the program reduces to some relations from which one finds the projections of the vectors during their rotation about any axis.

2. By kinematic analysis one finds the angular accelerations and the accelerations of the centers of gravity of the mechanism members. Hence the program is in the form of the corresponding recursive vector equations which enables one to determine the necessary accelerations without differentiating the position vectors.

3. The forces and the moments of the inertial forces are calculated for all the mechanism members. The linear and the angular accelerations needed for this calculation, are determined in the second phase. The program runs according to the corresponding dynamic equations.

4. Equations of kinetostatic equilibrium are formulated by applying D´Alambert´s principle. In order to do that, the sum of the external forces and driving torques, together with the inertial forces and

moments of inertial forces, are determined for each member. The resulting vectors are reduced to the members´ centers of gravity. The kinematic chain of the mechanism is stepwise "ruptured" at each joint and the equilibrium conditions are set for the free chain end with respect to the axis of the "ruptured" joint. The program runs according to the algorithm for reducing all forces and moments to the corresponding axes.

5. The fourth phase yields a system of linear algebraic equations with respect to unknown accelerations $\ddot{q}(t)$. By solving this system of equations the real acceleration values are obtained. The formulation of the equations is then complete. Under the supposition that the accelerations obtained are constant during the short time interval Δt, the equations can be integrated and the values of the coordinates and velocities in the time instant $t_2 = t + \Delta t$ determined. These coordinates and velocities can be regarded as initial conditions in the time instant t_2. Using these initial conditions the construction of the differential equations for time t_2 can be performed and the process repeated*).

Each calculation therefore consists of two parts: the formulation of the differential motion equations and their integration. This procedure can be illustrated by the general algorithm block-scheme in Fig. 1.12.

It should be emphasized that in the course of applying this method to active spatial mechanisms, the complex kinematic pairs must be reduced to simple fifth class pairs using equivalent kinematic chains (Fig. 1.1). For example, a kinematic pair of the third class (spherical) can be presented in the form of a three-joint chain, connected by zero-length members (Fig. 1.2). Hence, the method always can consider members with simple joints, always supposing that for joints of other the substitution has already been performed. The simplest kinematic pair is illustrated in Fig. 1.1. Here two members of an anthropomorphic mechanism (i-1 and i) are connected by a rotational joint, the axis of which is determined by means of the unit vector $\vec{e}_i$. By $\vec{r}_{i-1,i}$ and $\vec{r}_{i,i}$ we denote the vectors from some point on the joint axis (conditionally, the axis is taken to pass through the center of the joint) to the center of gravity of the (i-1)st and i-th member, respectively. Fig. 1.1. illustrates one pair of that type. The joint center is marked by a

---

*) This concerns Euler´s integration method.

black circle and the centers of gravity by empty circles which contain the member numbers. A generalized coordinate q^i is chosen for the angle of relative rotation for a rotational kinematic pair. This angle is defined as the angle between the projections of the vectors $-\vec{r}_{i-1,i}$, $\vec{r}_{i,i}$ onto the plane perpendicular to the joint axis $\vec{e}_i$ (Fig. 1.1). The positive direction is counterclockwise for an observer from the $\vec{e}_i$ end. The angle is calculated from $-\vec{r}_{i-1,i}$. A special case can arise when one of the vectors $\vec{r}_{ij}$ defining the angle is parallel to the $\vec{e}_j$ joint axis.

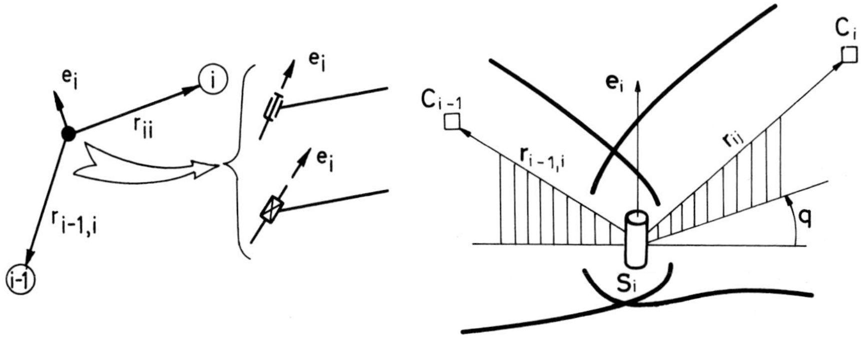

Fig. 1.1. Kinematic pair of the fifth class

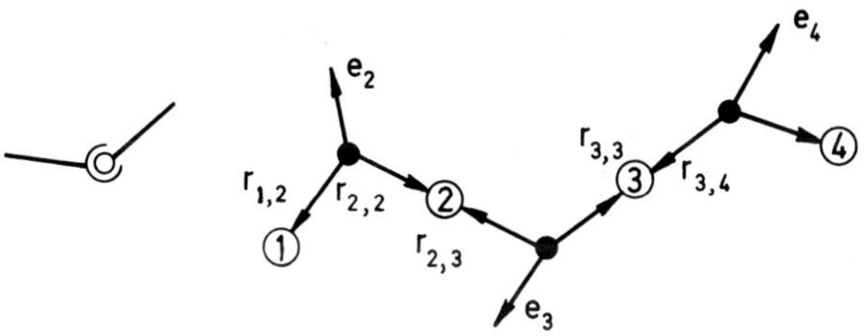

Fig. 1.2. Representation of the spherical pair by means of an equivalent kinematic chain

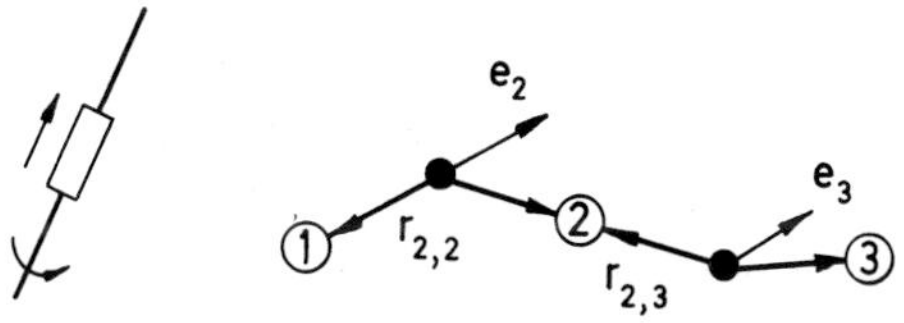

Fig. 1.3. Kinematic chain, equivalent to the forth class cylindrical type pair

Then its projection onto the plane, perpendicular to the joint axis coincides with the projection of this axis onto that plane, so determining the angle by means of the method described becomes impossible. Some unit vector $\vec{e}_{ij} \perp \vec{e}_j$ is therefore positioned at the i-th member. This vector $\vec{e}_{ij}$ is used instead of vector $(-\vec{r}_{ij})$ in the determination of the rotation angle. With the translatory kinematic pair (Fig. 1.3) one chooses as a generalized coordinate the linear displacement of the i-th member with respect to the (i-1)-st member in the direction of the unit vector $\vec{e}_i$.

1.1.1. Mechanism parameters

The mechanism members can be considered as rigid homogenous bodies of masses m_i and moments of inertia J_{i1}, J_{i2}, J_{i3}. The position of each member in the absolute coordinate system is determined by the projections of the vectors $\vec{r}_{ij}$ and $\vec{e}_i$ (index i denoting the member number). With the mechanism members is associated an internal coordinate system with normalized coordinate base $\vec{q}_{ij}$ (j=1,2,3). At each joint we add unit vector $\vec{e}_{ij}$ directed along the joint axis.

The vectors $\vec{e}_{ij}$ and $\vec{r}_{ij}$ (the latter is the vector from the joint center to the member center of gravity) can be written in the form of projections onto the axes of the internal coordinate system, i.e.,

$$\begin{aligned} \vec{\tilde{e}}_{ij} &= \tilde{e}^1_{ij}\vec{q}_{i1} + \tilde{e}^2_{ij}\vec{q}_{i2} + \tilde{e}^3_{ij}\vec{q}_{i3}, \\ \vec{\tilde{r}}_{ij} &= \tilde{r}^1_{ij}\vec{q}_{i1} + \tilde{r}^2_{ij}\vec{q}_{i2} + \tilde{r}^3_{ij}\vec{q}_{i3}, \end{aligned} \tag{1.1.1}$$

where $\tilde{e}^k_{ij}$ and $\tilde{r}^k_{ij}$ are the projections, the first index of which is the member number, the second index is the joint number and the upper index denotes the number of the projection onto the corresponding axis of the relative system. Accordingly, while the vectors $\vec{e}_j$ and $\vec{r}_{ij}$ determine the orientation of the joint axes and the distances of the members´ centers of gravity and the joint centers in the fixed coordinate system, (after "assembly" of the mechanism vectors) the vectors $\vec{\tilde{e}}_{ij}$ and $\vec{\tilde{r}}_{ij}$ represent the internal member parameters (Fig. 1.4). One can see that after the "mechanism assembly" the corresponding vectors $\vec{e}_{ij}$ and $\vec{\tilde{e}}_{ij}$ coincide.

1.1.2. Mechanism "assembly"

The "assembly" consists in uniting all the mechanism members, i.e., ensuring that the corresponding joint axes and their center coincide in space. After "assembly" it is necessary to determine the position of the mechanism in the absolute coordinate system, i.e., to calculate the following:

- the directions of the joint axes and the main axes of inertia of all the members ($\vec{e}_i$ and $\vec{q}_{ij}$),
- all the vectors $\vec{r}_{ij}$.

In the "assembly" of the mechanism it is supposed that all the relative coordinates (q^i) are zero. Let the position of the (i-1)-st member in the absolute system be given, i.e., the projections of vectors $\vec{r}_{i-1,i}$ and $\vec{e}_i$. That is, suppose the mechanism part up to the (i-1)-st member is already "assembled". It is now necessary to join to it the i-th member and determine the projections of vectors $\vec{r}_{ij}$ and $\vec{e}_j$ connected to this member. "Uniting" will consist in making the vectors $\vec{e}_i$ and $\vec{\tilde{e}}_{ii}$ coincide and determining the projections of the vectors $\vec{r}_{ij}$ and $\vec{q}_{ij}$ in the absolute coordinate system (Fig. 1.5).

Consider the case where the i-th joint is rotational. When the axes of both joints coincide, member i must be positioned in space in such way that angle $q^i = 0$. To determine q^i it is necessary to find the projections $\vec{r}_{i-1,i}$ and $\vec{r}_{ii}$ onto the plane perpendicular to $\vec{e}_i$. If $\vec{r}_i$ and $\vec{\tilde{r}}_i$ are reduced to their components perpendicular and parallel to $\vec{\tilde{e}}_i$, $\vec{\tilde{r}}_{iN}$ will be

$$\vec{r}_N = \vec{e} \times (\vec{r} \times \vec{e}),$$
$$\vec{\tilde{r}}_N = \vec{\tilde{e}} \times (\vec{\tilde{r}} \times \vec{\tilde{e}}). \tag{1.1.2}$$

The lower indexes are omitted for simplicity. After normalization of the components $\vec{r}_N$ and $\vec{\tilde{r}}_N$, and by changing the sign of $\vec{r}_N$, the unit vectors are obtained as $\vec{a}$ and $\vec{\tilde{a}}$:

$$\vec{a} = \frac{-\vec{r}_N}{|\vec{r}_N|}, \qquad \vec{\tilde{a}} = \frac{\vec{\tilde{r}}_N}{|\vec{\tilde{r}}_N|} \tag{1.1.3}$$

Since, after "assembly", q = 0, the coincident condition is

$$\vec{e} = A\vec{\tilde{e}}, \qquad \vec{a} = A\vec{\tilde{a}}, \tag{1.1.4}$$

where A is the transformation matrix (to be determined) from the internal to the absolute system. The matrix A has nine unknown elements, so a new vector, non-collinear with $\vec{e}$ and $\vec{a}$, is defined as follows

$$\vec{\tilde{b}} = \vec{\tilde{e}} \times \vec{\tilde{a}}; \qquad \vec{b} = \vec{e} \times \vec{a}; \qquad \vec{b} = A\vec{\tilde{b}} \tag{1.1.5}$$

If (1.1.4) and (1.1.5) are written in the form of projections:

$$\begin{aligned} &A^{11}\tilde{a}^1 + A^{12}\tilde{a}^2 + A^{13}\tilde{a}^3 = a^x \\ &A^{21}\tilde{a}^1 + A^{22}\tilde{a}^2 + A^{23}\tilde{a}^3 = a^y \\ &A^{31}\tilde{a}^1 + A^{32}\tilde{a}^2 + A^{33}\tilde{a}^3 = a^z \\ &A^{11}\tilde{e}^1 + A^{12}\tilde{e}^2 + A^{13}\tilde{e}^3 = e^x \\ &- - - - - - - - - - - - - - - - \\ &A^{31}\tilde{b}^1 + A^{32}\tilde{b}^2 + A^{33}\tilde{b}^3 = b^z \end{aligned} \tag{1.1.6}$$

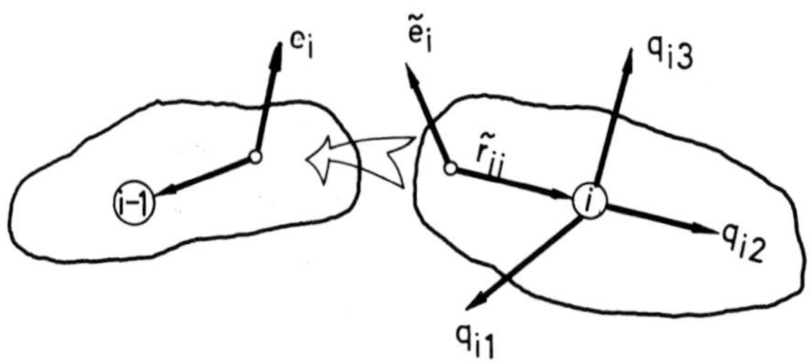

Fig. 1.4. Parameters of the internal and fixed coordinate systems

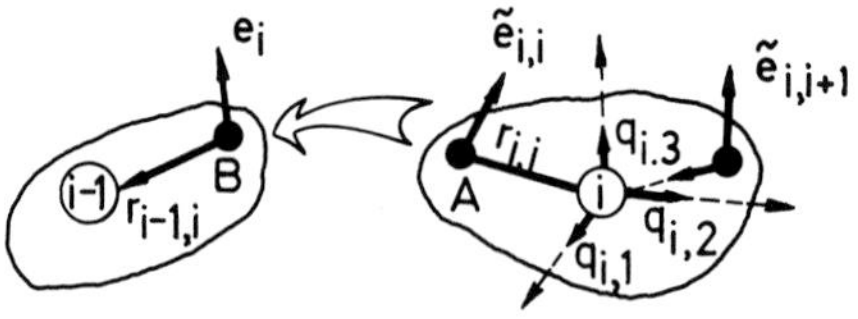

Fig. 1.5. Assembly the chain members

Solving this system with respect to A^{ij}, the expanded matrix B takes the form

$$B = \begin{bmatrix} \tilde{a}^1\ \tilde{a}^2\ \tilde{a}^3 & 0 & 0 & a^x \\ 0 & \tilde{a}^1\ \tilde{a}^2\ \tilde{a}^3 & 0 & a^y \\ 0 & 0 & \tilde{a}^1\ \tilde{a}^2\ \tilde{a}^3 & a^z \\ \tilde{e}^1\ \tilde{e}^2\ \tilde{e}^3 & 0 & 0 & e^x \\ 0 & \tilde{e}^1\ \tilde{e}^2\ \tilde{e}^3 & 0 & e^y \\ 0 & 0 & \tilde{e}^1\ \tilde{e}^2\ \tilde{e}^3 & e^z \\ \tilde{b}^1\ \tilde{b}^2\ \tilde{b}^3 & 0 & 0 & b^x \\ 0 & \tilde{b}^1\ \tilde{b}^2\ \tilde{b}^3 & 0 & b^y \\ 0 & 0 & \tilde{b}^1\ \tilde{b}^2\ \tilde{b}^3 & b^z \end{bmatrix} \qquad (1.1.7)$$

According to the non-collinearity condition, the system determinant is not equal to zero. The vectors $\vec{r}$ and $\vec{e}$ are obtained from (1.1.5). The projections of the unit vectors of the internal coordinate system ($\vec{q}_{ij}$) are the columns of matrix A. The block-scheme of the "assembly" algorithm of the i-th kinematic pair is given in Fig. 1.6. It is supposed that the (i-1)-st mechanism member has already been added to the mechanism.

1.1.3. Determining the position of the open chain

By means of the mechanism "assembly" we have determined the initial position of the kinematic chain, where all relative coordinates $q^i = 0$. To determine the new mechanism positions after a change of q^i, the so-called formula of finite rotations (Rodrigue´s formula) is used:

$$\vec{r}\,' = \vec{r}\cos q + (1-\cos q)(\vec{e}\cdot\vec{r})\vec{e} + \vec{e}\times\vec{r}\sin q, \qquad (1.1.8)$$

where $\vec{e}$ is the unit vector of axis, q is the turning angle, $\vec{r}\,'$ and $\vec{r}$ are the vectors after and before rotation.

Applying this formula to the unit vectors $\vec{q}_{ij}$, one obtains

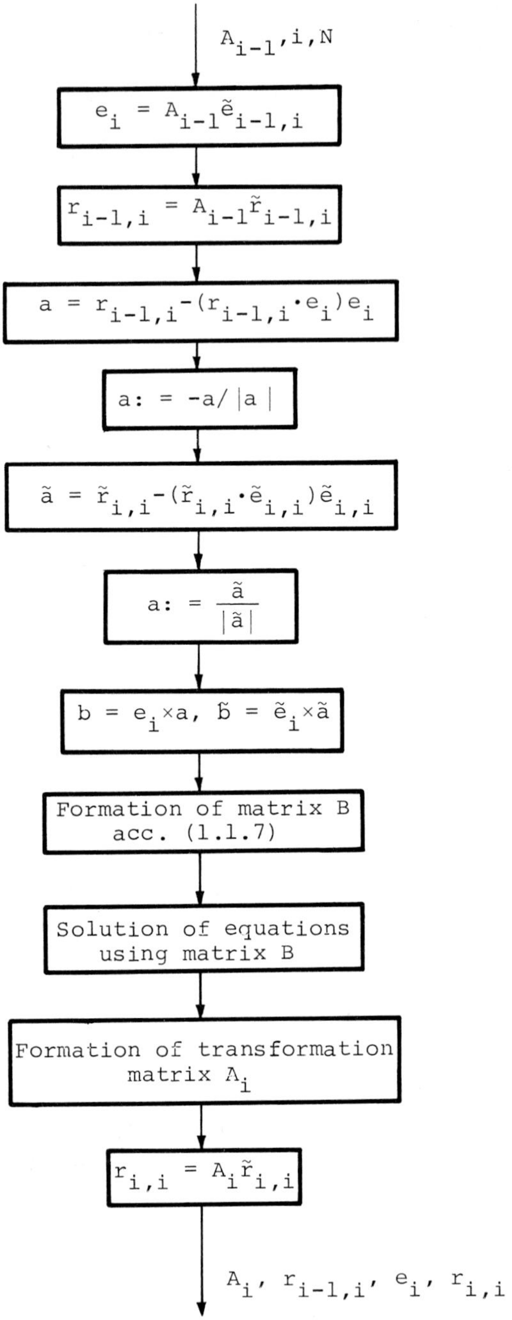

Fig. 1.6. Block-scheme for assembly kinematic pairs

$$\vec{q}'_{ij} = \vec{q}_{ij}\cos q^i + (1-\cos q^i)(\vec{e}_i\cdot\vec{q}_{ij})\vec{e}_i + \vec{e}_i \times \vec{q}_{ij}\sin q^i, \quad j=1,2,3$$

To determine the i-th member position one still has to determine $\vec{e}_i$ and $\vec{r}_{i,i}$ with respect to the absolute system, either by using the transformation matrix A_i or using Rodrigue's formula. Matrix A_i has the form

$$A_i = [\vec{q}'_{i1}\ \vec{q}'_{i2}\ \vec{q}'_{i3}]. \tag{1.1.9}$$

The new position of vector $\tilde{r}_{ii}$ with respect to the absolute system will be

$$\vec{r}'_{ii} = A_i\vec{\tilde{r}}_{ii}.$$

If a translatory kinematic pair is being considered, then, to determine the position, only the projections of vector $\vec{r}_{ii}$ change. For the case of rotation of the vector $\vec{r}$ about the unit vector $\vec{e}_i$ through the angle q, one uses the modified formula

$$\vec{r}' = \vec{r} + \frac{2\theta}{1+\theta^2}(\vec{r}+\vec{\theta}\times\vec{r}),$$

where $\vec{\theta} = \vec{e}\ \mathrm{tg}\ \frac{q}{2}$, and $\vec{r}'$ is the vector $\vec{r}$ after turning.

1.1.4. Determining velocities and accelerations

Suppose that the mechanism position is determined and that all vectors $\vec{e}$ and $\vec{r}$ are known. The angular velocities and accelerations $\vec{\omega}_i$ and $\vec{\varepsilon}_i$ and the translatory velocities and accelerations $\vec{v}_i$ and $\vec{w}_i$ must be determined for all members. Using the rule of general rigid body motion, which states that the angular velocity equals the sum of the transfer and relative angular velocity,

$$\vec{\omega} = \vec{\omega}_p + \vec{\omega}_r,$$

and that the translatory velocity of some point on the rigid body is the sum of the transfer (pole) velocity and relative rotation around that pole,

$$\vec{v}_B = \vec{v}_A + \vec{\omega} \times \vec{r}_{AB},$$

all the members' velocities can be obtained. Here, the motion of each preceding member is considered as transferred and the motion of the member considered as relative. For a linear kinematic pair (Fig. 1.7),

$$\vec{\omega}_i = \vec{\omega}_{i-1},$$
$$\vec{v}_i = \vec{v}_{i-1} - \vec{\omega}_{i-1} \times \vec{r}_{i-1,i} + \vec{\omega}_i \times \vec{r}_{ii} + \dot{q}^i \vec{e}_i. \tag{1.1.10}$$

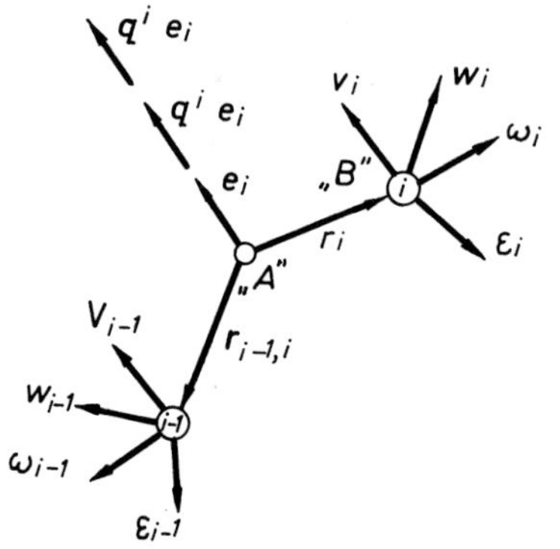

Fig. 1.7. Velocities and accelerations of the kinematic pair

For a rotational kinematic pair,

$$\vec{\omega}_i = \vec{\omega}_{i-1} + \dot{q}^i \vec{e}_i,$$
$$\vec{v}_i = \vec{v}_{i-1} - \vec{\omega}_{i-1} \times \vec{r}_{i-1,i} + \vec{\omega}_i \times \vec{r}_{i,i}.$$

For the accelerations of points during free motion of the rigid body the following relations hold

$$\vec{\varepsilon} = \vec{\varepsilon}_p + \vec{\varepsilon}_r + \vec{\omega}_p \times \vec{\omega}_r,$$
$$\vec{w}_B = \vec{w}_A + \vec{\varepsilon} \times \vec{r}_{AB} + \vec{\omega} \times (\vec{\omega} \times \vec{r}_{AB})$$

If the kinematic pair between the (i-1)-st and i-th member is linear,

$$\vec{\varepsilon}_i = \vec{\varepsilon}_{i-1}, \quad \vec{w}_i = \vec{w}_{i-1} - \vec{\varepsilon}_{i-1} \times \vec{r}_{i-1,i} - \vec{\omega}_{i-1} \times (\vec{\omega}_{i-1} \times \vec{r}_{i-1,i}) + \vec{\varepsilon}_i \times \vec{r}_{ii} + \vec{\omega}_i \times (\vec{\omega}_i \times \vec{r}_{ii}) + \ddot{q}^i \vec{e}_i + 2\dot{q}^i \vec{\omega}_{i-1} \times \vec{e}_i. \tag{1.1.11}$$

For a rotational kinematic pair the angular and translatory acceleration of the i-th member center of gravity are

$$\vec{\varepsilon}_i = \vec{\varepsilon}_{i-1} + \ddot{q}^i\vec{e}_i + \dot{q}^i\vec{\omega}_{i-1} \times \vec{e}_i,$$

$$\vec{w}_i = \vec{w}_{i-1} - \vec{\varepsilon}_{i-1}\times\vec{r}_{i-1,i} - \vec{\omega}_{i-1}\times(\vec{\omega}_{i-1}\times\vec{r}_{i-1,i}) + \vec{\varepsilon}_i\times\vec{r}_{ii} + \vec{\omega}_i\times(\vec{\omega}_i\times\vec{r}_{i,i}). \tag{1.1.12}$$

The expressions for velocities and accelerations of the mechanism members´ centers of gravity are recursive. By increasing the index number of the member by one, one can calculate the velocities and accelerations of all members. As we said, the relative displacements q^i and relative velocities $\dot{q}^i$ are known, while the relative accelerations are still to be determined. The expressions for accelerations (1.1.11) and (1.1.12), which are functions of the second-order derivatives of the relative coordinates, should therefore be rearranged without changing their recursive character. Thus,

$$\vec{\varepsilon}_i = \sum_{j=1}^{i} \vec{\alpha}_{ij}\ddot{q}^j + \vec{\theta}_i + \vec{\varepsilon}_o,$$

$$\vec{w}_i = \sum_{j=1}^{i} \vec{\beta}_{ij}\ddot{q}^j + \vec{\eta}_i + \vec{w}_o, \tag{1.1.13}$$

where $\vec{\varepsilon}_o$ and $\vec{w}_o$ are the angular and linear accelerations of the basic member, while the vector coefficients $\vec{\alpha}_{ij}$ and $\vec{\beta}_{ij}$ are functions of the relative coordinates, and vectors $\vec{\theta}_i$ and $\vec{\eta}_i$ are functions of relative coordinates and velocities. By writing $\vec{\varepsilon}_i$ and $\vec{\varepsilon}_{i-1}$ in modified form in the expression for angular acceleration of the rotational kinematic pair (1.1.12), one obtains

$$\sum_{j=1}^{i} \vec{\alpha}_{ij}\ddot{q}^j + \vec{\theta}_i = \sum_{j=1}^{i-1} \vec{\alpha}_{i-1,j}\ddot{q}^j + \ddot{q}^i\vec{e}_i + \dot{q}^i\vec{\omega}_{i-1} \times \vec{e}_i + \vec{\theta}_{i-1}.$$

Equating the coefficients with index j one obtains

$$\vec{\alpha}_{ij} = \vec{\alpha}_{i-1,j}, \qquad \text{for} \quad j \neq i,$$

$$\vec{\alpha}_{ii} = \vec{e}_i, \tag{1.1.14}$$

$$\vec{\theta}_i = \vec{\theta}_{i-1} + \dot{q}^i\vec{\omega}_{i-1} \times \vec{e}_i.$$

Likewise, for a linear kinematic pair,

$$\vec{\alpha}_{ij} = \vec{\alpha}_{i-1,j}, \qquad \text{for} \quad j \neq i,$$

$$\vec{\alpha}_{ii} = 0, \tag{1.1.15}$$

$$\vec{\theta}_i = \vec{\theta}_{i-1}.$$

Substituting the modified expressions for $\vec{w}_{i-1}$ and $\vec{w}_i$ into the expressions for linear accelerations (1.1.11) and (1.1.12), we find the following:

- for a rotational kinematic pair

$$\vec{\beta}_{ij} = \vec{\beta}_{i-1,j} + \vec{\alpha}_{i-1,j} \times (\vec{r}_{i,i}-\vec{r}_{i-1,i}), \qquad \text{for} \quad j \neq i,$$

$$\vec{\beta}_{ii} = \vec{e}_i \times \vec{r}_{i,i}, \tag{1.1.16}$$

$$\vec{\eta}_i = \vec{\eta}_{i-1} + \vec{\theta}_{i-1} \times (\vec{r}_{ii}-\vec{r}_{i-1,i}) + \dot{q}^i(\vec{\omega}_{i-1}\times\vec{e}_i) \times \vec{r}_{ii} +$$

$$+ \vec{\omega}_i \times (\vec{\omega}_i\times\vec{r}_{ii}) - \vec{\omega}_{i-1} \times (\vec{\omega}_{i-1}\times\vec{r}_{i-1,i}).$$

- for a linear kinematic pair,

$$\vec{\beta}_{ij} = \vec{\beta}_{i-1,j} + \vec{\alpha}_{i-1,j} \times (\vec{r}_{ii}-\vec{r}_{i-1,i}), \qquad \text{for} \quad j \neq i,$$

$$\vec{\beta}_{ii} = \vec{e}_i \tag{1.1.17}$$

$$\vec{\eta}_i = \vec{\eta}_{i-1} + \vec{\theta}_{i-1} \times (\vec{r}_{ii}-\vec{r}_{i-1,i}) + 2\dot{q}^i\vec{\omega}_{i-1} \times \vec{e}_i +$$

$$+ \vec{\omega}_i \times (\vec{\omega}_i\times\vec{r}_{ii}) - \vec{\omega}_{i-1} \times (\vec{\omega}_{i-1}\times\vec{r}_{i-1,i}).$$

One thus obtains the recursive relations for determining all the coefficients.

1.1.5. Determining the forces and moments of inertial forces

The main vector $\vec{F}_i$ and main moment of inertial forces $\vec{M}_i$ are reduced to the i-th member center of gravity. The inertial force of the i-th member, according to relation $\vec{F}_i = -m_i\vec{w}_i$, can be written as

$$\vec{F}_i = \sum_{j=1}^{i} \vec{a}_{ij}\ddot{q}_j + \vec{a}_i^o - m_i\vec{w}_o, \tag{1.1.18}$$

where $\vec{a}_{ij} = -m_i\vec{\beta}_{ij}$, $\vec{a}_i^o = -m_i\vec{\eta}_i$.

The moment of inertial forces can be determined by means of Euler's dynamic equations:

$$\begin{aligned} \tilde{M}_i^1 &= -J_{i1}\tilde{\varepsilon}_i^1 + (J_{i2}-J_{i3})\tilde{\omega}_i^2\tilde{\omega}_i^3 , \\ \tilde{M}_i^2 &= -J_{i2}\tilde{\varepsilon}_i^2 + (J_{i3}-J_{i1})\tilde{\omega}_i^3\tilde{\omega}_i^1 , \\ \tilde{M}_i^3 &= -J_{i3}\tilde{\varepsilon}_i^3 + (J_{i1}-J_{i2})\tilde{\omega}_i^1\tilde{\omega}_i^2 , \end{aligned} \tag{1.1.19}$$

where the tilde indicates that the vectors $\vec{M}$, $\vec{\varepsilon}$ and $\vec{\omega}$ are taken as projections onto the internal (moving) coordinate system, the upper index showing the projection number, i.e.,

$$\tilde{\varepsilon}_i^j = \vec{\varepsilon}_i \cdot \vec{q}_{ij}.$$

For transfer to the fixed system, the transformation matrix of the i-th member, A_i, is used:

$$\vec{M}_i = A_i \begin{bmatrix} \tilde{M}_i^1 \\ \tilde{M}_i^2 \\ \tilde{M}_i^3 \end{bmatrix} \tag{1.1.20}$$

By multiplying the left and right sides of expressions (1.1.19) with A_i and substituting the projections of accelerations $\tilde{\varepsilon}_i^j$ by the above expressions, the projections of the moments of the inertial forces onto the axes of the absolute system are obtained:

$$\begin{aligned} M_i^j = -[&(A_i^{j1}J_{i1}q_{i1}^1 + A_i^{j2}J_{i2}q_{i2}^1 + A_i^{j3}J_{i3}q_{i3}^1)\varepsilon_i^1 + \\ &+ (A_i^{j1}J_{i1}q_{i1}^2 + A_i^{j2}J_{i2}q_{i2}^2 + A_i^{j3}J_{i3}q_{i3}^2)\varepsilon_i^2 + \\ &+ (A_i^{j1}J_{i1}q_{i1}^3 + A_i^{j2}J_{i2}q_{i2}^3 + A_i^{j3}J_{i3}q_{i3}^3)\varepsilon_i^3] + \lambda_i^j, \quad j=1,2,3, \end{aligned} \tag{1.1.21}$$

where

$$\vec{\lambda}_i \equiv \begin{bmatrix} \lambda_i^1 \\ \lambda_i^2 \\ \lambda_i^3 \end{bmatrix} = A_i \begin{bmatrix} (J_{i2}-J_{i3})(\vec{\omega}_i\vec{q}_{i2})(\vec{\omega}_i\vec{q}_{i3}) \\ (J_{i3}-J_{i1})(\vec{\omega}_i\vec{q}_{i3})(\vec{\omega}_i\vec{q}_{i1}) \\ (J_{i1}-J_{i2})(\vec{\omega}_i\vec{q}_{i1})(\vec{\omega}_i\vec{q}_{i2}) \end{bmatrix}. \tag{1.1.22}$$

Since $A_i^{jk} = q_{ik}^j$, the expression for the moment of inertial forces of the i-th member can be written as

$$\vec{M}_i = -T_i\vec{\varepsilon}_i + \vec{\lambda}_i \ , \tag{1.1.23}$$

where matrix T_i (3×3) has the components

$$T_i^{jk} = \sum_{\ell=1}^{3} A_i^{j\ell} J_{i\ell} q_{i\ell}^k = \sum_{\ell=1}^{3} q_{i\ell}^j q_{i\ell}^k J_{i\ell} \tag{1.1.24}$$

Substituting the preceding expressions for angular accelerations, one obtains:

$$\vec{M}_i = \sum_{j=1}^{i} \vec{b}_{ij}\ddot{q}^j + \vec{b}_i^o - T_i\vec{\varepsilon}_o, \tag{1.1.25}$$

where $\vec{b}_{ij} = -T_i\vec{\alpha}_{ij}$, $\vec{b}_i^o = -T_i\vec{\theta}_i + \vec{\lambda}_i$.

Fig. 1.8. illustrates the block-scheme of the algorithm for calculating forces and moments of inertial forces. The algorithm calculates the vectors $\vec{a}_{ij}$, $\vec{b}_{ij}$, $\vec{a}_i^o$, $\vec{b}_i^o$, T_i by means of formulae (1.1.18) - (1.1.25).

When the length of members is five or more times their diameter, instead of equations (1.1.25) the moment of inertial forces can be written directly in the fixed coordinate system, in which the projections of the angular velocity onto the fixed axes do not figure. For this purpose, the notion of equivalent angular acceleration $\vec{\tau}$ is introduced. This has the property that the moment of inertial forces due to its action is equal to the moment of inertial forces due to angular velocity $\vec{\omega}$. Thus,

$$\vec{\tau}_i = (\vec{\omega}_i \cdot \vec{S}_i)(\vec{S}_i \times \vec{\omega}_i), \tag{1.1.26}$$

where $\vec{S}_i$ is the unit vector directed along the cane´s axis. The moment of inertial forces is

$$\vec{M}_i = -J_i(\vec{\varepsilon}_i + \vec{\tau}_i),$$

where J_i is the tensor of inertia. By reducing $\vec{M}_i$ and $\vec{\omega}_i$ to components perpendicular to the cane´s axis (index N) and parallel to it (index S) one finds that

$$\vec{M}_{iN} = (\vec{S}_i \times \vec{M}_i) \times \vec{S}_i; \quad \vec{M}_{iS} = (\vec{M}_i \cdot \vec{S}_i)\vec{S}_i,$$

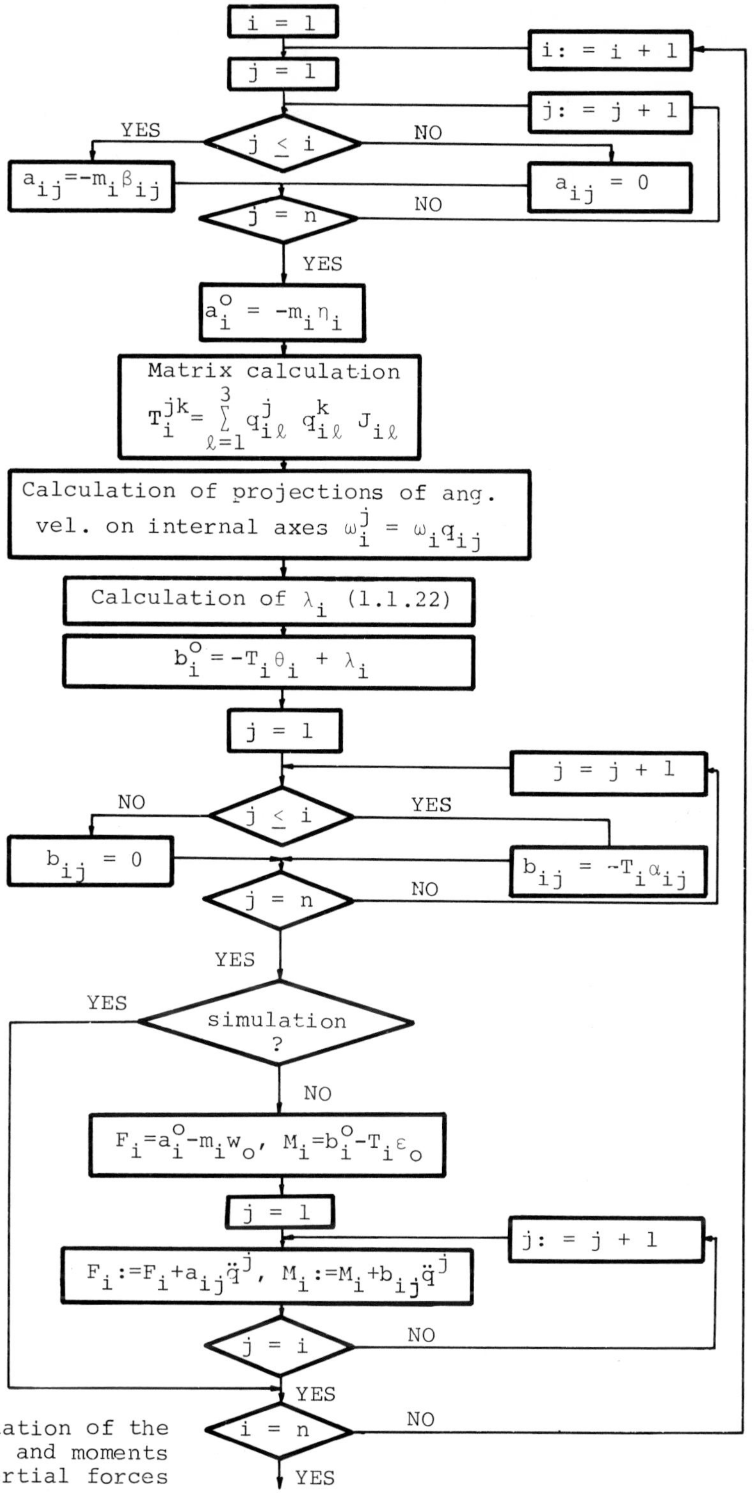

Fig. 1.8. Calculation of the forces and moments of inertial forces

$$\vec{\varepsilon}_{iN} = (\vec{S}_i \times \vec{\varepsilon}_i) \times \vec{S}_i ; \qquad \vec{\varepsilon}_{iS} = (\vec{\varepsilon}_i \cdot \vec{S}_i)\vec{S}_i .$$

Since $\vec{\tau}_i$ is perpendicular to the i-th member axis (1.1.26),

$$\vec{M}_i = -J_{iN}(\vec{\varepsilon}_{iN}+\vec{\tau}_i) - J_{iS}\vec{\varepsilon}_{iS}.$$

When the i-th member can be considered as a cane, then using the preceding equation instead of (1.1.25), the computation speed is increased two to three times. Introducing (1.1.26) into the expression for the cane moment, the vector of the inertial forces moment can be written in the form

$$\vec{M}_i = \sum_{j=1}^{i} \vec{C}_{ij}\ddot{q}^j + \vec{C}_i^o ,$$

where $\vec{C}_{ij}$ and $\vec{C}_i^o$ are the vector coefficients, which are being calculated according to the preceding expressions for $\vec{M}_{iN}$, $\vec{\varepsilon}_{iN}$, $\vec{M}_i$ and (1.1.26).

1.1.6. Kinetostatic procedure for obtaining differential motion equations

Let the kinematic chain be fictively ruptured at the i-th joint and consider the equilibrium of the mechanism free end (Fig. 1.9). The action of the rejected mechanism part is substituted by forces and reaction moments $\vec{M}_i^e$, $\vec{M}_i^N$, $\vec{R}_i^e$, $\vec{R}_i^N$. The driving torque in the "ruptured" joint is divided into the collinear and perpendicular component to the joint axis: $\vec{M}_i^e$ and $\vec{M}_i^N$. The force components are collinear $\vec{R}_i^e$ and perpendicular $\vec{R}_i^N$ to the joint axis $\vec{e}_i$. When considering the kinetostatic equilibrium the external forces and moments of inertial forces of the rejected chain members are reduced to the center of the i-th ("ruptured") joint. According to D´Alambert´s principle, the force and moment after the reduction are balanced by the reactions. The resulting force and moment of the j-th member are given by

$$\vec{\Gamma}_j' = \vec{F}_j + \vec{G}_j , \qquad \vec{L}_j' = \vec{M}_j + \vec{M}_j^G , \qquad \vec{G}_j = \vec{\Gamma}_j , \qquad \vec{M}_j^G = \vec{L}_j ,$$

where $\vec{G}_j$ is the external force and $\vec{M}_j^G$ the external moment. Introducing (1.1.18) and (1.1.25) one obtains

$$\vec{\Gamma}_j' = \sum_{k=1}^{j} \vec{a}_{jk}\ddot{q}^k + \vec{a}_j^o - m_j\vec{w}_o + \vec{G}_j ,$$

$$\vec{L}'_j = \sum_{k=1}^{j} \vec{b}_{jk}\ddot{q}^k + \vec{b}^o_j - T_j\vec{\varepsilon}_o + \vec{M}^G_j.$$

To obtain the reactions, it is appropriate to establish the recursive relations between the reactions of two adjacent joints. From Fig. 1.10a, where one mechanism member is illustrated, it is evident that

$$\vec{R}_i = \vec{R}_{i+1} - \vec{\Gamma}'_i,$$
$$\vec{M}_i = \vec{M}_{i+1} + (\vec{r}_{ii}-\vec{r}_{i,i+1}) \times \vec{R}_{i+1} - \vec{r}_{ii} \times \vec{\Gamma}'_i - \vec{L}'_i \qquad (1.1.27)$$

Circling the chain from the last to the first member according to (1.1.27), all reactions can be determined. The chain is "ruptured" gradually, progressing from the last to the first (basic) member. The corresponding forces $\vec{\Gamma}'_j$ and moments $\vec{L}'_j$ are therefore reduced to the joints illustrated in Fig. 1.9. Reducing $\vec{\Gamma}'_j$ and $\vec{L}'_j$ to the center of the i-th joint and to their components parallel and perpendicular to $\vec{e}_i$, one determines the values of reactions (Fig. 1.10b)

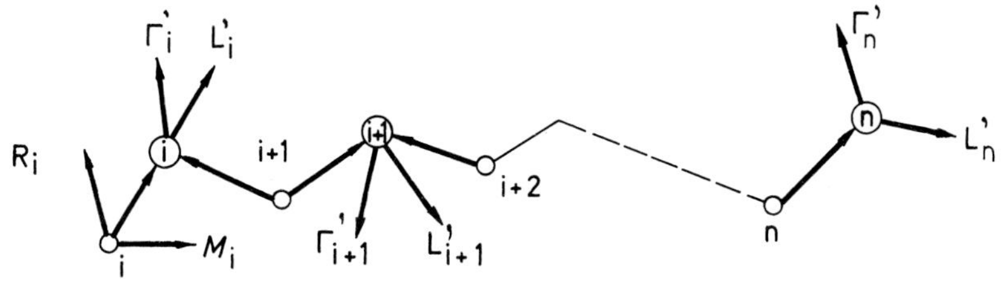

Fig. 1.9. Equilibrium of the free chain

$$\vec{R}_i = -\sum_{j=i}^{n} \vec{\Gamma}'_j = -\sum_{(j)} \left(\sum_{k=1}^{j} \vec{a}_{jk}\ddot{q}^k + \vec{a}^o_j - m_j\vec{w}_o + \vec{G}_j \right)$$

$$\vec{R}^e_i = (\vec{R}_i\vec{e}_i)\vec{e}_i; \quad \vec{R}^N_i = \vec{R}_i - \vec{R}^e_i \qquad (1.1.28)$$

$$\vec{M}_i = -\sum_{j=i}^{n} (\vec{L}'_j+\vec{r}_{ji}\times\vec{\Gamma}'_j) = -\sum_{(j)} \Big[\sum_{k=1}^{j} (\vec{b}_{jk}+\vec{r}_{ji}\times\vec{a}_{jk})\ddot{q}^k + \vec{r}_{ji} \times \vec{a}^o_j -$$
$$- m_j\vec{r}_{ji} \times \vec{w}_o + \vec{b}^o_j - T_j\vec{\varepsilon}_o + \vec{M}^G_j + \vec{r}_{ji} \times \vec{G}_j \Big]$$

$$\vec{M}^e_i = (\vec{M}_i\vec{e}_i)\vec{e}_i; \quad \vec{M}^N_i = \vec{M}_i - \vec{M}^e_i$$

These expressions enable us to determine the drives in the i-th joint. Here, the normal components cannot produce mechanism motion. Forces and driving torques depend on the collinear components. If the i-th joint is rotational,

$$\vec{P}_i^M = M_i^e \, \vec{e}_i = P_i^M \vec{e}_i, \tag{1.1.29}$$

and for the linear kinematic pair,

$$\vec{P}_i^F = R_i^e \, \vec{e}_i = P_i^F \vec{e}_i.$$

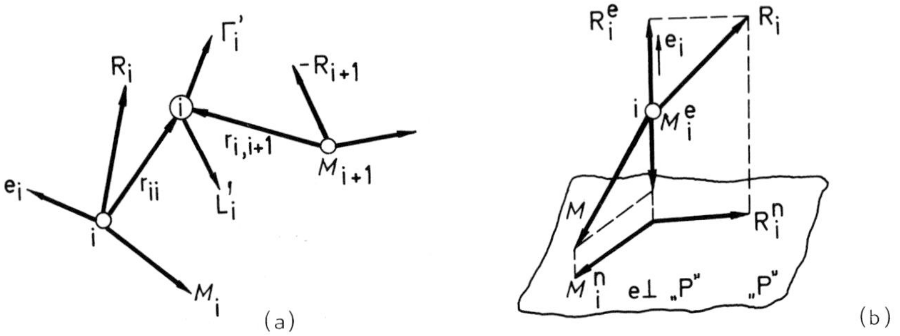

Fig. 1.10. Reactions of the i-th member

The block-scheme for determining reactions is given in Fig. 1.11. The algorithm calculates the reactions at all joints in terms of two components (parallel and perpendicular to the axis of the kinematic pair). The algorithm is based on relations (1.1.27) and (1.1.28). The value of the cycle counter i corresponds to the joint number and circling of all joints starts from the chain end.

The preceding expression for the equilibrium of moments in the i-th joint represents a differential equation of motion. Writing this equation for all mechanism joints, one finds that

$$H\ddot{q} + h = P, \tag{1.1.30}$$

where $H = [H_{ik}]$; $h = [h_1 \ldots h_n]^T$; $P = [P_1 \ldots P_{n-1} P_n]^T$; $\ddot{q} = [\ddot{q}^1 \ldots \ddot{q}^n]^T$

$$H_{ikj} = -\vec{e}_i(\vec{b}_{jk} + \vec{r}_{ji} \times \vec{a}_{jk}), \quad \text{where } P_i \text{ denotes either } P_i^F \text{ or } P_i^M,$$

$$h_{ij} = -\vec{e}_i(\vec{r}_{ji} \times \vec{a}_j^o - m_j \vec{r}_{ji} \times \vec{w}_o + \vec{b}_j^o - T_j \vec{\varepsilon}_o + \vec{M}_j^G + \vec{r}_{ji} \times \vec{G}_j),$$

$$H_{ik} = \sum_{(j)} H_{ikj}; \quad h_i = \sum_{(j)} h_{ij},$$

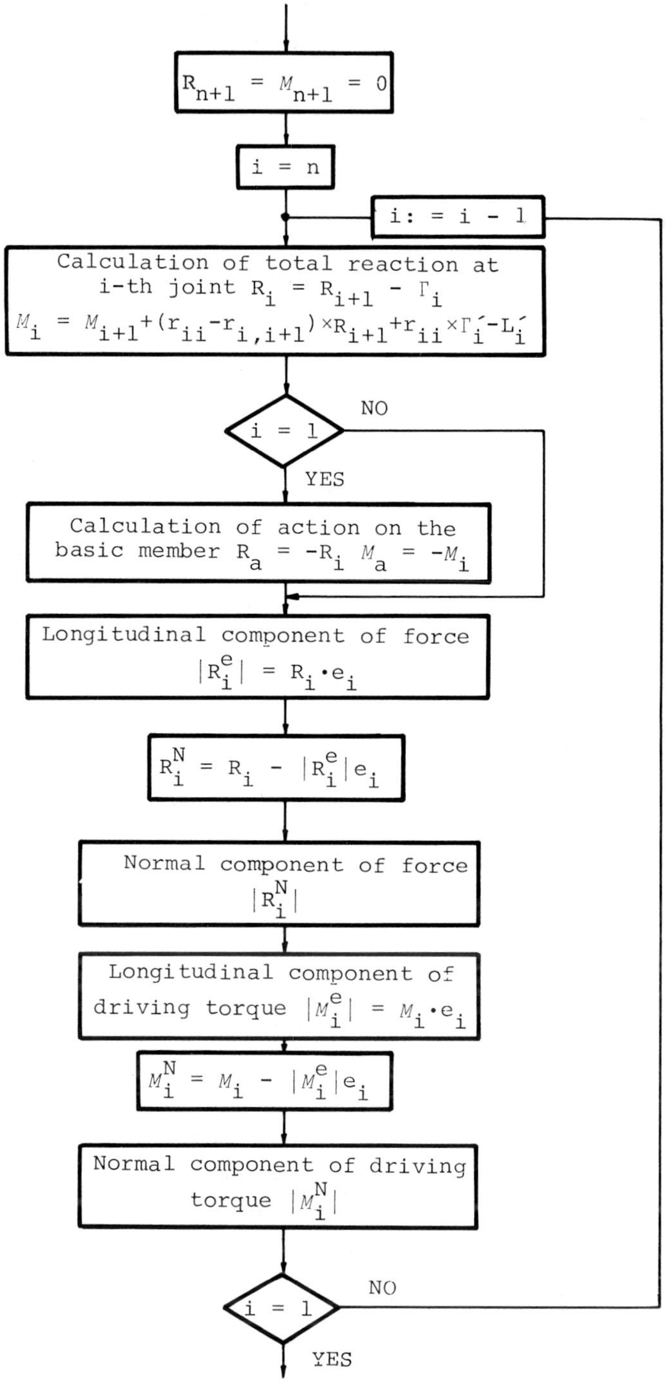

Fig. 1.11. Block-scheme of algorithm for calculating reactions

The motion equations can be differently derived in a slightly modified form. Using the virtual displacement principle, one obtains the analogous equations

$$\sum_{i=1}^{n} (\vec{\Gamma}_i'\vec{v}_{ij} + \vec{L}_i'\vec{\omega}_{ij} + \vec{P}_i\dot{\vec{q}}_{ij}) = 0, \qquad (1.1.31)$$

where the index j of the variables $\vec{v}_i$, $\vec{\omega}_i$ and $\dot{\vec{q}}_i$ indicates that they have been produced by the virtual velocity $\dot{q}^j$.

Equation (1.1.31) holds both for an open and closed kinematic chain. For the open chain, relative motion is realized in the i-th joint only and the preceding member stays fixed. Taking $\dot{q}^i=1$ for simplicity, one finds that for the open kinematic chain

$$\sum_{i=1}^{n} (\vec{\Gamma}_i'\vec{v}_{ij} + \vec{L}_i'\vec{\omega}_{ij}) = -\vec{P}_j\vec{e}_i; \quad (j=1,2,\ldots,n). \qquad (1.1.32)$$

Substituting expressions (1.1.27) and (1.1.28) into (1.1.31) and collecting the terms containing $\ddot{q}$, the matrix differential equation

$$H\ddot{q} = H_1 P + H_2 w_o + H_3 \varepsilon_o + \zeta \qquad (1.1.33)$$

is obtained, where H, H_1, H_2, H_3 are n×n, n×n, n×3 and n×3 matrices, respectively. The elements of these matrices are:

$$H^{ij} = \sum_{k=1}^{n} (\vec{v}_{ki}\vec{a}_{kj} + \vec{\omega}_{ki}\vec{b}_{kj}),$$

$$H_1^{ij} = -\dot{q}_{ij} \text{ (the virtual velocity in direction of } \vec{e}_j), \qquad (1.1.34)$$

$$H_2^{ij} = \sum_{k=1}^{n} m_k v_{ki}^j, \quad H_3^{ij} = -\sum_{k=1}^{n} (T_k\vec{\omega}_{ki})^j,$$

where index j indicates that the j-th component of these vectors is taken into account. The components of vector ζ from (1.1.33) are

$$\zeta_i = -\sum_{k=1}^{n} (\vec{v}_{ki}\vec{a}_k + \vec{\omega}_{ki}\vec{b}_k), \qquad (1.1.35)$$

where

$$\vec{a}_k = \vec{a}_k^o + \vec{G}_k, \quad \vec{b}_k = \vec{b}_k^o + \vec{M}_k^G.$$

H_1 is a diagonal matrix with components equal to the virtual velocities.

If the virtual velocities are equal to unity, H_1 becomes a unit matrix, so (1.1.33) becomes

$$\begin{aligned} H\ddot{q} &= P + H_2 w_o + H_3 \varepsilon_o + \zeta, \quad \text{i.e.,} \\ \ddot{q} &= AP + Bw_o + C\varepsilon_o + D, \end{aligned} \tag{1.1.36}$$

where $A=H^{-1}$, $B=H^{-1}H_2$, $C=H^{-1}H_3$, $D=H^{-1}\zeta$.

Fig. 1.12. illustrates the block-scheme of the algorithm for mechanism modelling on a digital computer. Here it is supposed that the motion law ($\vec{v}_o$, $\vec{w}_o$, $\vec{\omega}_o$, $\vec{\varepsilon}_o$, $\vec{R}_1$) of the basic mechanism members is known. If the chain is connected to the base, the first four vectors ($\vec{v}_o,\ldots,\vec{\varepsilon}_o$) are equal to zero. The block-scheme illustrated contains two algorithms previous considered for "assembling" the chain and forming differential equations of motion.

To put the model into a working system, it is necessary to form a few auxiliary blocks, "serving" the model. By means of these blocks information is prepared and introduced about the following.

1. Mechanism structure and parameters (block 1),
2. Type of task (block 2),
3. Working mode (block 3 - 5).

The algorithm illustrated is sufficiently general because it has not been prepared for any particular mechanism. Beside the basic laws of mechanics it contains no specific properties. The model acquires the feature of a particular mechanism after data is introduced about the mechanism structure and its parameters. In the next phase the algorithm "assembles" the selected mechanism, etc. (see block-sheme in Fig. 1.12).

1.2. The Complete Dynamical Model of Manipulation Systems

Manipulation systems, and robots in general, consist of the mechanical part of the system S^M and actuators S^i driving the degrees of freedom of the mechanism. The previous paragraph described one method for producing a mathematical model of dynamics of open kinematic chains, representing the mechanical part of the manipulator system. As shown, the mathematical model of dynamics of active mechanism S^M with n d.o.f. can

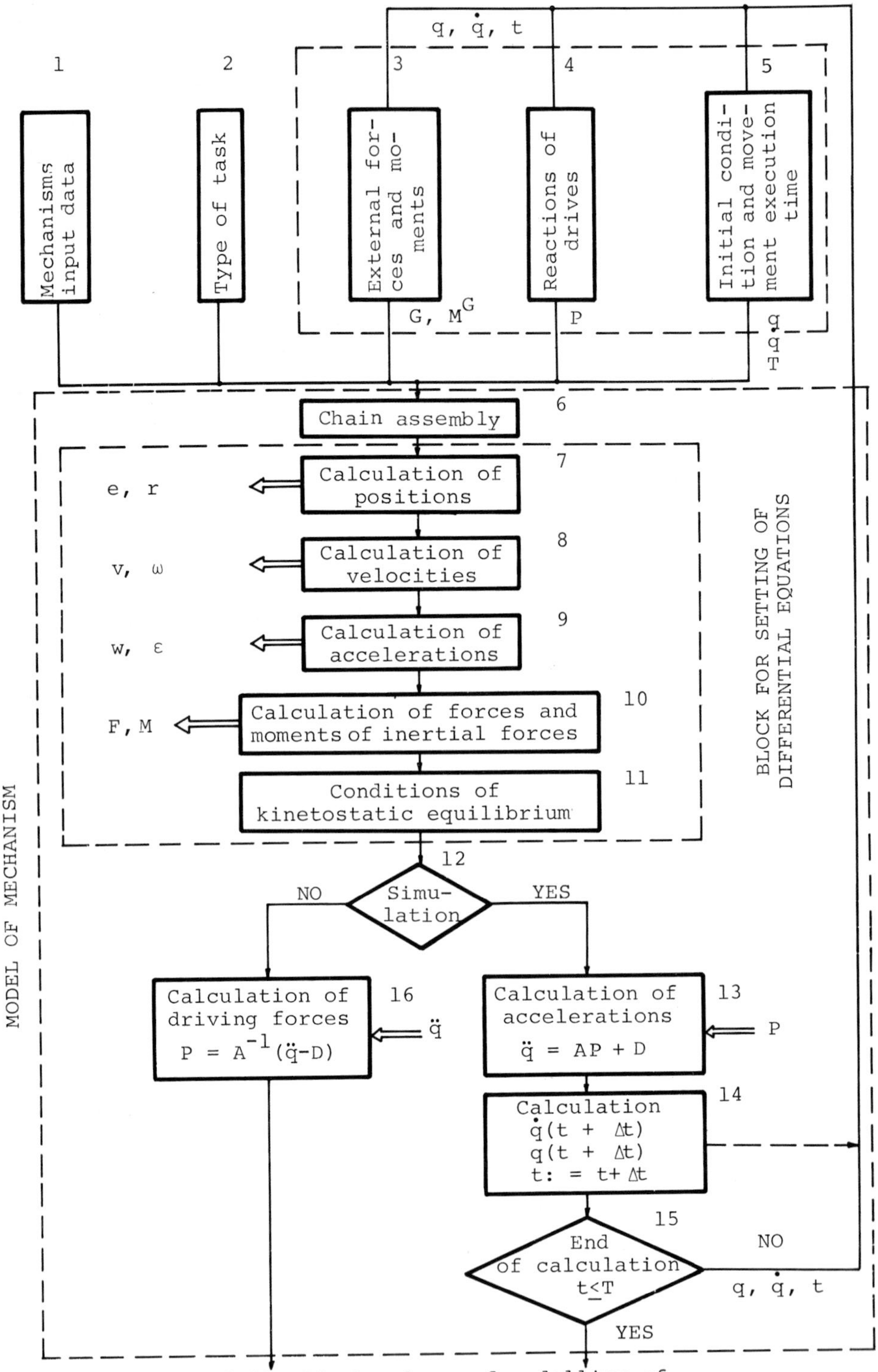

Fig. 1.12. Block scheme of modelling of spatial open-chain configuration

be written in the form (1.1.30).

In the general case all d.o.f. of manipulation systems need not be powered by separate actuators. However, with industrial manipulators the most common situation is that all d.o.f. are powered by separate actuators, so in this book we shall only consider this case. The general case, when the number of actuators is less then the number of d.o.f., is presented in [17].

In the general case, the models of actuators can be given in the form of nonlinear systems. Here, we shall consider the particular case, which is most frequently encountered in practice, when the actuators' dynamics can be presented by linear, time-independent systems. Thus,

$$S^i : \dot{x}^i = A^i x^i + b^i N(u^i) + f^i P_i, \qquad \forall i \varepsilon I, \tag{1.2.1}$$

where $x^i \varepsilon R^{n_i}$ is the state vector of the i-th actuator model, n_i is the order of the i-th actuator (which drives the i-th d.o.f. of the mechanism), $u^i \varepsilon R^1$ is the input of the i-th actuator, $P_i \varepsilon R^1$ is the driving force (or torque) acting at the i-th d.o.f., $A^i \varepsilon R^{n_i \times n_i}$ is the matrix of subsystem S^i, $b^i \varepsilon R^{n_i}$ is the input distribution vector, $f^i \varepsilon R^{n_i}$ is the load distribution vector, $N(u^i)$ is the nonlinearity of the amplitude saturation type:

$$N(u^i) = \begin{cases} -u_m^i & \text{for} \quad u^i < -u_m^i \\ u^i & \text{for} \quad -u_m^i \leq u^i \leq u_m^i \\ u_m^i & \text{for} \quad u^i > u_m^i \end{cases} \tag{1.2.2}$$

and $I = \{i: i=i,2,...,n\}$.

The mathematical model of the system S is given by the model of the mechanism S^M(1.1.30) and the models of actuators S^i(1.2.1), for all $i\varepsilon I$. Let us consider the case when, in each state vector of S^i - x^i, two coordinates coincide with q^i and $\dot{q}^i$, the angle (displacement) and angular (linear) velocity respectively of the i-th d.o.f. The state vector of the complete system S is given by $x = (x^{1T}, x^{2T}, ..., x^{nT})^T$, while the order of the system S is $N = \sum_{i=1}^{n} n_i$, $x\varepsilon R^N$; i.e., $R^N = R^{n_1} \times R^{n_2} \times ... R^{n_n}$. The system input is given by $u = (u^1, u^2, ..., u^n)^T$, $u\varepsilon R^n$.

The model of the complete system S can be obtained by uniting the model of the system S^M and the models of the actuators S^i. Since in each vector x^i one coordinate coincides with $\dot{q}^i$, the vector P (1.1.30) can be expressed by the state vector of the system x as:

$$S^M : P = (I_n - HTF)^{-1}[HT(Ax + BN(u)) + h], \tag{1.2.3}$$

where $T \in R^{n \times N}$, $T = \mathrm{diag}(T^i)$, $T^i \in R^{1 \times n_i}$, $\ddot{q}^i = T^i \dot{x}^i$, $A \in R^{N \times N}$, $A = \mathrm{diag}(A^i)$, $B \in R^{N \times n}$, $B = \mathrm{diag}(b^i)$, $F \in R^{N \times n}$, $F = \mathrm{diag}(f^i)$, and $N(u) = (N(u^1), N(u^2), \ldots, N(u^n))^T$. Introducing (1.2.3) in (1.2.1) and by uniting the models, we obtain the model of the systems S in the form

$$S : \dot{x} = \hat{A}(x) + \hat{B}(x)N(u), \tag{1.2.4}$$

where $\hat{A} : R^N \to R^N$, $\hat{A}(x) = Ax + F(I_n - HTF)^{-1}[HTAx + h]$, $\hat{B} : R^N \to R^{N \times n}$, $\hat{B} = B + F(I_n - HTF)^{-1}HTB$.

In this way we obtain the mathematical model of manipulation and robot systems which are considered in this book. Utilizing the procedure for automatic formulation of the mathematical model of active mechanisms described in the previous paragraph, based on the adopted models of the actuators (1.2.1), the mathematical model of the complete manipulation system S dynamics (1.2.4) can be set on a computer and used to analyse system performance and control synthesis for solving particular manipulation tasks.

It should be mentioned that in the general case the relation between the coordinates q^i, $\dot{q}^i$ of the mechanical part of the system S^M and the coordinates of the state vector x^i need not be linear or such that the two state coordinates in x^i coincide with q^i, $\dot{q}^i$, as we assumed above. In the general case, the relation between q^i, $\dot{q}^i$ and x^i is given by the nonlinear functions

$$q^i = g_1^i(x^i), \quad \dot{q}^i = g_2^i(x^i), \qquad i \in I \tag{1.2.5}$$

where $g_1^i : R^{n_i} \to R^1$, $g_2^i : R^{n_i} \to R^1$. If as before we take the state vector in the form $x = (x^{1T}, x^{2T}, \ldots, x^{nT})^T$, then the model of the mechanical part of the system S can be expressed by the state vector x as

$$S^M : P = (I_n - HG(x)F)^{-1}[HG(x)(Ax + BN(u)) + h], \tag{1.2.6}$$

where the notation is the same as in (1.2.3) and $G(x):R^N \to R^{n\times N}$, $G(x) = \mathrm{diag}(G^i(x^i))$, $G^i(x^i):R^{n_i} \to R^{1\times n_i}$, $G^i(x^i) = \partial g_2^i(x^i)/\partial x^i$. Introducing (1.2.6) into (1.2.1) and by uniting these models we obtain the model of the system S in the form analogous to (1.2.4), where the system matrices are now given by $\hat{A} = Ax + F(I_n - HG(x)F)^{-1}[HG(x)Ax + h]$, $\hat{B} = B + F(I_n - HG(x)F)^{-1} \cdot HG(x)B$. Obviously, the general case can be treated like the particular case when $g_1^i = T_1^i x^i$ and $g_2^i = T^i x^i$, where $T_1^i \varepsilon R^{1\times n_i}$. In what follows we shall restrict ourselves to this particular case, which is encountered whenever actuators are applied which have the same character of motion as the corresponding d.o.f. (for example, the application of the D.C. motors for rotational d.o.f., or application of translational hydraulic actuators for linear d.o.f., etc.). We shall generalize the results in ch. 3. and apply them to a particular manipulation system (which uses hydraulic actuators to drive the rotational joints).

1.3. Computer Oriented Method for Linearization of Dynamic Models of Open Kinematic Chains

In the previous paragraph the mathematical model of the dynamics of the open kinematic chain was given in the case when all d.o.f. are powered by appropriate actuators (1.2.4). This general model of the system dynamics can be utilized to investigate performances of different configurations, control synthesis, system design and so on. For the system performance analysis, and for the control synthesis, it is often convenient to have a linearized model of the system dynamics, where the linearization is performed around some point in the system state space or around a particular trajectory $x^o(t)$ of the system. As will be shown in ch. 2, the control synthesis for the manipulation systems is performed in two stages. The first stage synthesizes the nominal trajectory $x^o(t)$, which the manipulation system is supposed to realize. In the second stage, the model of the deviation of the system from the nominal trajectory $x^o(t)$ is considered. In order to apply various well-known methods of control synthesis of linear systems as well as linear theory to the system analysis, it is desirable to obtain the linearized model of the system dynamics around the nominal trajectory.

The linearized model of the manipulation systems dynamics can be determined in several ways. If the exact nonlinear model of the system dynamics (1.2.4) is set on the computer, then it is possible, applying various procedures, to identify the linearized model of the system. However, the procedures for identification of complex large-scale systems require lengthy computing time and the results obtained are not very reliable. It is therefore useful to linearize the models of the manipulation systems dynamics analytically. However, as we said in previous paragraphs the mathematical models of active mechanisms might be very complex, so it is convenient to develop some computer - oriented algorithms for constructing the linearized model of the active mechanism. Such an approach, based on "composite matrices" was given by E.P.Popov and associates [14].

We briefly present a procedure for computer construction of linearized models of open-chain active mechanisms. According to our adopted approach, the linearized model is constructed in parallel to the dynamic equations for nominal mode. The algorithm developed is based on the method of general theorems of mechanics but the same procedure may be carried out for algorithms based on other methods [18].

The nominal trajectory for the mechanical part of the system must satisfy the equation

$$S^M : P^o = H(q^o)\ddot{q}^o + h(q^o, \dot{q}^o), \quad q^o(0), \dot{q}^o(0) \text{ are given} \qquad (1.3.1)$$

where $P^o \varepsilon R^n$ is the vector of the nominal driving torques (forces), $q^o \varepsilon R^n$ is the vector of the nominal generalized coordinates, $H(q^o): R^n \to R^{n\times n}$ and $h(q^o, \dot{q}^o): R^n \times R^n \to R^n$ are matrices of the system in nominal dynamic mode.

The mathematical model of the deviation of the system state from the nominal trajectory is given in the form

$$S^M : \Delta P = H^o(t, \Delta q)\Delta\ddot{q} + h(t, \Delta q, \Delta\dot{q}), \quad \Delta q(0), \Delta\dot{q}(0) \text{ are given} \qquad (1.3.2)$$

where $\Delta q \varepsilon R^n$, $\Delta q = q(t) - q^o(t)$ is the deviation vector of the generalized coordinates, $\Delta P \varepsilon R^n$, $\Delta P = P(t) - P^o(t)$ is the deviation vector of torques (forces), $H^o: T\times R^n \to R^{n\times n}$ and $h^o: T\times R^n \times R^n \to R^n$ are matrices of the deviation model. However, if only small perturbations are present (the case most frequently encountered in industrial manipulation) the nonlinear model (1.3.2) may be replaced by its first-order approximation

$$S_1^M : \delta P = H(q^o)\delta\ddot{q} + h_v(q^o, \dot{q}^o)\delta\dot{q} + h_p(q^o, \dot{q}^o, \ddot{q}^o)\delta q, \qquad (1.3.3)$$

where $\delta q \varepsilon R^n$ is a small deviation of the vector of the generalized coordinates from the nominal trajectory, $\delta P \varepsilon R^n$ is the deviation of the vector of driving torques from $P^o(t)$, $h_v : R^n \times R^n \to R^{n\times n}$, $h_p : R^n \times R^n \times R^n \to R^{n\times n}$ are matrices of the linearized system given by

$$h_v(q^o, \dot{q}^o) = \frac{\partial h(q^o, \dot{q}^o)}{\partial \dot{q}} \qquad (1.3.4)$$

$$h_p(q^o, \dot{q}^o, \ddot{q}^o) = \frac{\partial H(q^o)}{\partial q}\ddot{q}^o + \frac{\partial h(q^o, \dot{q}^o)}{\partial q} \qquad (1.3.5)$$

By solving the model dynamic equations (1.3.3) with respect to $\delta\ddot{q}$, which is possible on account of the regularity of matrix $H(q^o)$, and by introducing the state vector $\xi = [\delta q^T \; \delta\dot{q}^T]^T$ of the mechanical part of the system, the model (1.3.3) is obtained in the form

$$S_1^M : \dot{\xi} = \begin{bmatrix} 0 & I_n \\ -H^{-1}h_p & -H^{-1}h_v \end{bmatrix}\xi + \begin{bmatrix} 0 \\ H^{-1} \end{bmatrix}\delta P, \quad \xi(0) \text{ is given} \qquad (1.3.6)$$

where I_n is unit $n\times n$ matrix.

It is assumed that an actuator S^i is associated with each joint (degree of freedom) of the mechanism. The model of deviation from the nominal for each actuator can be written in the form

$$S_1^i : \delta\dot{x}^i = A^i\delta x^i + f^i\delta P^i + b^i N(t, \delta u^i), \; \delta x^i(0) \text{ is given} \qquad (1.3.7)$$

where δx^i, δP^i and δu^i are deviations of the corresponding vectors.

The linearized model of the complete system S is obtained by uniting the linearized model of the mechanical part S_1^M and the model of the actuators S_1^i into an integral mathematical model. This is done by a simple mathematical operation,

$$S_1 : \delta\dot{x} = \tilde{A}^o(x^o(t), u^o(t))\delta x + \tilde{B}^o(x^o(t))N(t, \delta u), \; x(0) \text{ is given} \qquad (1.3.8)$$

where $\tilde{A}^o : R^N \times R^n \to R^{N\times N}$, $\tilde{B}^o : R^N \to R^{N\times n}$ are matrices of the linearized model.

The models (1.3.6) and (1.3.7), as well as the integral model (1.3.8), represent a basis for the control synthesis of open-chain active mechanisms in the presence of perturbations (see ch. 2).

The problem considered is the development of a computer-oriented procedure for evaluation of the matrices of the linearized model (1.3.3) of open-chain active spatial mechanisms: $h_p(q^o, \dot{q}^o)$ and $h_v(q^o, \dot{q}^o, \ddot{q}^o)$. In addition, the algorithm should not involve operations which might produce cumulative errors (for instance, numerical differentation). In order to form the matrices h_p and h_v, it is necessary, according to (1.3.4) and (1.3.5), to form matrices of partial derivatives

$$\frac{\partial H(q^o)}{\partial q}, \quad \frac{\partial h(q^o, \dot{q}^o)}{\partial q}, \quad \frac{\partial h(q^o, \dot{q}^o)}{\partial \dot{q}}, \tag{1.3.9}$$

where $q^o(t):T\rightarrow R^n$ is a given continuous nominal trajectory.

The algorithm should construct the matrices (1.3.9) in parallel with the matrices $H(q^o)$ and $h(q^o, \dot{q}^o)$ of the basic model (1.1.30).

The following notation will be used: $J = \{j:j=1,\ldots,i\}$ denotes a set of indices,

$$\eta_{\ell i} = \sum_{k=1}^{i} \delta_{\ell k}, = \begin{cases} 1, & \text{if } i \geq \ell, \\ 0, & \text{otherwise,} \end{cases}$$

where $\delta_{\ell k}$ is the Kronecker delta ($\delta_{\ell k} = 1$ if $\ell = k$, and 0 otherwise), ξ_i is a symbol whose value is 1 if the i-th kinematic pair is translatory, and 0, if it is rotational, and $\Delta_\ell A = \frac{\partial A}{\partial q^\ell}$, $\Delta_{\dot{\ell}} A = \frac{\partial A}{\partial \dot{q}^\ell}$ are partial derivatives of the variable A with respect to the coordinates q^ℓ and $\dot{q}^\ell$.

The following relations will also be used.

The partial derivative of a noncomposite vector $\vec{a}_i$ *), $i\in I$, with respect to the generalized coordinate q^ℓ, $\ell \in I$, is given by

$$\Delta_\ell \vec{a}_i = (\vec{e}_\ell \times \vec{a}_i)(1 - \xi_\ell)\eta_{\ell i}, \tag{1.3.10}$$

where $\vec{e}_\ell$ denotes the unit vector of the joint axis of ℓ-th kinematic pair.

---

*) Noncomposite vectors $\vec{a}_i$ are vectors that are composed of vectors related to the i-th member of a mechanism.

When the i-th kinematic pair is rotational ($\xi_\ell = 0$), $i \geq \ell$ and if $\eta_{\ell i}=1$ (Fig. 1.4), the problem of determining

$$\Delta_\ell \vec{a}_i = \lim_{\Delta q^\ell \to 0} \frac{\Delta \vec{a}_i}{\Delta q^\ell} \Bigg|_{\substack{\Delta q^j = 0 \\ (j \neq \ell)}}$$

reduces to the well-known theorem of elementary rotations in theoretical mechanics. According to this theorem we obtain $\lim\limits_{\Delta q^\ell \to 0} \frac{\Delta \vec{a}_i}{\Delta q^\ell} = \vec{e}_\ell \times \vec{a}_i$. This proves (1.3.10) for $\xi_\ell = 0$ and $\ell \leq i$. It is obvious from Fig. 1.4. that the value of $\Delta_\ell \vec{a}_i$ is 0 for $\ell > i$. In addition, for a translatory pair ℓ ($\xi_\ell = 1$), the value of $\vec{a}_i$ does not change, i.e., $\Delta_\ell \vec{a}_i = 0$, which completes the proof of (1.3.10).

The partial derivative of vector $\vec{r}_{ij}$, $i \varepsilon I$, $j \varepsilon J$, connecting the joint of j-th kinematic pair and the center of gravity of i-th mechanism link, with respect to the generalized coordinate q^ℓ, $\ell \varepsilon I$, is given by

$$\Delta_\ell \vec{r}_{ij} = [(\vec{e}_\ell \times \vec{r}_{ij})\eta_{\ell,j-1} + (\vec{e}_\ell \times \vec{r}_{i\ell})(\eta_{\ell i} - \eta_{\ell,j-1})](1-\xi_\ell) +$$

$$+ \vec{e}_\ell(\eta_{\ell i} - \eta_{\ell,j-1})\xi_\ell \qquad (1.3.11)$$

Using the relation $\vec{r}_{ij} = \vec{r}_{i\ell} + \vec{R}_{\ell j}$, where $\vec{R}_{\ell j}$ is the vector from the joint j to the joint ℓ, according to (1.3.10), $\Delta_\ell \vec{r}_{ij}$ is given by (1.3.11).

The following relation will also be used.

The partial derivative of vector or scalar product of two vectors (which may be composite) $\vec{a}_i$ and $\vec{b}_i$ is given by

$$\Delta_\ell(\vec{a}_i \times \vec{b}_i) = (\Delta_\ell \vec{a}_i) \times \vec{b}_i + \vec{a}_i \times (\Delta_\ell \vec{b}_i). \qquad (1.3.12)$$

This is one of the basic formulae in vector analysis.

According to the method of general theorems, the joint angles are generalized coordinates q^i, $i \varepsilon I$, of the mechanism. As we mentioned in para. 1.1, a Cartesian coordinate system $S_i = (\vec{q}_{i1}, \vec{q}_{i2}, \vec{q}_{i3})$ is associated with each link of the mechanism, with the origin in its center of gravity (Fig. 1.4).

Let us briefly describe the principal stages of the general theorems approach to the construction of the nonlinear model. At the same time, the stages will be expanded by relations corresponding to the developed algorithm. A complete flow-chart of the primary and developed algorithm is given in Fig. 1.13.

The first stage of the general theorems approach envolves the "assembly" of the links of i-th kinematic pair. It is assumed that the first link of the i-th pair is already "assembled" i.e., that its initial position ($q^j = 0$, $j \in J$) in the absolute coordinate system is determined. The i-th pair is assembled by making the vectors $\vec{e}_i$ and $\vec{\tilde{e}}_i$ coincide, assuming $q^i = 0$. As shown in the flow-chart (Fig. 1.13), the matrix $A_i^o = [\vec{q}_{i1}^{\,o}, \vec{q}_{i2}^{\,o}, \vec{q}_{i3}^{\,o}]$ is easily determined. Its coloumns are vectors $\vec{q}_{ij}^{\,o} = \vec{q}_{ij}$, in the initial position of the link i*).

According to the algorithm developed, instead of partial derivation we use (1.3.10). Since the vectors $\vec{q}_{ij}^{\,o}$ are noncomposite, it follows that

$$\Delta_\ell \vec{q}_{ij}^{\,o} = (\vec{e}_\ell \times \vec{q}_{ij}^{\,o})(1-\xi_\ell)\eta_{\ell i}, \tag{1.3.13}$$

$$\Delta_\ell A_i^o = [\Delta_\ell \vec{q}_{i1}^{\,o} \; \Delta_\ell \vec{q}_{i2}^{\,o} \; \Delta_\ell \vec{q}_{i3}^{\,o}]^T, \tag{1.3.14}$$

where $i, \ell \in I$, $j \in \{1, 2, 3\}$. According to the definition of $\eta_{\ell i}$, the relation (1.3.13) may be written in the form

$$\Delta_\ell \vec{q}_{ij}^{\,o} = (\vec{e}_\ell \times \vec{q}_{ij}^{\,o})(1-\xi_\ell), \qquad \ell \in J \tag{1.3.15}$$

while for all the remaining indices $\ell \in \{i+1, \ldots, n\}$, $\Delta_\ell \vec{q}_{ij} = 0$.

In the next stage of the primary algorithm the position of the i-th mechanism link is determined when $q^i = q_i^o$, where q_i^o is the given value of the i-th generalized coordinate. The position of the i-th link is determined using the projections of the vectors $\vec{r}_{ii}$, $\vec{r}_{i,i+1}$, $\vec{r}_{ij}$, $i \in I$, $j \in J$ in the absolute coordinate system. These vectors, as well as the vector $\vec{e}_{i+1}$, are determined by applying Rodrigue´s finite rotations formula.

Since all the vectors (except $\vec{r}_{ij}$) formed at this stage are noncomposite, their partial derivatives are determined by (1.3.10). The partial

---

*) By $\vec{q}_{ij}^{\,o}$ we denote the vector $\vec{q}_{ij}$ in the phase of mechanism assembly ($q^i = 0$).

INPUT PARAMETERS

$i = 1$

(1)

block 1, "assembly"

$r_{Ni} = e_i\times(r_{i-1,i}\times e_i)$, $\tilde r_{Ni} = \tilde e_i\times(\tilde r_{ii}\times\tilde e_i)$, $a_i = -r_{Ni}/|r_{Ni}|$, $\tilde a_i = \tilde r_{Ni}/|\tilde r_{Ni}|$

$b_i = e_i\times a_i$, $\tilde b_i = \tilde e_i\times\tilde a_i$, $A_i^0 = [q_{i1}^0\ q_{i2}^0\ q_{i3}^0] = [a_i\ e_i\ b_i][\tilde a_i\ \tilde e_i\ \tilde b_i]^T$

block 1´

$\Delta_\ell q_{ij}^0 = (e_\ell\times q_{ij}^0)(1-\xi_\ell)\eta_{\ell i}$, $\Delta_\ell A_i^0 = [\Delta_\ell q_{i1}^0\ \Delta_\ell q_{i2}^0\ \Delta_\ell q_{i3}^0]$

block 2, "positions"

$q_{ij} = [q_{ij}^0\cos q^i + (1-\cos q^i)(e_i\ q_{ij}^0)e_i + e_i\ q_{ij}^0\sin q^i](1-\xi_i) + q_{ij}^0\xi_i$, $(j=1,2,3)$

$A_i = [q_{i1}\ q_{i2}\ q_{i3}]$, $r_{ii} = A_i\tilde r_{ii} + q^i e_i\xi_i$, $r_{i,i+1} = A_i\tilde r_{i,i+1}$, $e_{i+1} = A_i\tilde e_{i+1}$

$r_{ij} = r_{i-1,j} - r_{i-1,i} + r_{ii}$, $j=1,\ldots,i-1$

block 2´

$\Delta_\ell q_{ij} = (e_\ell\times q_{ij})(1-\xi_\ell)\eta_{\ell i}$, $\Delta_\ell A_i = [\Delta_\ell q_{i1}\Delta_\ell q_{i2}\Delta_\ell q_{i3}]$, $\Delta_\ell r_{ii} = (e_\ell\times r_{ii})(1-\xi_\ell)\eta_{\ell i} + e_i\xi_i$

$\Delta_\ell r_{i,i+1} = (e_\ell\times r_{i,i+1})(1-\xi_\ell)\eta_{\ell i}$, $\Delta_\ell e_{i+1} = (e_\ell\times e_{i+1})(1-\xi_\ell)\eta_{\ell i}$

$\Delta_\ell r_{ij} = [(e_\ell\times r_{ij})\eta_{\ell,j-1} + (e_\ell\times r_{i\ell})(\eta_{\ell i} - \eta_{\ell,j-1})](1-\xi_\ell) + e_\ell(\eta_{\ell i} - \eta_{\ell,j-1})\xi_\ell$

block 3, "velocities and accelerations"

$\omega_i = \omega_{i-1} + \dot q^i e_i(1-\xi_i)$, $v_i = v_{i-1} - \omega_{i-1}\times r_{i-1,i} + \omega_i\times r_{ii} + \dot q^i e_i\xi_i$

$\varepsilon_i = \sum_{j=1}^{i}\alpha_{ij}\ddot q^j + \alpha_i^0$, $w_i = \sum_{j=1}^{i}\beta_{ij}\ddot q^j + \beta_i^0$

$\alpha_{ii} = e_i(1-\xi_i)$, $\alpha_{ij} = \alpha_{i-1,j}$, $\alpha_i^0 = \alpha_{i-1}^0 + \dot q^i(\omega_{i-1}\times e_i)(1-\xi_i)$

$\beta_{ii} = (e_i\times r_{ii})(1-\xi_i) + e_i\xi_i$, $\beta_{ij} = \beta_{i-1,j} + \alpha_{i-1,j}\times(r_{ii} - r_{i-1,i})$

$\beta_i^0 = \beta_{i-1}^0 + \alpha_{i-1}^0\times(r_{ii} - r_{i-1,i}) + \dot q^i(\omega_{i-1}\times e_i)\times r_{ii}(1-\xi_i) + 2\dot q^i(\omega_{i-1}\times e_i)\xi_i - \gamma_{i-1,i} + \gamma_{ii}$

where $\gamma_{ij} = \omega_i\times(\omega_i\times r_{ij})$

block 3´

$$\Delta_\ell\omega_i = \Delta_\ell\omega_{i-1} + \dot{q}^i(e_\ell\times e_i)(1-\xi_i), \quad \Delta_\ell\alpha_{ij} = \Delta_\ell\alpha_{i-1,j},$$

$$\Delta_\ell\alpha_i^0 = \Delta_\ell\alpha_{i-1}^0+\dot{q}^i(\Delta_\ell\omega_{i-1}\times e_i+\omega_{i-1}\times\Delta_\ell e_i)(1-\xi_i), \quad \Delta_\ell\alpha_{ii} = \Delta_\ell e_i(1-\xi_i)$$

$$\Delta_{\dot\ell}\omega_i = \Delta_{\dot\ell}\omega_{i-1}+e_i(1-\xi_i)\delta_{\ell i}, \; \Delta_{\dot\ell}\alpha_i^0 = \Delta_{\dot\ell}\alpha_{i-1}^0+[(\omega_{i-1}\times e_i)\delta_{\ell i}+\dot{q}^i(\Delta_{\dot\ell}\omega_{i-1}\times e_i)](1-\xi_i)$$

$$\Delta_\ell\beta_{ij} = \Delta_\ell\beta_{i-1,j}+\Delta_\ell\alpha_{i-1,j}\times(r_{ii}-r_{i-1,i})+\alpha_{i-1,j}\times(e_\ell\times(r_{ii}-r_{i-1,i}))$$

$$\Delta_\ell\beta_{ii} = (\Delta_\ell e_i\times r_{ii}+e_i\times\Delta_\ell r_{ii})(1-\xi_i)+\Delta_\ell e_i\xi_i, \; \Delta_\ell\beta_i^0 = \Delta_\ell\beta_{i-1}^0+\Delta_\ell\alpha_{i-1}^0\times(r_{ii}-r_{i-1,i})+$$
$$+ \alpha_{i-1}^0\times(e_\ell\times(r_{ii}-r_{i-1,i}))+\dot{q}^i[(\Delta_\ell\omega_{i-1}\times e_i)\times r_{ii}+(\omega_{i-1}\times\Delta_\ell e_i)\times r_{ii}+(\omega_{i-1}\times e_i)\times$$
$$\times \Delta_\ell r_{ii}](1-\xi_i)+2\dot{q}^i(\Delta_\ell\omega_{i-1}\times e_i+\omega_{i-1}\times\Delta_\ell e_i)\xi_i-\Delta_\ell\gamma_{i-1,i}+\Delta_\ell\gamma_{ii}$$

where $\Delta_\ell\gamma_{ij} = \Delta_\ell\omega_i\times(\omega_i\times r_{ij})+\omega_i\times(\Delta_\ell\omega_i\times r_{ij})+\omega_i\times(\omega_i\times\Delta_\ell r_{ij})$

$$\Delta_{\dot\ell}\beta_i^0 = \Delta_{\dot\ell}\beta_{i-1}^0+\Delta_{\dot\ell}\alpha_{i-1}^0\times(r_{ii}-r_{i-1,i})-\Delta_{\dot\ell}\omega_{i-1}\times(\omega_{i-1}\times r_{i-1,i})-\omega_{i-1}\times(\Delta_{\dot\ell}\omega_{i-1}\times r_{i-1,i})+$$
$$+ \Delta_{\dot\ell}\omega_i\times(\omega_i\times r_{ii})+\omega_i\times(\Delta_{\dot\ell}\omega_i\times r_{ii})+[(\omega_{i-1}\times e_i)\times r_{ii}(1-\xi_i)+2(\omega_{i-1}\times e_i)\xi_i]\delta_{\ell i} +$$
$$+ [\dot{q}^i(\Delta_{\dot\ell}\omega_{i-1}\times e_i)\times r_{ii}(1-\xi_i)+2\dot{q}^i(\Delta_{\dot\ell}\omega_{i-1}\times e_i)\xi_i]n_{\ell,i-1}$$

block 4, "inertial forces and torques"

$$F_i = -m_i w_i = \sum_{j=1}^{i} a_{ij}\ddot{q}^j + a_i^0, \quad a_{ij} = -m_i\beta_{ij}, \quad a_i^0 = -m_i\beta_i^0$$

$$M_i = \sum_{j=1}^{i} b_{ij}\ddot{q}^j + b_i^0, \quad b_{ij} = -J_{Ni}(s_i\times\alpha_{ij})\times s_i-J_{si}(\alpha_{ij}\cdot s_i)s_i$$

$$b_i^0 = -J_{Ni}((s_i\times\alpha_i^0)\times s_i+\tau_i)-J_{si}(\alpha_i^0\cdot s_i)s_i, \quad \tau_i = (\omega_i\cdot s_i)(s_i\times\omega_i), \quad s_i = r_{ii}/|r_{ii}|$$

block 4´

$$\Delta_\ell a_{ij} = -m_i\Delta_\ell\beta_{ij}, \quad \Delta_\ell a_i^0 = -m_i\Delta_\ell\beta_i^0, \quad \Delta_{\dot\ell}a_i^0 = -m_i\Delta_{\dot\ell}\beta_i^0$$

$$\Delta_\ell b_{ij} = -J_{Ni}[(\Delta_\ell s_i\times\alpha_{ij})\times s_i+(s_i\times\Delta\alpha_{ij})\times s_i+(s_i\times\alpha_{ij})\times\Delta_\ell s_i] -$$
$$-J_{si}[(\Delta_\ell\alpha_{ij}\cdot s_i)s_i +(\alpha_{ij}\cdot\Delta_\ell s_i)s_i+(\alpha_{ij}\cdot s_i)\Delta_\ell s_i], \quad \Delta_\ell s_i=e_\ell\times s_i$$

$$\Delta_\ell\tau_i = (\Delta_\ell\omega_i s_i+\omega_i\Delta_\ell s_i)\cdot(s_i\times\omega_i)+(\omega_i\cdot s_i)\cdot(\Delta_\ell s_i\times\omega_i+s_i\times\Delta_\ell\omega_i)$$

$$\Delta_\ell b_{io} = -J_{Ni}[(\Delta_\ell s_i\times\alpha_{io})\times s_i+(s_i\times\Delta_\ell\alpha_{io})\times s_i+(s_i\times\alpha_{io})\times\Delta_\ell s_i+\Delta_\ell\tau_i]$$
$$-J_{si}[(\Delta_\ell\alpha_{io}\cdot s_i)s_i+(\alpha_{io}\cdot\Delta_\ell s_i)s_i+(\alpha_{io}\cdot s_i)\Delta_\ell s_i]$$

$$\dot{\Delta}_\ell b_{io} = -J_{Ni}[(s_i \times \dot{\Delta}_\ell \alpha_{io}) \times s_i + \dot{\Delta}_\ell \tau_i] - J_{si}(\dot{\Delta}_\ell \alpha_i^0 \cdot s_i)s_i$$

$$\dot{\Delta}_\ell \tau_i = (\dot{\Delta}_\ell \omega_i \cdot s_i)(s_i \times \omega_i) + (\omega_i \cdot s_i)(s_i \times \dot{\Delta}_\ell \omega_i)$$

$i = n$ — no: $n+1 \rightarrow 1$; yes: continue.

block 5, "dynamic model matrices"

$$\xi_i = 0:\ H_{ik} = -e_i \cdot \sum_{j=k}^{n} (b_{jk} + r_{ji} \times a_{jk}), \quad h_i = -e_i \cdot \sum_{j=i}^{n} (r_{ji} \times (a_j^0 + G_j) + b_j^0)$$

$$\xi_i = 1:\ H_{ik} = -e_i \cdot \sum_{j=k}^{n} a_{jk}, \quad h_i = -e_i \cdot \sum_{j=i}^{n} (a_j^0 + G_j), \quad G_j = (0,\ 0,\ m_j g)^T$$

block 5´

$$\xi_i = 0:\ \Delta_\ell H_{ik} = -\Delta_\ell e_i \cdot \sum_{j=k}^{n} (b_{jk} + r_{ji} \times a_{jk}) - e_i \cdot \sum_{j=k}^{n} (\Delta_\ell b_{jk} + \Delta_\ell r_{ji} \times a_{jk} + r_{ji} \times \Delta_\ell a_{jk})$$

$$\Delta_\ell h_i = -\Delta_\ell e_i \cdot \sum_{j=i}^{n} (r_{ji} \times (a_j^0 + G_j) + b_j^0) - e_i \cdot \sum_{j=i}^{n} [\Delta_\ell r_{ji} \times (a_j^0 + G_j) + r_{ji} \times \Delta_\ell a_j^0 + \Delta_\ell b_j^0]$$

$$\dot{\Delta}_\ell h_i = -e_i \cdot \sum_{j=i}^{n} (r_{ji} \times \dot{\Delta}_\ell a_j^0 + \dot{\Delta}_\ell b_j^0)$$

$$\xi_i = 1:\ \Delta_\ell H_{ik} = -\Delta_\ell e_i \cdot \sum_{j=k}^{n} a_{jk} - e_i \cdot \sum_{j=k}^{n} \Delta_\ell a_{jk}$$

$$\Delta_\ell h_i = -\Delta_\ell e_i \cdot \sum_{j=i}^{n} (a_j^0 + G_j) - e_i \cdot \sum_{j=i}^{n} \Delta_\ell a_j^0, \quad \dot{\Delta}_\ell h_i = -e_i \cdot \sum_{j=i}^{n} \dot{\Delta}_\ell a_j^0$$

Fig. 1.13. Flow-chart of the algorithm for nonlinear and linearized model construction

derivatives of $\Delta_\ell \vec{r}_{ij}$, $i \varepsilon I$, j, $\ell \varepsilon J$ are determined by (1.3.11). All the operations are thus reduced to the calculation of the vectorial products of the vectors produced by the primary algorithm (Fig. 1.13, block 2´).

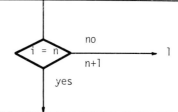

In the next step the angular and linear velocities, $\vec{\omega}_i$ and $\vec{v}_i$, and the angular and linear accelerations, $\vec{\varepsilon}_i$ and $\vec{w}_i$, of the i-th link are calculated. The superposition of the transmission and relative velocities and accelerations is made according to the known relations from rigid body dynamics. In order to prepare a computer-oriented algorithm suitable for the dynamic analysis of the mechanism the expressions for $\vec{\varepsilon}_i$ and $\vec{w}_i$ are written in a slightly modified form using the coefficients $\vec{\alpha}_{ij}$ and $\vec{\beta}_{ij}$, respectively. These coefficients are determined from the recursive relations of Fig. 1.13, block 3.

The velocities and accelerations of the i-th link of the mechanism are composite vectors. For example, the angular velocity $\vec{\omega}_i$ may be expressed by the following sum

$$\vec{\omega}_i = \sum_{k=1}^{i} \dot{q}^k \vec{e}_k (1-\xi_k), \qquad \forall i \varepsilon I \tag{1.3.16}$$

Applying (1.3.10) we get

$$\Delta_\ell \vec{\omega}_i = \sum_{k=1}^{i} \dot{q}^k (\vec{e}_\ell \times \vec{e}_k)(1-\xi_k), \qquad \forall \ell \varepsilon J \tag{1.3.17}$$

This relation may be expressed recursively

$$\Delta_\ell \vec{\omega}_i = \Delta_\ell \vec{\omega}_{i-1} + \dot{q}^i (\vec{e}_\ell \times \vec{e}_i)(1-\xi_i) \tag{1.3.18}$$

In the developed algorithm (Fig. 1.13, block 3´), the relation (1.3.18) is used to calculate the partial derivatives $\Delta_\ell \vec{\omega}_i$, $\ell \varepsilon J$, in order to minimize the number of numerical operations. Using (1.3.10) - (1.3.12), recursive relations for the partial derivatives of the coefficients $\vec{\alpha}_{ij}$ and $\vec{\beta}_{ij}$, $i \varepsilon I$, $j \varepsilon J$ can be found. The partial derivatives of the coefficients $\vec{\alpha}_i^o$ and $\vec{\beta}_i^o$ with respect to generalized coordinates and velocities q^ℓ and $\dot{q}^\ell$ are formed in a similar way.

The next step in the primary algorithm is the calculation of the inertial force $\vec{F}_i$ and the moment $\vec{M}_i$ of the inertial force of the i-th link by using the angular and linear accelerations $\vec{\varepsilon}_i$ and $\vec{w}_i$. It is assumed that all links can be described by the transversal (J_{N_i}) and longitudinal (J_{S_i}) moment of inertia, i.e. can be considered as canes. This results in simpler calculation of the manipulator dynamics. Similar to the procedure in the preceding stage, the coefficients $\vec{a}_{ij}$ and $\vec{b}_{ij}$, $i \varepsilon I$, $j \varepsilon J$, are introduced. These are proportional to $\vec{\alpha}_{ij}$ and $\vec{\beta}_{ij}$ (Fig. 1.13, block 4´). The values of $\Delta_\ell \vec{a}_{ij}$, $\Delta_\ell \vec{b}_{ij}$, $i \varepsilon I$; j, $\ell \varepsilon J$ are determined using (1.3.10), (1.3.12) (Fig. 1.13, block 4´). All values relevant to the calculation of the matrices of the models (1.3.1) and (1.3.3) are formed during the "assembly" of the mechanism links, with the index i changing from 1 to n. The matrices of the nonlinear model (1.3.1) are formed by using the general theorems which describe the translation of the center of mass of a rigid body and rotation about it (para. 1.1.). Depending on the type of the i-th kinetic pair (linear or rotational), the first or the second theorem is applied. In this way all terms of the dynamic model matrices $H_{ik}(q)$ and $h_i(q, \dot{q})$, i, $k\varepsilon I$ (Fig. 1.13, block 5´) are determined by (1.1.30). By applying (1.3.10) and (1.3.12) one obtains the following

$$\Delta_\ell H_{ik}=-(\vec{e}_\ell\times\vec{e}_i)\sum_{j=k}^{n}(\vec{b}_{jk}+\vec{r}_{ji}\times\vec{a}_{jk})-\vec{e}_i\cdot\sum_{j=k}^{n}(\Delta_\ell\vec{b}_{jk}+\Delta_\ell\vec{r}_{ji}\times\vec{a}_{jk}+\vec{r}_{ji}\times\Delta_\ell\vec{a}_{jk}), \tag{1.3.19}$$

where $\vec{b}_{ij}$, $\vec{r}_{ij}$, $\vec{a}_{ij}$ and $\vec{e}_j$ are formed in the primary algorithm and $\Delta_\ell\vec{a}_{ij}$, $\Delta_\ell\vec{b}_{ij}$ and $\Delta_\ell\vec{r}_{ij}$ are formed in the expanded part of the algorithm. A similar method is applied to obtain the partial derivatives of h_i, $i\varepsilon I$. This completes the process of forming the matrices H and h, as well as their partial derivatives with respect to generalized coordinates and velocities. In order to estimate the computing time needed for simultaneous implementation of the basic and expanded algorithm (basic and linearized model), let us consider the dimensions of matrices of the models (1.3.1) and (1.3.3). The matrices $H(q)$ and $h(q, \dot{q})$ contain n^2+n terms in all and the matrices of the partial derivatives (1.3.9) contain n^3+2n^2 terms, which are determined in the expanded part of the algorithm. This means that there are $n(n+2)/(n+1)$ times more terms to be calculated in the expanded part of the algorithm then in the primary algorithm. On the other hand, the number of algebraic operations of the basic and expanded algorithms may be determined from the flow-chart in Fig. 1.13. Here only the operations of the three dimensional vector dot and cross product may be taken into account since they require the greatest amount of computer time. In the expanded part of the algorithm there are about 2n times more of these operations than in the basic algorithm, which implies that the necessary computer time for performing the expanded algorithm part is in fact 2n times longer than for the basic algorithm.

References

[1] Vukobratović K.M., V.Potkonjak, Dynamics of Active Spatial Mechanisms and Manipulation Robots, Monograph, Springer-Verlag, Berlin, 1982.

[2] Stepanenko A.Y., Dynamics of Spatial Mechanisms, (in Russian), Monograph, Mathematical Institute, Beograd, 1974.

[3] Vukobratović K.M., Y.Stepanenko, "Mathematical Models of General Anthropomorphic Systems", Math. Biosc., Vol. 17, pp 191-242, 1973

[4] Vukobratović K.M., Legged Locomotion Robots and Anthropomorphic Mechanisms, Monograph, Institute "M.Pupin", Beograd, (in English) 1974, also published by Mir, Moscow, (in Russian), 1976.

[5] Fletcher H.J., L.Rongved, E.Y.Au, "Dynamics Analysis of a Two-Body Gravitational Oriented Satellite", The Bel System Technical Journal, September, 2339-2366, 1963.

[6] Hooker W.W. and Margulies G., "The Dynamical Attitude Equations for a n-Body Satellite", The Journal of the Astronautical Sciences, Vol. XII, No. 4, pp. 123-128, Winder, 1965.

[7] Hooker W.W., "A Set of n Dynamical Attitude Equations for an Arbitrary n-Body Satellite Having r Rotational Degrees of Freedom", AIAA Journal, Vol. 8, No 7, 1970.

[8] Juričić D., M.K.Vukobratović, "Mathematical Modelling of a Bipedal Walking System", ASME Publ. 72-WA/BHF-13, 1972.

[9] Roberson, R.E., J.Wittenburg, "A Dynamical Formalism for an Arbitrary Number of Interconnected Rigid Bodies, with Reference to the Problem of Satellite Attitude Control", 3rd. IFAC-CONGRESS, Session 46, Paper 46D, 24th June, 1966.

[10] Wittenburg J., "Automatic Setting of Nonlinear Equations of Motion for Systems with Many Degrees of Freedom", Euromech 38, Colloquium,Louvain La Neuve-Belgium, 1973.

[11] Ossenberg - Franzes, Equations of Motion of Multiple Body Systems with Translational and Rotations of Motion of Freedom, Euromech 38 Colloquium, Louvain La Neuve, Belgium, 3-5 September, 1973.

[12] Uicker J.J., "Dynamic Behaviour of Spatial Linkages", Trans. of the ASME, No 68, March - 5, 1-15, 1968.

[13] Liegeois A., Renaud M., "Modèle Mathématique des Systèms Mechaniques Articulés en vue de la Commande Automatique de leurs Mouvements", C.R.A.S.T, 278 - Serie B-29, April, 1974.

[14] Popov E.P., Vereschagin A.F., Ivkin A.M., Leskov A.S., Medvedov V.S., "Synthesis of Control Systems of Robots Using Dynamic Models of Manipulation Mechanisms", (in Russian), Proc. VI IFAC Symp. on Autom. Contr. in Space, Erevan, USSR, 1974.

[15] Vukobratović K.M., V.Potkonjak, "Contribution to Automatic Forming of Active Chain Models via Lagrangian Form", Journal of Applied Mechanics, No 1, 181-185, 1979.

[16] Vukobratović K.M., V.Potkonjak, "Contribution to the Forming of Computer Methods for Automatic Modelling of Active Spatial Mechanisms Motion. Part I: Method of Basic Theorems of Mechanics", Journal of Mechanisms and Machine Theory, Vol. 14, No 3, 1979.

[17] Vukobratović K.M., D.M.Stokić, "Contribution to the Decoupled Control of Large-Scale Mechanical Systems", Automatica, Jan., No 1, 1980.

[18] Vukobratović K.M., N.M.Kirćanski, "Computer-Oriented Method for Linearization of Dynamic Models of Active Spatial Mechanisms", Journal of Mechanism and Machine Theory, Vol. 17, No 1, 1982.

Chapter 2
Synthesis of Manipulation Control

2.1. Introduction

In this chapter we present the theoretical basis of the control synthesis of complex mechanical systems, such as robots and manipulators. As we have said, we shall treat only the problem of the control synthesis of the realization of prescribed mechanism motion, i.e., we treat the two lowest hierarchical levels of robot and manipulator control under the supposition that, at the higher control levels, the task (the motion to be realized by the robot) has been defined. In this chapter systems of robots and manipulators will be considered without the action of external forces, so that the system behaves as an open kinematic chain.

Chapter 1. considered the mathematical model of the system (manipulators consisting of the mechanical part of the system and of the actuators). In this chapter consideration will be restricted to manipulation systems of invariant structure and chapter 3. will consider systems of variable structure, systems changing their structure as a result of contact with external obstacles and the action of reaction forces. In para. 2.2. we define the task which the synthesized control has to realize and the various methods of its solution are considered. As a result, we adopt the two-stage concept of the constrol synthesis, so that para. 2.3. considers the control synthesis at the stage of nominal dynamics. Central to this chapter is the synthesis of control at the stage of the perturbed dynamics. In order to visualize the various possibilities, provided by systems theory for the control of complex systems, we first consider the synthesis of the centralized optimal regulator for the stabilization of the system about the nominal trajectory (para. 2.5.). In para. 2.6. the synthesis of the decentralized control is given via the synthesis of local control of the particular subsystems corresponding to the separate mechanism degrees of freedom (para. 2.6.1.). Two habitual methods for the control synthesis are therefore considered: one by means of the incomplete local feedback loops and the other by minimizing the local criteria. The selected stability analysis of the total system is then presented using the aggregation - decomposition method (para. 2.6.2.). In addition we consider the "suboptimality" analysis of this decentralized control with respect to the standard quadratic criterion (para. 2.6.3.). The basic aim of this is to present possible ways of examining the validity of the decentralized control we select. In order to achieve satisfactory behaviour of the complete system, i.e., ensure the stability of the complete system and reduce the suboptimality of local control, global control is introduced. In para. 2.6.4. we consider various forms of global control and synthesis of the global control parameters at the basis of the "suboptimality" reduction of the control type. Due to arbitrariness in the choice of the part of the system model which is assigned to the coupling, and the part of the model assigned to the subsystems, and since determination the parameters of the local and global control depends directly on the subsystem models, i.e., on the coupling, the partition of the model is also considered (para. 2.6.5.). For the sake of demonstrating the possibilities provided by the theory of optimal regulators for the synthesis of manipulator decoupled control, we consider the synthesis of the asymptotic regulator (para. 2.6.6.).

In order to implement the selected control laws on a microcomputer, we briefly consider the transfer to the time-discrete domain. That is, we consider the synthesis of the linear discrete regulator (para. 2.7.1.), of the decentralized regulator and observer and of the global control in the discrete time domain (para. 2.7.2.). In appendix 2.A we derive the expression for estimating the maximal value of the quadratic criterion when decentralized and global control is applied to the complete system. In appendix 2.B we consider the partition of the model into the part assigned to coupling and the part assigned to subsystems for the special case of first order subsystems.

In appendix 2.C we consider the stability analysis of the system when decentralized regulator and decentralized observer are applied.

Thus, the aim of the chapter is to present the postulates of various concepts in the synthesis of robot and manipulator control in unobstructed working environments. Applications of these concepts to the synthesis of control for concrete systems will be considered in chapter 3. where we also consider all the aspects of the adopted approach to control synthesis. Implementation of control systems will also be considered in ch. 3.

2.2. Definition of the Control Task

Models of manipulation systems have been considered in ch. 1. The model of the complete system (mechanical model part given by (1.1.30) and the actuators, given by (1.2.1)) can be written in centralized form, as was shown in para. 1.2.

$$S : \dot{x} = \hat{A}(x) + \hat{B}(x)N(u), \tag{2.2.1}$$

where $\hat{A}:R^N \to R^N$, $\hat{B}:R^N \to R^{N\times n}$, $x \in R^N$, $u \in R^n$, $N(u) = (N(u^1), N(u^2),\ldots,N(u^n))$.

In this chapter we consider two types of system output: (La) when the system output of the order, lower than the order N of the system, is measurable, i.e., when not all the state coordinates are being measured but only those whose measurement is technically feasible and cheap, and (Lb), when the whole state vector is measured, so that the system output y coincides with the state vector. In this chapter the observer is introduced because it is necessary to reconstruct the whole system state vector in order to apply the synthesis of the optimal decentralized regulator in the case when the measurable system output is of lower order than that of the system. That is, with robots and manipulators it is usually possible to realize the measurement of the whole system state vector but this is not feasible in practice because of cost and complexity. Hence, either one applies the synthesis of the observer for reconstructing the system state vector or control is synthesized with incomplete feedback with respect to the system output.

We suppose that in all the regions of the state space considered the conditions of existence and uniqueness of solution (2.2.1) are satisfied during the time interval $T = \{t : t \in (0, \tau)\}$, $\tau > 0$, during which the system is observed. That means, for each particular control type $u(t)$ there exists a unique solution $x(0) \in X^t$, $\forall t \in T$, where X^t is a bounded region $X^t \subset R^N$ and $x(t) \in X^t$, $\forall t \in T$. Let suppose that the functions $\hat{A}$ and $\hat{B}$ are continuous and differentiable with respect to x in the considered region X^t.

In this chapter we precisely define the task to be performed and consider the various ways of accomplishing it. As we said, we consider the control synthesis at the two lowest control levels. It is therefore necessary to define the general task imposed at the lowest levels.

In general this can be defined as the task of transferring the system state from an arbitrary initial state to a defined point in the state space during a set time interval. The higher control levels define this point in the state space, as well the time interval for performing the task. However, the most frequent initial state can belong only to a bounded region in the system state space, $X^I \subset R^N$, and it is not necessary to transfer the system state to a point. It sufficies to transfer the state into a bounded region in the state space around the desired point $X^F \subset R^N$. The system is considered in a given time interval τ and it is required that the state is transferred from the region of initial conditions X^I into a region X^F in prescribed time interval $\tau_s \leq \tau$. It is also required that during the transfer the system state remains within a bounded region $X^t \subseteq R^N$. Accordingly the system is considered to be practically stable if $\forall x(0) \in X^I$ implies $x(t) \in X^F, \forall t \in T_s$, where $T_s = \{t : t \in (\tau_s, \tau)\}$, and $x(t) \in X^t$, $\forall t \in T$, where $X^I \subseteq X^t$ and $X^F \subseteq X^t$. This definition of practical stability is sufficiently broad to include all other feasible definitions of practical stability.

All manipulation tasks in practice can be regarded as having to satisfy the stipulated practical stability. It is usually required that the manipulator tip is conducted to some particular region in absolute space. For instance, in the assembly process it is necessary to transfer the manipulator gripper from various initial positions to a point near the element where the assembly is being performed. It is also necessary to ensure that the working object, carried by the manipulator, has some particular orientation in space. The final manipulator position need not be strictly attained but the working object should be brought into a certain region around the point of assembly with only a certain deviation from the desired orientation. The deviation depends on the region of initial conditions of the assembly process itself, which follows the transfer process. In industrial practice, the initial manipulator positions are mostly well defined in advance, but still some tolerance in the initial conditions must be allowed. Moreover, one often defines several initial conditions from which the manipulator should be transferred into one region around the point of assembly. The perturbations acting on manipulators produce certain deviations from the desired motion and, in principle, can be reduced to perturbations of the initial conditions. It is usually required that the manipulator be in a certain region in space, this defined by obstacles and various kinematic and other constraints. Often the region X^t, which must not be left by the manipulator during motion, imposes strict limitations on manipulator motion during the transfer.

Taking as a second typical task in practical manipulator application the task of transferring an open vessel with liquid, the dynamical conditions of liquid transfer define the region X^t even more strictly. The functional requirements of the orientation of the vessel during the transfer impose conditions which must be satisfied by the manipulator motion.

The task of assembling mechanical elements by means of manipulator, which can be reduced to the process of inserting an element into a fixed opening or hole, can also be treated as the task of transferring the system state from region X^I to the region X^F in a finite time interval. The region X^F is defined by the tolerances determined by the particular conditions. The region of initial conditions X^I represents the region of deviations of the manipulator tip from the exact position of the hole and the deviation of the working object with respect to the hole axis orientation. The region X^t, in which the manipulator state must remain during the insertion process, is very narrow and is defined by the hole configuration and the permissible manipulator velocity in order to avoid hard impacts on the hole walls. However, perturbations occur due to the reaction forces of the hole, so that not only impulse perturbations occur but the manipulator behaves as a system of variable structure. This case will be treated separately in ch. 3.

These manipulation tasks can be regarded as tasks of transferring the system state from one bounded region to another during a defined time interval. Complex manipulation tasks are reduced at higher hierarchical levels to those simple tasks which are realized at the lowest control levels by satisfying concrete conditions of practical stability. That is, higher control levels define the regions X^I, X^F and X^t and the settling time τ_s, as well the time τ in which the task is observed, and the lowest control levels should ensure the satisfaction of these concrete conditions.

Similarly, the control task with locomotion systems can be reduced to that of satisfying the practical stability. For instance, the locomotion system should be transferred from the present state to another posture or position in the absolute space [1, 2].

It should be mentioned that in some concrete tasks the regions X^I and X^t need not be connected as is the case when the manipulator is to be transferred to a defined region X^F from various positions with large distances between them. Then $X^I = X_1^I \cup X_2^I \cup \dots \cup X_L^I$, whereby $X_i^I \cap X_j^I = \phi$, $i, j \in L_1$, where $L_1 = \{\ell : \ell = 1, 2, \dots, L\}$. Also $X^t = X_1^t \cup X_2^t \cup \dots \cup X_L^t$ and $X_1^t \cap X_2^t \dots X_L^t = X_F^t \subseteq X^F$, since the conditions $X_\ell^F \subseteq X_\ell^t$ and $X_\ell^I \subseteq X_\ell^t$, $\forall \ell \in L_1$ must be satisfied. We henceforth restrict ourselves to considering the case when the regions X^I and X^t are connected [3].

It should be remembered that in this chapter only perturbations of the initial conditions are considered to act on the system S. Other perturbation will be treated in ch. 3. For instance, perturbations due to parameter variations and inexact models will be regarded as the problem of reliability and system sensitivity. It should also be understood that the synthesized control should be able to withstand white noise perturbations.

The set task can be accomplished by synthesizing the control which transfers the system state from a defined initial condition $x(0) \in X^I$ to some state $x(\tau_S, x(0)) \in X^F$ during time $\tau_S \leq \tau$ with the minimization of some criterion (e.g. minimum energy consumption). Thus,

$$J = \int_0^{t_S} L(x(t), u(t))dt + g(x(t_S)), \tag{2.2.2}$$

where time t_S is not fixed (but it is $\leq \tau$), $L: R^N \times R^n \to R^1$ is a scalar, continuous and differentiable function of state and control and $g: R^N \to R^1$ is the state function at the terminal time instant. Synthesis of such control by Pontryagin's maximum principle or by dynamic programming [4] results, on account of the non-linearity (2.2.1), in open-loop control. That is, the solution of the problem is programmed control and depends on the initial state $x(0)$. The set control task is usually impossible to solve in the form of a fixed structure and closed feedback with respect to the state, (i.e. in the form of a function of the momentary state $x(t)$ only). It is obtained in numerical form, so the synthesis of optimal control, in the sense of (2.2.2) must be repeated for each initial state $x(0)$. This type of synthesis of optimal control poses a number of numerical problems, so its on-line implementation would require the use of powerful processors and a very complex control structure. This would influence the price of the control system and its reliability and robustness (see para. 2.3.). It is obvious that optimal control synthesis would be sensitive to the variation of the system S parameters. On the other hand, optimal control would be centralized.

In order to facilitate the synthesis of a control ensuring practical stability of the system (2.2.1) and minimization of criterion (2.2.2), one usually considers various approximate models of the system, at the basis of which the feasible control is synthesized. The behaviour of the real system (exact model) with synthesized control [5] is then examined. The most common procedure is the linearization of the system model around a pre-selected system trajectory (satisfying the set conditions) and a control synthesis using the methods of linear systems theory. One first determines the trajectory satisfying the conditions of practical stability of the system. Linearization is then carried out (e.g. according to the procedure described in para. 1.3.). At the basis of the linearized model and minimization of criterion (2.2.2) it is often possible to obtain a control in suitable form. However, it is necessary to examine whether such control satisfies the exact, non-linear system model. It is obvious that such control is limited by the accuracy of the linear model approximation, so in the case of manipulation systems which are distinctly non-linear, it can happen that the control, synthesized at the basis of the linearized model, is not satisfactory.

It is clear, that the definition of criterion (2.2.2) is not unique, so the system optimality is also conditional. In addition, such centralized control does not enable one to follow sufficiently closely the physical system image when, for instance, it involves variation of control parameters and their influence on the required system performance. This results in a formal control synthesis.

Our aim in solving the problem considered above is to synthesize as simple a control as possible, this being the most insensitive to parameter variations. Thus, when solving the above control task with minimization of criterion (2.2.2), it is possible to impose on the control (in addition to amplitude constraint) constraint on its information structure. It is required that the control minimizing criterion (2.2.2), is restricted to the form

$$u^i(t) = \phi_i(x^{i'}(t)), \qquad \forall i \varepsilon I, \tag{2.2.3}$$

where ϕ_i, $\forall i \varepsilon I$ are the subsystem vector state functions, $x^{i'} \varepsilon R^{n_i}$, where $n_i \leq N$ and

$$x^{i'}(t) = \{x^{i'}(\tau''), \quad o<\tau''<t\} \tag{2.2.4}$$

Accordingly, optimal control with respect to (2.2.2) is required, where none of the inputs u^i possesses information about all state coordinates x of the system S but only about the state coordinates corresponding to the particular subsystems. By applying constraints on the information structure, an important simplification of the control structure is achieved. It is clear, that when once the control laws ϕ_i in (2.2.3) are chosen for $i \in I$, the $u^i(t)$ can be eliminated from (2.2.1) and (2.2.2). The optimization problem thus becomes only a function of the state and is well-defined. It is evident that the value of criterion (2.2.2) depends on the choice of control laws (2.2.3), i.e., on the choice of the information structure $x^{i'}$, (which information x^i will joint the input $u^i(t)$). It has been shown [6] that such an optimization problem is even more complex than in the centralized case, both from the numerical aspect and with respect to implementation. That is in the case of a linear, time invariant system, and for a quadratic criterion with respect to state and control, the optimal solution in the case of a decentralized information pattern requires a nonlinear control law, the implementation of which could be rather complex and the system behaviour very sensitive to parameter variations. However, it is possible, to restrict oneself á priori to linear control laws if one tolerates a certain increase in the value of the quadratic functional. However, even if it is required at the outset that the control laws be arbitrary linear functions of all past observations of the subsystem state, this will not necessarily result in a readily implemented optimal control because it cannot be guaranteed that the control is of finite dimensions. It is possible to introduce supplementary constraints on the control so that it be not only linear but of fixed structure, and then to optimize over the parameters of this fixed structure. However, the choice of the control structure is totaly heuristic and at present there are no known procedures for the choice of the optimal structure. All this makes the optimality of such a solution conditional.

From this discussion about the difficulties in synthesizing "optimal" control in decentralized form for linear, time invariant systems, it follows that in the case of the class of nonlinear systems considered (2.2.1), the difficulties with optimization would even be increased. If one decided on a fixed control structure in advance, i.e., for decoupling of the system S to subsystems, it would be possible to solve the above problem of transferring the system S from the region X^I to a bounded region X^F in a finite time τ_s by optimizing the criterion (2.2.2) over the parameters of the chosen control structure. Some of the "optimality" of the control would certainly be lost compared with the general centralized case.

If the control law is not specified in advance, then certainly in the general case a numerical solution of the problem would be obtained, not some analytical control form $u^i(t)$, $\forall i \in I$, as a function of the instantaneous subsystem $x^{i'}(t)$ state. So, the difficulties in implementating such a control in real time would be still present.

If the functions (2.2.3) were constrained to previously defined control laws applicable in practice, along with the already chosen control structure, it is questionable whether optimization problem (2.2.3) would be solvable in the general case. In any case, such constraints on control would still further increase the conditional optimality of the solution obtained. So it is uncertain how much would be lost by an "optimal" decentralized control using a fixed, arbitrarily chosen control structure and an arbitrarily chosen control law, compared with the centralized case of control.

It follows that if we want to apply decentralized, reliable and robust control, the general optimization procedure does not yield broader possibilities for synthesizing of practically attainable control. In addition, the optimality of the solution is extremely conditional. Optimization procedures do not offer a choice of optimal structure. Hence, it is sensible to choose the structure of decentralized control on the basis of experience.

In the case of concrete mechanical systems, such as robots and manipulators, engineering experience requires that the decoupling of system S to subsystems, corresponds to the separate degrees of freedom or their groups. Such decentralization of control structure is necessitated by the mechanical characteristics and functional requirements of mechanical systems and the fact that the mechanical degrees of freedom with some classes of mechanical system are weakly interconnected. With some mechanical systems such as active mechanisms, mechanical degrees of freedom are strongly coupled, so the usual application of decentralized control would be inadequate. If with such systems one applied decentralized control structure (for strongly coupled subsystems) and chosen control law, and then attempted to synthesize the given structure parameters by minimizing criterion (2.2.2), the optimization procedure would certainly involve a number of numerical difficulties.

In order to apply decentralized control to strongly coupled systems, one must move away from control synthesis which minimizes the criterion (2.2.2).

If it is not required that the control minimizes the conditionally chosen criterion (2.2.2), system S can be regarded as a set of subsystems adjoined to the separate degrees of freedom or their groups and one can synthesize independent local control for each subsystem, stabilizing each subsystem. If the complete system is controllable with the chosen decentralized information control structure, then local controls can be chosen, ensuring asymptotic stability of the global system [7]. Accordingly, if sufficiently high local gains are chosen [8], one can ensure the stabilization of the system with decentralized control. However, if the influence of the coupling among the subsystems is strong, decentralized control might require that local gains be too high, which would result in unsatisfactory system performance. Although we have abandoned the strict minimization of the criterion during the control synthesis, it is necessary to ensure that the control does not result in unacceptably high criterion values (since it does not consider the complete system dynamics of coupling). For that reason, when decentralized control is chosen, it should be considered whether or not it results in unnecessarily high criterion values.

In the case of the synthesis of decentralized or centralized control by the approximate model, it is necessary to ensure that the approximate model,be close to the exact system model and that the control synthesized using this model be the simplest possible. In order to ensure, that the approximate system model be as close as possible to the exact one, a two stage control synthesis is introduced [9 - 12]. The stages of control synthesis are as follows:

(a) a nominal dynamics stage

(b) a perturbed dynamics stage.

At the stage of nominal dynamics the synthesis of programmed control $u^o(t)$ is performed, which transfers the system S state from a defined initial state $x^o(0) \varepsilon X^I$ into some point in the region X^F in finite time $\tau_s < \tau$. Synthesis is performed at the basis of the complete centralized model S without any approximations. Since such programmed control is essentialy centralized, at this stage one part of the coupling is taken into account. Thus, the coupling at the next stage is reduced. Similarly, if we consider the control synthesis using the linearized system model around the nominal trajectory, then introducing the programmed nominal control ensures that the approximate model is "closer" to the exact model. That is, the nonlinear system model is expanded into a Taylor series, and the programmed control compensates the 0-th member, so the linearized model (first member of the series) is a better approximation of the system. Thus, control synthesized at the basis of the linearized model is more adequate when applied to the nonlinear system [13].

Such an approach is justified, since, as we said, the solutions of the mathematical system model are very close to the behaviour of the real systems. Thus, programmed control is synthesized using the exact mathematical model and when applied to the real system a good correspondence between the system response and the desired trajectory, $x^o(t)$, $\forall t \varepsilon T$, $x^o(0) \varepsilon X^I$, $x^o(t) \varepsilon X^t$, $\forall t \varepsilon T$ and $x^o(t) \varepsilon X^F$, $\forall t \varepsilon T_s$, can be expected.

The control synthesized at the next stage should transfer system S state to the region X^F in the case of deviations of the initial conditions from the programmed trajectories synthesized at the previous stage. At this stage a control is synthesized which should ensure tracking of the nominal trajectories in the case of the perturbations of the class considered. If, by using the control synthesized at this stage one can ensure that $\forall x(0) \varepsilon X^I$, that $x(t, x(0)) \varepsilon X^t$, $\forall t \varepsilon T$ and $x(t, x(0)) \varepsilon X^F$ $\forall t \varepsilon T_s$, the problem of the control synthesis has been solved.

Thus, the control is chosen in the form

$$u(t) = u^o(t) + \Delta u(t), \qquad (2.2.5)$$

where $\Delta u \varepsilon R^n$ is the control synthesized at the second stage of control synthesis.

As we said in the introduction, in the case of robots and manipulators the working conditions are often known in advance, so the nominal trajectories and the corresponding programmed control can be synthesized in advance, off-line, and stored in the computer memory. Since in industrial practice one central computer can control several manipulators, and each manipulator is controlled by a separate microcomputer, for example, both the generation of nominal trajectories and programmed control can be accomplished in the central computer, while the microcomputers realize the level of trajectory tracking. Generation of the nominal states is thereby performed at longer time intervals. The calculated nominal values are sent to the microcomputers, which use them up to the next change of the nominal states by the central computer.

The control at the second stage can be synthesized either in the centralized or in the decentralized form. In the first case, the centralized model of deviations from the nominal states is considered and a control synthesized, minimizing the corresponding criterion. The linearized system model is usually used. The second possibility is to consider the deviation model of the system as a set of decoupled subsystems and to synthesize the local control at the level of subsystems.

However, although in both cases introducing the nominal control reduces the difference between the behaviour of the exact and approximate model, the problem remains as to how accurate is the control synthesized at the basis of the approximate model when applied to the nonlinear system. The question in now raised, as to whether the nonlinear model is stable and to what extent the control "optimality" is reduced when the control, synthesized by minimizing the criterion based on the approximate model, is applied to the exact model.

Although by introducing the nominal, programmed control the influence of the coupling at the stage of perturbed dynamics (realization level of the nominal trajectories) has been reduced, with active mechanisms the influence of coupling among the subsystems (degrees of freedom) can still be rather strong. That is, local control only (which does not consider the real character of coupling among the subsystems) can either be insufficient to ensure the desired system behaviour (practical system stability around the nominal trajectory) or too "suboptimal". Hence, it is sometime necessary to introduce global control. We propose two versions of global control:

(Ga) At the stage of perturbed dynamics, another way of solving the problem of strong coupling among subsystems is to use a specificness of the mechanical systems of active mechanisms type. If the separate degrees of freedom are considered as subsystems, coupling among the subsystems is presented in the form of mechanical forces, or moments. These forces can be measured directly, so feedback with respect to them is introduced.

By introducing the global control in the form of force feedback it is possible to minimize the negative influence of the coupling on the stability of the complete system and to reduce the suboptimality of the applied decentralized control [14]. Thus, the simplicity of the control structure is not lost since the number of the feedback loops does not significantly increase. That is, by introducing the force feedbacks acting on the mechanical degrees of freedom (subsystems), the information control basis is increased, so the complete dynamics of the system can be taken into account in a suitable way. Moreover, by introducing force feedbacks, in some cases the control synthesis at the nominal dynamics stage can be simplified (see ch. 3). However, the convinience of introducing the force feedbacks is offset by the cost of the force sensors and certain technical difficulties in introducing them.

(Gb) A second possibility of introducing global control for mastering the destabilizing influence of the coupling among subsystems, is in on-line calculation of the forces acting in each degree of freedom by means of the mathematical system model. However, since the coupling is usually a function of all the system state coordinates, the basic advantage of decentralized control (its simplicity) would then be lost. On the other hand, this calculation is often so complex that a powerful computing system and long computation time are needed, so that it cannot be realized on-line. In some cases, forces in the degrees of freedom of the mechanical system can be calculated by the approximation formulae but this problem can best be illustrated with concrete systems [15]. We shall consider the general case when coupling can be approximated by means of a Taylor series as far its second term. We also consider concrete approximations for coupling with manipulators having 6 degrees of freedom (ch. 3).

Two-stage control synthesis and force feedback should both make possible the synthesis of a simple and reliable suboptimal control structure.

Later, both stages of the control synthesis will be considered. First we consider the possibility of synthesizing nominal, programmed control and nominal trajectories with complex mechanical systems, such as robots and manipulators.

At the stage of perturbed dynamics both versions of the control synthesis, centralized and decentralized, will be considered. As an example of centralized control we consider the synthesis of the linear optimal (centralized) regulator at the basis of the linearized system model. However, central to this chapter is the synthesis of decentralized control at the basis of the decoupled system model. In this case we adopt the decentralized information structure of control, in such a way that onto the input of each actuator are fed the system state coordinates corresponding to the degree of freedom being activated by the given actuator; i.e., the system is considered as a set of subsystems corresponding to the separate degrees of freedom. Thus, we will not consider the choice of the decentralized structure imposed on the control because it is supposed that above choice of structure is the most suitable one for mechanical systems and because it is often used in practice.

With this choice of control structure, i.e., subsystems, local control is synthesized separately for each subsystem. Two types of local control will be considered: (La) when the local control stabilizes the subsystems by means of an incomplete feedback with respect to the subsystem output and (Lb) when it is supposed that all state coordinates of the subsystems are measurable and the synthesis of local control is performed by minimizing the local quadratic criteria. Consideration of the two types of local control is intended to indicate the various possibilities concerning the choice of the decentralized control structure. Version (La) is technically simpler and closer to practice but version (Lb) provides the possibility of greater profit from decentralized control and makes possible a more precise analysis of the systems with local and global control. It is evident that system behaviour depends to a great extent on the choice of the form of local control. Since linear subsystems are concerned, it is logical to choose linear control laws for local control. This is also the method most acceptable in practice.

When synthesis of local control is performed either in the form (La), or (Lb), it is necessary to examine the stability of the complete system, since local control is chosen at the level of subsystems and not at the level of the complete system model. In case (Lb), the suboptimality of the decentralized control is analyzed, when this is applied to the complete system, on the basis of the criterion provided by the sum of the chosen local criteria for subsystems. This makes possible an iterative improvement of the choice of local control until stabilization of the system is achieved, i.e., until minimization of the decentralized control suboptimality is attained.

By means of the complete system analysis (together with the coupling) around the nominal trajectories, it is found that the decentralized control alone is sufficient to ensure the stability of the complete system, i.e., it can be determined whether or not the local gains are sufficient to overcome the coupling among the subsystems. Thus, one determines whether it is necessary to consider the complete system dynamics or whether the chosen local gains are sufficient to impose a behaviour on the system which "suits" the individual subsystems. By analyzing the suboptimality of the chosen decentralized control one can estimate the extent to which the "optimality" of control is lost, for decentralized control is applied, which does not take into account the complete system dynamics but only the dynamics of the individual actuators. By repeated choice of local control (with higher values of the feedback gains) it is possible in certain cases to attain, under the above conditions of controllability and observability, the stability of the complete system using decentralized control; so the system dynamics is not of prime influence.

If these analyses show that local control (apart from the nominal, programmed control) is not sufficient, it is necessary to introduce global control in one of the two forms (Ga), (Gb). We analyze the following problems: choice of global control and separation of local and global control from the standpoint of the complete system stability and the suboptimality of the chosen control. In addition, we consider the various approximations of the coupling among the subsystems. We then consider the redistribution between the model part assigned to local subsystems and the part assigned to the coupling among subsystems. This redistribution is considered from the standpoint of control suboptimality resulting from the separation. That is, although the decentralized control structure has been chosen in advance, the question of the separation of the mathematical model into the part assigned to the subsystems and the part assigned to the coupling, is still open. The coupling acting on the subsystems can be a function not only of the state coordinates of the other subsystems, but also of the state coordinates of the subsystem itself. Part of the coupling acting on the i-th subsystem can be assigned to the subsystem itself or can be left as the part of the coupling. Depending on how this separation is made, the distribution of local and global control is determined. According to the control suboptimality and the particular model separation, one can estimate the most favourable distribution between the subsystems and coupling.

It should be kept in mind that the coupling among the subsystems need not always be destabilizing and the system suboptimality need not be increased relative to the decoupled case [16]. Hence global control does not always diminish any influence of the coupling but it should overcome the destabilizing influence of the coupling. Accordingly, coupling is not treated as an external perturbation on the system [17] but global control tries to overcome only its destabilizing influence. Global control is aimed at reducing the maximal value of the criterion, appearing for at least one initial state in the bounded region in the state space.

The advantage of such a synthesis of decentralized control is that in the case of linear subsystems (which are considered here) the choice of local control does not depend on the nominal trajectories or the task to be performed. Thus, the same local gains can be used in various control tasks since the nominal stage takes into account the dynamics in the region X^t in the state space, in which the task is realized. Certainly the conditions arising from the task determine whether or not the chosen decentralized control is satisfactory. That is, depending on the desired settling time and shrinkage of the region X^F relative to X^I, the acceptability of the local gains is determined. However, in principle it is possible to choose gains satisfying all control tasks. Anyway, it should be kept in mind, that such a choice of a unique decentralized control for various tasks might be suboptimal.

The second characteristic of local and global control thus synthesized is that decoupling of system with respect to subsystems is achieved, which is significant both from the standpoint of system reliability and its sensitivity to variations of the system parameters, action of perturbations, and the like, as well as from the standpoint of implementation and simplicity of system maintenance. However, it should be remembered, that decoupling to subsystems can be achieved also by control synthesis based on the centralized system model. To illustrate such possibilities of synthesizing decoupled control of manipulators, we consider the so-called synthesis of the asymptotic optimal regulator, which is performed at the basis of the linearized system model. In this approach, the properties of the asymptotic regulator are used to achieve decoupling of the system to selected subsystems.

2.3. Stage of Nominal Dynamics

At the stage of nominal dynamics, the task of control synthesis is the following: there should be synthesized programmed control $u^o(t)$, $\forall t \varepsilon T$, which should transfer the system S state from some defined initial state $x^o(0) \varepsilon X^I$ into some desired state $x^o(\tau_s) \varepsilon X^F$ during the time interval $\tau_s < \tau$, and the nominal trajectories of the system state should satisfy the conditions $x^o(t, x^o(0)) \varepsilon X^t$, $\forall t \varepsilon T$ and $x^o(t, x^o(0)) \varepsilon X^F$, $\forall t \varepsilon T_s$.

Later, for the sake of brevity, the nominal trajectories of the state coordinates will be denoted by $x^o(t)$ instead of $x^o(t, x^o(0))$.

Basically, two approaches to the synthesis of the nominal trajectories are distinguished: (a) kinematical synthesis and (b) dynamical synthesis. The kinematical synthesis of the nominal trajectories includes determining the laws of the change of the internal coordinates of the manipulator based on the kinematical model, where the dynamics of the active mechanism is not taken into account. In the Introduction we surveyed the kinematical methods of synthesizing the nominal trajectories. With calculated kinematical trajectories, it is necessary to check against the dynamical system model (2.2.1) whether the desired motion is attainable during the projected time interval, i.e., it is necessary to synthesize the programmed control $u^o(t)$ (if possible) which would produce the calculated kinematical trajectories in ideal conditions. In the synthesis of nominal trajectories, using the complete dynamical model of the system, one takes into account all dynamical properties of the mechanism. Synthesis can be performed either by minimizing some chosen criterion or suboptimally, according to the functional demands of a particular task. Let us first consider the possibility of synthesizing "optimal" nominal trajectories.

In the general case, the synthesis of the nominal trajectory can be performed according to minimization of the criterion (2.2.2) and the constraint (2.2.1). As optimization task, the problem is to synthesize the control $u^o(t)$, $\forall t \varepsilon T$, which transfers the system from the state $x(0) = x^o(0)$ to state $x^o(t_s) \varepsilon X^F$, and minimizes the corresponding criterion $J(u, x^o(0))$.

Optimal control $u^o(t)$ and the corresponding optimal trajectory $x^o(t)$ and the adjoined vector $p^o(t)$ must satisfy the following system of equations and boundary conditions [18] *):

$$\dot{x}^o(t) = \frac{\partial H^T}{\partial p}(x^o(t),\ p^o(t),\ u^o(t)),\quad x^o(0) \text{ is given}$$

$$x^o(t_S) \varepsilon X^F$$

$$\dot{p}^o(t) = \frac{\partial H^T}{\partial x}(x^o(t),\ p^o(t),\ u^o(t)),\quad p_i^o(t_S) = \frac{\partial g^T(x(t_S))}{\partial x_i} \tag{2.3.1}$$

$$i = i_1 + 1, \dots, N$$

$$\frac{\partial H^T}{\partial u}(x^o(t),\ p^o(t),\ u^o(t)) = 0$$

$$H(x,\ p,\ u) = L(x,\ u) + p^T(\hat{A}(x) + \hat{B}(x)u),\quad H: R^N \times R^N \times R^n \to R^1.$$

The order of this system is relatively high, namely, 2N+n, which for a manipulator with 6 degrees of freedom for example, and an actuator 3-rd order model, is 42.

For the case of the quadratic criterion

$$J(u,\ x^o(0)) = \int_0^{t_S} (\frac{1}{2}\ u^T \underline{R} u + x^T S u)\, dt,$$

where $\underline{R}$ is a symmetrical positively definite matrix, and S is a symmetric positive semidefinite matrix, Eq. (2.3.1) becomes

$$\dot{x}^o(t) = \hat{A}(x^o) + \hat{B}(x^o)u^o(t),\qquad x^o(0) \text{ is given}$$

$$x^o(t_S) \varepsilon X^F$$

$$\dot{p}^o(t) = -Su^o(t) - (\frac{\partial \hat{A}^T(x^o)}{\partial x} + u^{oT}(t)\ \frac{\partial \hat{B}^T(x^o)}{\partial x})\, p^o(t), \tag{2.3.2}$$

$$p_i^o(t_S) = 0,\quad i = i_1 + 1, \dots, N$$

$$\underline{R} u^o(t) + S^T x^o(t) + \hat{B}^T(x^o) p^o(t) = 0$$

Solving of this system of equations requires, apart from determining the nonlinear functions $\hat{A}(x)$ and $\hat{B}(x)$ in each integration step, calculating the matrices of partial derivatives of these functions $\frac{\partial \hat{A}(x)}{\partial x}$ and $\frac{\partial \hat{B}(x)}{\partial x}$.

---

*) Here is supposed that constraint $x^o(t_S) \varepsilon X^F$ imposes conditions on the i_1-th state vector coordinate, while $N-i_1$ coordinates are free.

These matrices can be found in the following way. In ch. 1, it was shown that

$$\hat{A}(x) = Ax + F(I_n-HTF)^{-1}[HTAx+h],$$
$$\hat{B}(x) = B + F(I_n-HTF)^{-1}HTB, \qquad (2.3.3)$$

where $A\varepsilon R^{N\times N}$, $B\varepsilon R^{N\times n}$, $F\varepsilon R^{N\times n}$ are constant matrices of the actuator model, $h(q, \dot{q})$ is the vector and $H(q)$ is the matrix of the system's mechanical part S^M (1.3.30).

Taking into account (2.3.3),

$$\frac{\partial \hat{A}(x)}{\partial x} = A + F(I_n-HTF)^{-1}\left[\frac{\partial H}{\partial x} TF(I_n-HTF)^{-1}(HTAx+h) + \frac{\partial H}{\partial x} TAx+HTA +\frac{\partial h}{\partial x}\right],$$

$$\frac{\partial \hat{B}(x)}{\partial x} = F(I_n-HTF)^{-1}\left[\frac{\partial H}{\partial x} TF(I_n-HTF)^{-1}HTB + \frac{\partial H}{\partial x} TB\right].$$

Since matrix H is a function only of the manipulator generalized coordinates q,

$$\frac{\partial H}{\partial x^j} = [H_p^j \quad 0 \quad 0], \qquad j\varepsilon I,$$

where the elements of matrix $H_p^j \varepsilon R^{n\times n}$ are given by $H_p^j = \frac{\partial H}{\partial q^j}$.

The partial deviatives of matrix h can be written in the form

$$\frac{\partial h}{\partial x^j} = [h_p^j \quad h_v^j \quad 0], \qquad j\varepsilon I,$$

where the vectors $h_p^j \varepsilon R^n$ and $h_v^j \varepsilon R^n$ are given by

$$h_p^j = \frac{\partial h}{\partial q^j}, \quad h_v^j = \frac{\partial h}{\partial \dot{q}^j}.$$

Determining H_p^j, h_p^j, h_v^j is a very complex procedure with industrial manipulation systems having many degrees of freedom, so it is reasonable to do so with a computer. In the automatic linearization of the manipulator dynamic model, presented in para. 1.3, these matrices are determined parallel to determining matrices H and h (without the use of numerical differentation).

The system of equations (2.3.2) can be solved in two ways using first order gradient procedures, (the algorithms at the basis of Haysian's matrices cannot be applied in this case, because of the complexity of determining these matrices). The first of these algorithms is based on eliminating the control vector from the first two equations of system (2.3.2), using the third equation. Thus,

$$u^o(t) = -\underline{R}^{-1}(S^T x^o(t) + \hat{B}^T(x^o)p^o(t)),$$

$$\dot{x}^o(t) = \hat{A}(x^o) - \hat{B}(x^o)\underline{R}^{-1}(S^T x^o(t) + \hat{B}^T(x^o)p^o(t)),$$

$$x^o(0) \text{ is given, } x^o(t_S)\varepsilon X^F, \qquad (2.3.4)$$

$$\dot{p}^o(t) = S\underline{R}^{-1}S^T x^o(t) + x^{oT}(t)S\underline{R}^{-1}\frac{\partial\hat{B}(x^o)}{\partial x}p^o(t) +$$

$$+ ((S\underline{R}^{-1}\hat{B}^T(x^o) - \frac{\partial\hat{A}(x^o)}{\partial x})p^o(t) + p^{oT}(t)\hat{B}(x^o)\underline{R}^{-1}\frac{\partial\hat{B}(x^o)}{\partial x}p^o(t)),$$

$$p_i^o(t_S) = 0, \; i=i_1+1,\dots,N.$$

To solve this system of equations of order 2N is not easy because the initial state of the adjoined vector p : p(0) is not known. The system can be solved by supposing some initial vector p(0). With this and the prescribed $x^o(0)$ one integrates the system (2.3.4). One then checks that the boundary conditions are satisfied, i.e., that the conditions $x^o(t_S)\varepsilon X^F$ and $p_i^o(t_S) = 0$, $i=i_1+1,\dots,N$, are satisfied.

If an error occurs, the correction $\Delta p(0)$ should be found which reduces it. This procedure should be repeated until the error is reduced to some value determined in advance. This procedure, however, is difficult to realize, because the adjoined system p behaves very unstably.

The second algorithm which can be used to solve the system (2.3.2) is an algorithm in the control space. It is based on an exact solution of the first two equations, while the error in the third is reduced by an iterative procedure. Let us suppose some control $u_o^o(t)$. With this control one calculates the corresponding nominal trajectory $x_o^o(t)$ based on the first equation of system (2.3.2). Thus,

$$\dot{x}_o^o(t) = \hat{A}(x_o^o) + \hat{B}(x_o^o)u_o^o(t).$$

The control $u_o^o(t)$ should ensure that $x_o^o(t_S)\varepsilon X^F$, which is not so difficult (for instance, $u_o^o(t)$ can be taken as a control which transfers the manipulator along a straight line with a velocity which changes according to some parabolic law). With the pair $x_o^o(t)$, $u_o^o(t)$ thus determined, it is necessary to solve the other equation of system (2.3.2). There are $N - i_1$ boundary conditions. The remaining i_1 boundary conditions $p_{io}^o(t_S)$, should be assumed, so the second equation should be solved starting at the instant t_S. Then one should check to what extent the third equation of system (2.3.2) is satisfied, i.e., choose the control correction

$$\delta u_o^o(t) = -k\frac{\partial H^T}{\partial u}(x_o^o(t),\ p_o^o(t),\ u_o^o(t)) = -k(\underline{R}u_o^o(t) +$$

$$+\ S^T x_o^o(t) + \hat{B}^T(x_o^o)p_o^o(t)),$$

where k is an arbitrary step size.

Now, the control becomes

$$u_1^o(t) = u_o^o(t) + \delta u_o^o(t).$$

However, there is the question of whether or not the chosen $u_1^o(t)$ ensures the satisfaction of the state vector boundary conditions in the terminal time instant t_S. If it does not, the starting supposition for the terminal value of the adjoined state vector part $p_i^o(t_S)$, $i=1,\ldots,i_1$, as well as the step-size k, should be changed. If $u_1^o(t)$ transfers the system to the terminal state in the region X^F, the minimization process continues in the same way until the following inequality is not satisfied:

$$\int_0^{t_S} \left|\left|\frac{\partial H^T}{\partial u}(t)\right|\right| dt \leq \varepsilon,$$

where ε is given in advance.

This algorithm in the control space does not create problems during integration of the equations. However, problems which are imposed by the choice of $p_i(t_S)$, $i=1,\ldots,i_1$ (ensuring the satisfaction of the state vector terminal conditions), are very great.

Hence, the solution of the canonical system of equations for the case of manipulator motion between two given states, is a very complex one, i.e., the synthesis of the optimal nominal trajectories by the "two boundary values" approach is a very tedious task. Apart from that, this method does not consider the control constraints of the real manipulation system or the constraints on the manipulator trajectories (region X^t).

That is, with many systems, notably with active mechanisms, the optimal solution need not satisfy the conditions of the system functionality. Often severe constraints are made on the system motion $x^o(t)$, $t \in T$, or the region X^t, to which the system state must belong, is very narrow. If these severe constraints were to be taken into account during optimization, numerous numerical problems would be encountered. If, on the other hand, these constraints are not considered in the first phase but one first synthesizes the optimal nominal trajectories and then checks that the constraints are satisfied, one can find it difficult to obtain an optimal solution with properties of system functionality. These severe constraints are the result of the natural demand that the nominal trajectories have the property of functionality. For instance, during the synthesis of nominal trajectories with biped systems, one can impose the condition of solution functionality-realization of anthropomorphic gait. The condition of anthropomorphic gait imposes very severe constraints on the dynamics of lower extremities. These are difficult to formulate precisely, so the synthesis of the optimal dynamically stable anthropomorphic gait is complicated by considerable numerical problems. Moreover, sometimes a particular type of anthropomorphic gait is required, which almost completely closes the region X^t in which the optimal solution is sought. Thus, the optimization of some conditionally chosen criterion becomes unnecessary, because a solution $x^o(t)$, $\forall t \in T$ is determined by the requirement of functionality.

As we said, with the class of mechanical systems considered, the condition of solution optimization, in the sense of some strictly defined criterion is usually not imposed at all, so the optimization of the nominal trajectories is not of prime significance. On the other hand, with the system class in question, in addition to the conditions of functionality, the condition of reliability is usually imposed, so it is necessary to synthesize a programmed control $u^o(t)$, $\forall t \in T$, which ensures at least a partial functioning of the system and satisfaction of the conditions of varying parameters or omission of some relevant components. In order to ensure this, as well to simplify the problem of the synthesis of programmed control from the point of view of calculation, it is desirable to perform, even at the stage of nominal dynamics, decoupling of system S to subsystems of lower order and thus reduce the problem to synthesis of programmed control for lower order subsystems (but taking care about the coupling among subsystems). It should be stressed that this decoupling is not aimed at simplifying the control structure because we are concerned with programmed control but at overcoming the numerical problems and obtaining a functional and reliable solution, ensuring the minimum of functionality even in conditions of parameter variations and fallouts. Hence, in this case it is not necessary to decouple system parts, which are weakly coupled, but those system parts should be identified, which separate themselves in the functional sense as separate entities.

As stated in para. 2.2. it is difficult to carry out a procedure for optimal partition of the system into lower order subsystems. Thus, in the choice of functional subsystems in the concrete mechanical systems one must rely on engineering experience and separate the system parts which are functionally independent. Usually these functional subsystems are imposed by the characteristics of the system itself. Thus, for instance, with biped locomotion robots [3, 19] two functional subsystems can be separated. The subsystem formed by all the degrees of freedom of the legs should realize the chosen gait type, while the subsystem with the degrees of freedom of the upper body part has the task of ensuring the dynamical equilibrium of the complete system during the gait. With manipulators of anthropomorphic type with 6 degrees of freedom, for instance, the first three degrees of freedom (the so-called minimal manipulator configuration) form a subsystem which has to realize the translation of the gripper with the working object. The other subsystem (three degrees of freedom of the gripper) has to realize the necessary working object orientation [20]. Similarly with many systems, two or more functional subsystems can be separated [21]. It is evident that with all the systems mentioned the coupling among the subsystems is significant, so that separate synthesis of the nominal states for each subsystem is not possible. However, it is possible to synthesize nominal trajectories for one subsystem so that a certain functionality is obtained, while during the synthesis of the trajectories for the second subsystem, care must be taken about the system as a whole.

If a partition of the system to functional subsystems is adopted, the synthesis of programmed control by minimizaiton of criterion (2.2.2) is not simplified. If we accept the solution which is optimal for each functional subsystem, this solution would be suboptimal with respect to the complete system S. However, often the nature of the problem, the demands of functionality and reliability, and the partition to tasks for functional subsystems all lead to a unique determination of the nominal trajectories for the functional subsystems; so the possibility for optimization is anulled.

Consider the case when the system S can be divided into two functional subsystems $\tilde{S}_o$ and $\tilde{S}_x$. In this case the state vector x is divided into two state vectors of subsystems x_o and x_x. These are of order N_1 and N_2, respectively: $x_o \varepsilon R^{N_1}$, $x_x \varepsilon R^{N_2}$, $N_1 + N_2 = N$. The input vector u(t) is also divided in two input vectors u_o and u_x, of order m_1 and m_2, respectively: $u_o \varepsilon R^{m_1}$, $u_x \varepsilon R^{m_2}$, $m_1 + m_2 = n$. In this case the system (2.2.1) can be written in the form

$$\tilde{S}_o : \dot{x}_o = \hat{A}_o(x_o, x_x) + \hat{B}_{oo}(x_o, x_x)\cdot N(u_o) + \hat{B}_{ox}(x_o, x_x)\cdot N(u_x),$$

$$x_o(0),\ x_x(0) \text{ - given} \qquad (2.3.5)$$

$$\tilde{S}_x : \dot{x}_x = \hat{A}_x(x_o, x_x) + \hat{B}_{xo}(x_o, x_x)\cdot N(u_o) + \hat{B}_{xx}(x_o, x_x)\cdot N(u_x), \qquad (2.3.6)$$

where $\hat{A} = (\hat{A}_o^T, \hat{A}_x^T)^T$, $\hat{B} = \left[\begin{array}{c|c} \hat{B}_{oo} & \hat{B}_{ox} \\ \hline \hat{B}_{xo} & \hat{B}_{xx} \end{array}\right]$, $x = (x_o^T, x_x^T)^T$,

and $\hat{A}_o : R^{N_1}\times R^{N_2}\to R^{N_1}$, $\hat{A}_x : R^{N_1}\times R^{N_2}\to R^{N_2}$, $\hat{B}_{oo} : R^{N_1}\times R^{N_2}\to R^{N_1\times m_1}$, $\hat{B}_{ox}$: $R^{N_1}\times R^{N_2}\to R^{N_1\times m_2}$, $\hat{B}_{xo} : R^{N_1}\times R^{N_2}\to R^{N_2\times m_1}$, $\hat{B}_{xx} : R^{N_1}\times R^{N_2}\to R^{N_2\times m_2}$.

The control task can also be distributed to subsystems. Subsystem $\tilde{S}_o$ should be transferred from a given state $x_o^o(0)\varepsilon X_o^I \subset R^{N_1}$ into region $x_o^o(t, x^o(0))\varepsilon X_o^F$, $X_o^F \subset R^{N_1}$, $\forall t\varepsilon T_s$, where the subsystem state trajectory should satisfy*) condition $x_o^o(t, x^o(0))\varepsilon X_o^t \subset R^{N_1}$, $\forall t\varepsilon T$. Similarly, the state $x^o(0)\varepsilon X_x^I \subset R^{N_2}$ of subsystem $\tilde{S}_x$ should be transferred to the region $x_x^o(t, x^o(0))\varepsilon X_x^F \subset R^{N_2}$, $\forall t\varepsilon T_s$, where $x_x^o(t, x^o(0))\varepsilon X_x^t \subset R^{N_2}$, $\forall t\varepsilon T$. The following must be also satisfied: $X_x^I\times X_o^I = X^I$, $X_x^F\times X_o^F = X^F$ and $X_x^t\times X_o^t = X^t$.

When system S is thus separated into two subsystems $\tilde{S}_o$ and $\tilde{S}_x$ and when

---

*) We mean that the subsystem $x_o(t)$ and $x_x(t)$ states are functions of the chosen control $u_o^o(t)$ and $u_x^o(t)$. However, as the programmed control is synthesized, the nominal trajectories are explicit time functions and also depend on the choice of the initial states $x_o^o(0)$ and $x_x^o(0)$.

the task is defined in such a way that it can be distributed with respect to subsystems, definition of the nominal trajectories $x_o^o(t)$, $x_x^o(t)$ and of the programmed control u_o^o, $u_x^o(t)$, $\forall t\varepsilon T$ is unique. Such a case arises in two earlier examples. Thus, the biped system is a typical example of a redundant system if synthesis of arbitrary leg motion is concerned. However, if some particular gait type is required and if system S is divided into two functional subsystems $\tilde{S}_o$ (lower extremities) and $\tilde{S}_x$ (upper body part), the synthesis of the nominal dynamics becomes unique. Subsystem $\tilde{S}_o$ should realize the desired gait type so that, if the gait speed has also been defined (step period and step size), the leg trajectories are determined uniquely. When the nominal state $x_o^o(t)$, $\forall t\varepsilon T$ for $\tilde{S}_o$ is defined, the dynamics of the upper body part $\tilde{S}_x$ is defined from the conditions of the system equilibrium dynamics during gait in nominal (unperturbed) conditions [3, 9, 22].

One can likewise proceed in the case of a manipulator with six degrees of freedom. Consider the task of transferring the working object along a determined trajectory in absolute space with a determined orientation during the transfer. If this task is transferred to the problem of trajectory synthesis for two functional subsystems, the solution becomes unique. First, for subsystem $\tilde{S}_o$ (formed by the three degrees of freedom of the minimal configuration) one synthesizes the trajectories $x_o^o(t)$, $\forall t\varepsilon T$, which realize the motion of the manipulator tip along some desired trajectory. If the manipulator tip trajectory has been defined, all three mechanical degrees of freedom are uniquely defined. The only remaining question is about the distribution of the velocity and acceleration in the time interval T, which can be optimized at the level of subsystem $\tilde{S}_o$*). When $x_o^o(t)$ has been defined, the trajectories of the subsystem $\tilde{S}_x$ (gripper) state are defined from the conditions of working object orientation during motion. In this way one obtains the nominal trajectories $x_x^o(t)$, $\forall t\varepsilon T$ for the subsystem $\tilde{S}_x$. Finally, from the complete system S (2.2.1) the programmed inputs $u^o(t)$, $\forall t\varepsilon T$ are determined.

It is thus possible with some system classes (with which we primarily concerned in this book) to simplify the problem of the nominal trajectories synthesis from the numerical point of view. This can be significant if the stage of the nominal dynamics can also be realized on-line or if this level is left to the operator, who is included in the con-

---

*) Supposing a "weak" influence of $\tilde{S}_x$ dynamics on $\tilde{S}_o$.

trol of the concrete system. On the other hand, artificial partition to functional subsystems ensures the finding of solutions to the desired functionality (i.e. that the system functions in a way which is expected by experience). Such a synthesis, apart from ensuring the minimal functionality in the case of fallouts and parameter variations, also ensures much easier control, change and maintenance of the nominal dynamics than is the case with optimal solutions, where all state coordinates contribute to the realization of requirements of the particular task.

It is therefore clear that the synthesis of nominal trajectories with concrete manipulation systems and concrete manipulation tasks can be made by the synthesizing suboptimal nominal dynamics. It is supposed that the trajectories of the manipulator gripper are determined by the task itself and the corresponding trajectories of the internal manipulator coordinates, as well the corresponding programmed control (ch. 3) are also thus determined. However, this does not exclude the possibility of minimizing the chosen criterion in determining particular parameters of the chosen trajectory (e.g. velocity profile), where various optimization algorithms can be used. Such a procedure is dynamical programming, the use of which for optimizing the velocity distribution along the set trajectory of the manipulator tip, will be also considered in ch. 3.

It should be mentioned that various procedures for synthesizing nominal trajectories have also been developed based on the minimization of particular criteria. These were mentioned in the Introduction and will not be considered here in more detail [23].

Consider the optimality problem with redundant manipulators. By partitioning the system to subsystems and distributing the task over the subsystems, a unique definition of the trajectories is not obtained. In this case the problem is how to distribute the desired motion over the degrees of freedom of the manipulation system, even when the motion of the working object (or gripper) has been fully defined. This problem was treated in the Introduction and will not be persuaded further.

It should be kept in mind that the synthesis of nominal trajectories does not necessarily include calculation of programmed control $u^o(t)$, $\forall t \epsilon T$ which realizes $x^o(t)$. Often in applications the nominal trajectories are synthesized at the basis of kinematic models of manipulation systems and it is left to the control synthesized at the stage of perturbed dynamics to realize them. As we said in para. 2.2, however, the realization of trajectories by a control, synthesized at the basis of approximate models, is frequently inadequate. Hence, it is necessary to introduce programmed centralized control. This will be explicitly demonstrated by concrete examples in ch. 3. However, it should be said that the synthesis of the nominal trajectories at the basis of the kinematical system models, without calculating the programmed control, is much simpler and hence much more suitable for application, when the trajectories have to be synthesized on-line.

2.4. Stage of Perturbed Dynamics

At the stage of perturbed dynamics the control should be synthesized, which will satisfy the conditions of practical stability. At this stage the control synthesis is performed using the model of deviations of the system state from the nominal trajectories $x^o(t)$. The deviation model of the system S state from nominal trajectories can be written in the form

$$S : \Delta\dot{x} = \bar{A}^o(\Delta x,\ t) + \bar{B}^o(\Delta x,\ t)N(t,\ \Delta u), \qquad (2.4.1)$$
$$\Delta x(0) = x(0) - x^o(0) \text{ - given},$$

where $\Delta x \varepsilon R^N$ is the deviation vector of the system state from the nominal trajectory. That is, $\Delta x(t) = x(t) - x^o(t)$, $\Delta u \varepsilon R^n$, is the deviation vector of control from the programmed input $\Delta u(t) = u(t) - u^o(t)$, $\bar{A}^o(\Delta x,\ t) : R^N \times T \to R^N$ is the vector function of order N, $\bar{A}^o(\Delta x,\ t) = \hat{A}(x) - \hat{A}(x^o(t)) + [\hat{B}(x) - \hat{B}(x^o(t))]u^o(t)$ and $\bar{B}^o(\Delta x,\ t) : R^N \times T \to R^{N\times n}$ is the matrix function of order $N\times n$ given by $\bar{B}^o(\Delta x,\ t) = \hat{B}(x) = \hat{B}(\Delta x + x^o(t))$. The nonlinearity of the amplitude saturation type of inputs $\Delta u(t)$ is given by $N(t,\ \Delta u) = (N(t,\ \Delta u^1),\ldots,N(t,\ \Delta u^n))^T$, where

$$N(t,\ \Delta u^i) = \begin{cases} -u^i_m - u^{oi}(t) & \text{for } \Delta u^i \leq -u^i_m - u^{oi}(t) \\ \Delta u^i & \text{for } -u^i_m \leq u^i + u^{oi}(t) \leq u^i_m \\ u^i_m - u^{oi}(t) & \text{for } \Delta u^i \geq u^i_m - u^{oi}(t) \end{cases} \qquad (2.4.2)$$

and $\Delta u = (\Delta u^1,\ \Delta u^2,\ldots,\Delta u^n)^T$. From the nature of $\bar{A}^o(\Delta x,\ t)$ it follows that the condition $\bar{A}^o(0, t) = 0$, $\forall t \varepsilon T$ is satisfied, i.e., the coordinate origin for $\Delta x = 0$ is the equilibrium point ($\Delta u = 0$ for $\Delta x = 0$).

The stabilization problem of the system now reduces to the problem of synthesizing the control $\Delta u(t)$, ensuring that if the initial state of system S is in the region $\bar{X}^I = X^I - x^o(0)$, the system state also satisfies $\Delta x(t) \varepsilon \bar{X}^t = X^t - x^o(t)$, $\forall t \varepsilon T$ and $\Delta x(t) \varepsilon \bar{X}^F(t) = X^F - x^o(t)$, $\forall t \varepsilon T_s$. We supposed that the regions $\bar{X}^t(t)$, $\forall t \varepsilon T$, and $\bar{X}^F(t)$, $\forall t \varepsilon T_s$, are continually time-varying regions as usually defined in the literature.

In the course of control synthesis at the stage of perturbed states it is possible to minimize the chosen criterion (2.2.2) and thus ensure the "optimal" system stabilization around the nominal trajectory. However, as shown in paras. 2.2. and 2.3, the minimization of the criterion during the control synthesis for complex, nonlinear systems is fraught with difficulty. Hence, one usually considers approximate system models. At the basis of these control is synthesized and applied to the nonlinear system.

As we said in para. 2.2., we consider two versions of the control synthesis at the stage of perturbed dynamics, namely, centralized and decentralized control. As an example of centralized control, the synthesis of the linear optimal regulator will be demonstrated at the basis of the linearized system model. Decentralized control will be synthesized at the basis of the decoupled system model. In both cases synthesis will be based on minimization of the standard quadratic criterion.

2.5. Linear Optimal Regulator

In order to synthesize the linear regulator and ensure the above conditions of system practical stability, it is necessary to linearize the mathematical model of deviation of system (2.4.1) state around the nominal trajectories, i.e. around the coordinate origin in the state space for $\Delta x(t)$. Linearization of the deviation model (2.4.1) can be performed analytically or by means of various numerical procedures for model linearization on a digital computer, as shown in para. 1.3. Linearization can also be performed by means of various identification procedures. Regardless of how the linearization of the mathematical model is achieved, a linear mathematical model is obtained which approximately describes the deviation of the system state from the nominal trajectories $x^o(t)$ in the form:

$$S : \Delta\dot{x} = \tilde{A}^o(t)\Delta x + \tilde{B}^o(t)N(t, \Delta u), \ \Delta x(0) \text{ is given} \qquad (2.5.1)$$

where $\tilde{A}^o(t) : T \to R^{N\times N}$ is an $N\times N$ matrix, the elements of which are continuous functions of time, $\tilde{B}^o(t) : T \to R^{N\times n}$ is an $N\times n$ matrix, whose elements are piecewise continuous in the interval T. When considering the linear regulator we will suppose that it is possible to realize feedback with respect to the complete state vector Δx, i.e., that all state coordinates of system S are measured. If the order of the system output is lower than N then, under the supposition that system (2.5.1) is observable $\forall t \in T$, the state vector Δx can be estimated and the problem reduced to the upper case. These considerations will be presented in para. 2.7.

A square performance criterion is defined in the form

$$J(\Delta u) = \frac{1}{2} \int_0^{t_S} e^{2\Pi t}[\Delta x^T Q(t)\Delta x + \Delta u^T \underline{R}(t)\Delta u]dt + \\ + \frac{1}{2}\Delta x^T(t_S)Q_T x(t_S), \qquad (2.5.2)$$

where $Q(t) : T \to R^{N\times N}$, $Q_T : R^{N\times N}$ are positive semidefinite and $\underline{R}(t) : T \to R^{n\times n}$ is a positive definite matrix of corresponding dimensions. Interval t_S, prescribed stability degree Π and the weighting matrices $Q(t)$, Q_T, $\underline{R}(t)$ should be chosen in such way as to satisfy the conditions of practical system stability.

By minimizing the criterion (2.5.2), taking care of constraint (2.5.1) and supposing that the matrix pair $[\tilde{A}^o, \tilde{B}^o]$ is controllable in the interval $(0, t_S)$, optimal control is obtained in the form [24, 25]:

$$\Delta u(t) = -\underline{R}^{-1}(t)\tilde{B}^{oT}(t)K(t)\Delta x(t) = D(t)\Delta x(t), \qquad (2.5.3)$$

where $K(t) : T \to R^{N\times N}$ is a positive definite symmetrical matrix of order $N\times N$, which is the solution of the differential matrix equation of Riccati type, and $D(t) : T \to R^{n\times N}$ is a matrix of order $n\times N$, representing the feedback gains. The elements of matrices $K(t)$ and $D(t)$ are piecewise continuous functions of time in interval $(0, t_S)$. By the choice of t_S, Π, $Q(t)$, $\underline{R}(t)$ one can ensure, that the solution of the system with closed feedback

$$S : \Delta\dot{x} = [\tilde{A}^o(t) - \tilde{B}^o(t)\underline{R}^{-1}(t)\tilde{B}^{oT}(t)K(t)]\Delta x, \ \Delta x(0) \text{ is given} \qquad (2.5.4)$$

satisfies the stability conditions

$$\Delta x(t)\varepsilon \bar{X}^t(t), \quad \forall t\varepsilon T, \quad \Delta x(t)\varepsilon \bar{X}^F(t), \quad \forall t\varepsilon T_S \tag{2.5.5}$$

One should examine whether the solution of the nonlinear model of the deviation from the nominal state (2.4.1) satisfies the conditions (2.5.5) if control in the form (2.5.3) is used.

However, control in the form (2.5.3) requires time-varying gains in the feedback loops. The realization of such a control is difficult and this produces unreliability and more complex system maintenance. Hence, one simplifies the control by introducing constant gains in the feedback loops. A "time-average" of the mathematical model (2.5.1) should be taken in order to obtain a linear time-invariant model of the deviation from the nominal state:

$$S : \Delta\dot{x} = \bar{\tilde{A}}^o \Delta x + \bar{\tilde{B}}^o N(t, \Delta u), \ \Delta x(0) \text{ is given.} \tag{2.5.6}$$

If instead of (2.5.3) a criterion is introduced

$$J(\Delta u) = \int_0^\infty e^{2\Pi t}[\Delta x^T Q \Delta x + \Delta u^T \underline{R} \Delta u]dt, \tag{2.5.7}$$

where $Q_T = 0$ and $t_S \to \infty$, while $Q(t)$ and $\underline{R}(t)$ become time invariant matrices, control is obtained in the form [24, 25]

$$u(t) = -\underline{R}^{-1}\bar{\tilde{B}}^{oT} K \Delta x(t) = D\Delta x(t), \tag{2.5.8}$$

where $D\varepsilon R^{n\times N}$, $K\varepsilon R^{N\times N}$ is the solution of the matrix Riccati type equation

$$K(\bar{\tilde{A}}^o + \Pi I) + (\bar{\tilde{A}}^{oT} + \Pi I)K - K\bar{\tilde{B}}^o \underline{R}^{-1} \bar{\tilde{B}}^{oT} K + Q = 0. \tag{2.5.9}$$

Thus, one obtains a control with constant feedback gains. The right choice of Π, Q, R, ensures that the solution of the system with closed feedback

$$S : \Delta\dot{x} = (\bar{\tilde{A}}^o - \bar{\tilde{B}}^o \underline{R}^{-1} \bar{\tilde{B}}^{oT} K)\Delta x, \ \Delta x(0) \text{ is given,} \tag{2.5.10}$$

satisfies the stability conditions (2.5.5) for a finite region of initial conditions.

Up to now we have assumed that control is synthesized without constraints on the control amplitudes. However, as indicated in the deviation model (2.5.1) or (2.5.6), the inputs are limited with respect to amplitude. According to [26], the linear regulator ensures exponential stability of the linear system with the required stability degree Π in the region X^I_{LR} of the initial conditions defined by:

$$\Delta x(0) \in X^I_{LR}, \quad (Q + K\tilde{\bar{B}}^o \underline{R}^{-1}(2\lambda - I)\tilde{\bar{B}}^{oT}K) \text{ - positive definite matrix}$$

$$\lambda = \text{diag}\ (\lambda_i),\ \lambda_i = \frac{\dfrac{u^i_m - \max(u^{oi}(t))}{T}}{|(D\dot{\Delta}x(0))_i|}, \qquad |\lambda_i| \leq 1, \tag{2.5.11}$$

where $(\ )_i$ denotes the i-th element of the vector in parenthesis.

Hence, only for the initial states belonging to the finite region X^I_{LR}, will the linear system (2.5.1), or (2.5.6) satisfy the conditions (2.5.5). However, it is necessary to examine the behaviour of the nonlinear system (2.4.1) when the linear regulator (2.5.8) is used, i.e. it should be determined, whether the control, synthesized at the basis of the linearized model (2.5.6) satisfies the system S. One should analyze the stability of the nonlinear system (2.4.1) under the control (2.5.8) and one should compare with criterion (2.5.2) the "suboptimality" of the control (2.5.8) when applied to the system (2.4.1). For the sake of analysis, Liapunov's function for the system S is introduced in the form

$$v = (\Delta x^T K \Delta x)^{1/2}, \tag{2.5.12}$$

where $v : R^N \to R^1$, $v>0$. In order to examine the system S stability, the derivative of the function v is examined along the solution of system (2.4.1) and with control (2.5.8):

$$\dot{v} = (\text{grad } v)^T \Delta\dot{x} \leq -(\Pi + \frac{\lambda_m(W)}{2\lambda_M(K)})v + (\text{grad } v)^T[\bar{A}^o(\Delta x, t) -$$
$$- \tilde{\bar{A}}^o \Delta x + (\bar{B}^o(\Delta x, t) - \tilde{\bar{B}}^o) N(t, -\underline{R}^{-1}\tilde{\bar{B}}^{oT}K\Delta x(t))], \tag{2.5.13}$$

where $W = Q + K\tilde{\bar{B}}^o \underline{R}^{-1}(2\lambda - I)\tilde{\bar{B}}^{oT}K$, $W \in R^{N\times N}$, λ_m and λ_M denote the minimal and maximal eigenvalues of the corresponding matrix. The expression (2.5.13) is valid in the region X^I_{LR}, according to (2.5.11). The second member on the right side of (2.5.13) represents the influence of nonlinearity of the system S on its stability. In order that system S be asymptotically stable, the following sufficient condition must hold

$$\dot{v} \leq 0 \tag{2.5.14}$$

At the basis of (2.5.13) one can examine the region $\bar{X}_{LR}(t)$ in the state space R^N in which the condition of the asymptotic stability of system is satisfied. By examining the expression on the right side of (2.5.13) one can determine in the state space the region $\bar{X}_{LR}(t)$ for which this expression is negative. If the influence of the nonlinearity in (2.5.13) is strong, it might be that the nonlinear system is not practically stable even if the linear system itself is practically stable.

It should be mentioned that it is possible to linearize the nonlinear system in such a way that, with unlimited or limited control (2.5.8), one can achieve the required exponential stability of the original nonlinear system S in the same region X^I_{LR} in which the linear model is exponentially stable. However, the control designed in this way is not always purposeful from the point of view of energy since a linear model is sought which masters the "worst case" of the nonlinear system.

Since the criterion (2.5.7) has been minimized using the model (2.5.6) and not the exact model (2.4.1), it is evident that control (2.5.8) is suboptimal for system (2.4.1) with respect to criterion (2.5.7). That is, when the control (2.5.8) is applied to (2.4.1), the value of the criterion (2.5.7) will be different from the value which would be achieved if the control would be synthesized based on (2.4.1), with minimization of (2.5.7). It will also be different from the value taken by the chosen performance index when (2.5.8) is applied to the linearized model (2.5.6). One can examine how much is lost in "optimality" of control when (2.5.8) is applied to the nonlinear system just as one can examine the stability of system S [27, 28]. We shall not consider here the suboptimality of the linear regulator because this will come within the scope of the synthesis of decentralized control, which is of prime interest in this book.

It is evident that introducing the linear regulator requires a complex control structure with many feedback loops (n×N in the concrete case). For the sake of simplifying the control structure it should be determined, by varying the weighting matrices Q and $\underline{R}$, which of the feedback loops are essential [29]. One should then choose Q and $\underline{R}$ such that number of feedback loops is minimized and (2.5.5) remains satisfied. Nevertheless, the fact remains that the optimal regulator produces a complex and unsufficiently reliable control scheme. The stationary linear regulator in the form (2.5.8) is substantially simpler and more reliable than the time-varying regulator in the form (2.5.3). However, this simplification has been obtained at the expense of "optimality" loss in the solutions or an increase in energy requirements (since control by means of a stationary regulator covers the "worst case" of the time-varying regulator).

It should be mentioned that the choice of criterion, i.e. the weighting matrices in the criterion, is not unique, so the result of the "optimization" of such control is conditional. It is possible to examine the choice of the weighting matrices and their influence on the suboptimality of the linear regulator when applied to the nonlinear system S. By the choice of weighting matrices it is possible to ensure that the linear regulator decouples the linearized system model to the corresponding subsystems. Thus, by synthesizing centralized control, it is possible to decouple manipulation systems. This will be demonstrated in para. 2.6.6., where the asymptotic linear regulator for manipulation systems will be synthesized.

The synthesis of the linear regulator is, in the case of high dimensionality systems, linked with several problems of a numerical nature. The synthesis of the linear regulator overlooks the physical characteristics of systems. Thus, the use of engineering experience and the specific characteristics of the system (which can often simplify control structure and make control synthesis easier) is made more difficult. A fallout of one part of the system usually results, in the case of linear regulators, to instability of other system parts because of the coupling among them. Fallout of one feedback loop in the linear regulator usually results in instability of the complete system, this can be a serious drawback in such an approach to control system synthesis.

2.6. Synthesis of Decentralized and Global Control-Stage of Perturbed Dynamics

The major part of this book concerns decentralized control synthesis at the stage of perturbed dynamics. As we said in para. 2.2., the decentralized control will be synthesized using an approximate decoupled model of the system. Then we examine the performance of the exact model of the system when synthesized control is applied. We shall therefore consider the model of state deviation from nominal motion, not in the centralized form (2.4.1), but in terms of the model of the mechanical part of the system S^M and the models of the actuators S^i.

The model of deviation from the nominal state for the mechanical part of the system is given in the form

$$S^M : \Delta P = H(t, \Delta q)\Delta\ddot{q} + h(t, \Delta q, \Delta\dot{q}), \ \Delta q(0), \ \Delta\dot{q}(0) \text{ are given,} \tag{2.6.1}$$

where $\Delta P(t) = P(t) - P^o(t)$, $\Delta P \varepsilon R^n$ is the deviation of the driving torques in the joints of the mechanism from the nominal torques, $\Delta q(t) = q(t) - q^o(t)$, $\Delta q \varepsilon R^n$ is the deviation of the angles of the mechanism from their nominal values, $H(t, \Delta q) : T \times R^n \to R^{n\times n}$ is a matrix of dimensions $n\times n$ and $h(t, \Delta q, \Delta\dot{q}) : T\times R^n \times R^n \to R^n$ is a vector of dimension $n\times 1$. Here, q^o, $\dot{q}^o$, $\ddot{q}^o$ and $P^o(t)$ are the nominal angles, angular velocities, accelerations and driving torques, respectively (or linear displacements, linear velocities, accelerations and forces).

The model of deviation from the nominal trajectory for the actuator S^i is given in the form

$$S^i : \Delta\dot{x}^i = A^i\Delta x^i + b^i N(t, \Delta u^i) + f^i\Delta P_i, \ \Delta x^i(0) \text{ is given, } \forall i \varepsilon I, \tag{2.6.2}$$

where $\Delta x^i(t) = x^i(t) - x^{io}(t)$, $\Delta x^i \varepsilon R^{n_i}$ is the deviation of the actuator subsystem state vector from the nominal trajectory, $\Delta u^i(t) = u^i(t) - u^{io}(t)$, $\Delta u^i \varepsilon R^1$ is the deviation of the input from the nominal input.

The model of deviation from the nominal dynamics of the system S can be regarded as a set of subsystems S^i (2.6.2) interconnected through ΔP_i, which are given by (2.6.1). This "decoupling" of the system S into the set of subsystems corresponding to actuators in each d.o.f. of the mechanical part of the system, is very common in practice. However, the separation of the system S into the part which is to be associated with the subsystems S^i and the part which is to be associated with couplings ΔP_i may be done in various ways. Since a coupling ΔP_i is also the function of the state coordinates Δx^i of the subsystem S^i, it is possible to delete some terms from the expression for ΔP_i (those which depend on $\Delta x^i(t)$) and associate them with the free subsystem S^i. Let us consider ΔP_i in the form (2.6.1). The diagonal elements of the matrix $H^O(t, \Delta q)$ can be written in the form

$$H^O_{ii}(t, \Delta q) = \bar{H}_{ii} + (H^O_{ii}(t, \Delta q) - \bar{H}_{ii}), \quad \forall i \varepsilon I, \tag{2.6.3}$$

where the numbers $\bar{H}_{ii}>0$ will be chosen during the control synthesis. The i-th element of the vector $h^O(t, \Delta q, \Delta\dot{q})$ can be written in the form

$$h^O_i(t, \Delta q, \Delta\dot{q}) = \bar{h}_i \Delta q^i + (h^O_i(t, \Delta q, \Delta\dot{q}) - \bar{h}_i \Delta q^i), \quad \forall i \varepsilon I, \tag{2.6.4}$$

where the numbers $\bar{h}_i$ will be also chosen during the control synthesis. The subsystems models S^i can be written in the form

$$S^i: \Delta\dot{x}^i = \bar{A}^i \Delta x^i + \bar{b}^i N(t, \Delta u^i) + \bar{f}^i \Delta P^*_i, \quad \Delta x^i(0) \text{ is given}, \tag{2.6.5}$$

where $\bar{A}^i = \bar{C}^i(A^i + f^i \bar{h}_i T^i_1)$, $\bar{b}^i = \bar{C}^i b^i$, $\bar{f}^i = \bar{C}^i f^i$, $\bar{C}^i = (I_{n_i} - f^i \bar{H}_{ii} T^i)^{-1}$, $\Delta P^* = \Delta P_i - \bar{H}_{ii} T^i \Delta\dot{x}^i - \bar{h}_i T^i_1 \Delta x^i$, T^i, $T^i_1 \varepsilon R^{1\times n_i}$, $T^i \Delta\dot{x}^i = \Delta\ddot{q}^i$ and $T^i_1 \Delta x^i = \Delta q^i$.

Now, if the system is observed as a set of subsystems S^i (2.6.5) interconnected through ΔP^*_i instead through ΔP_i, as in the previous case, the model is distributed between the coupling and subsystems in a new way. The introduction of $\bar{H}_{ii}$ in the subsystem model S^i is very common: each actuator is considered to have an inertia (moment of inertia or equivalent mass) which includes not only its own inertia but also the inertia of that mechanical part of the system which is moved by the actuator. Introducing $\bar{h}_i$ into the subsystem model S^i means that the actuator model is considered with gravitational load, i.e., the gravitational moment, which is a function of Δq^i, is associated with the subsystem model S^i. The significance of such a distribution will be considered in para. 2.6.5.

By such decoupling the model of the system S one determines the information structure of the control. First, the system is stabilized by a control with decentralized structure, where each input Δu^i has information only about the state coordinates $\Delta x^i(t)$ of the corresponding subsystem, i.e. $\Delta u^i = \phi(\Delta x^i)$, $\forall i \varepsilon I$. In order to improve the stabilization of the system, global control is also introduced in the second step, so the input Δu^i has information about other state coordinates Δx. In order to ensure that the system S can be stabilized in the first step, applying only decentralized control, it should be assumed that the system S is controllable with the chosen control structure $\Delta u^i = \phi_i(\Delta x^i)$, $\forall i \in I$ [7, 30].

Thus, in order to solve the control task at the stage of perturbed dynamics, the control will be applied in the form [8, 12, 31]

$$\Delta u^i(t) = \Delta u_i^L(t) + \Delta u_i^G(t), \quad \forall i \in I, \tag{2.6.6}$$

where $\Delta u_i^L(t)$ will be chosen as a local control law for stabilizing the local subsystems S^i (free of coupling) which have information only about the state coordinates Δx^i, while $\Delta u_i^G(t)$ will be chosen as global control and a function of the coupling ΔP_i^*.

2.6.1. Local control synthesis

Local control of "free" subsystems S^i can be synthesized in various ways. Since linear, time-constant comparatively low order subsystems are in question, it is possible to use various classical methods of linear systems control synthesis, such as those concerning frequency domain the root-locus method, different pole assignment methods, methods based on minimizing different criteria is procedures for nonlinear control synthesis. Here, we shall only consider two common procedures for local control synthesis, namely, synthesis based on pole assignment by incomplete output feedbacks and synthesis based on local criterion minimization. The first method is a common approach to control synthesis, while the second permits a rigorous control synthesis and analysis of system performance (i.e. analysis of suboptimality). However, it should be noted that the procedure for the system analysis and the introduction of global control do not essentially depend on local control selection. Thus, the procedures for system analysis and global control synthesis that will be presented can be implemented when different local control laws are applied.

As already stated, we shall consider two procedures for local control synthesis:

(La) Let us assume that for each subsystem S^i the output Δy^i is measurable, where the output Δy^i is given by

$$\Delta y^i = C^i \Delta x^i, \quad \forall i \in I, \tag{2.6.7}$$

where $C^i \in R^{k_i \times n_i}$ is the output matrix and $\Delta y^i \in R^{k_i}$, $k_i < n_i$. That control should be synthesized which stabilize the free subsystems S^i (free of coupling)

$$S^i: \Delta \dot{x}^i = \bar{A}^i \Delta x^i + \bar{b}^i N(t, \Delta u_i^L), \; \Delta x^i(0) \text{ is given}, \quad \forall i \in I. \tag{2.6.8}$$

The local control will be synthesized in the form of output feedback:

$$\Delta u_i^L = -k_i^T \Delta y^i = -k_i^T C^i \Delta x^i, \quad \forall i \in I, \tag{2.6.9}$$

where $k_i \in R^{k_i}$ is the vector of local gains. Let us assume that the subsystems S^i are observable and controllable and that by incomplete output feedback the exponential stability of subsystem S^i with stability degree Π_i can be achieved, if the constraints on the input are not taken into account. That is, although the control task requires practical stability of the system, we synthesize local control to achieve local exponential stability of the subsystems. The synthesis of the control (2.6.9) is simple, since the subsystems S^i are of low order, linear, time-invariant and with single input. Thus, various classical methods for pole placement by incomplete feedback can be apllied. However, the choice of pole positions (i.e. degree of subsystems exponential stability) is not unique. But by using an iterative approach, by checking the practical stability of the global system, the poles of the free subsystems can be chosen, in some cases, so as to achieve the practical stability of the global system S (when local controls only are applied).

Since the inputs to the system are constrained (with respect to their amplitude), the local control in the form (2.6.9) guarantees the exponential stability (with stability degree Π_i) of the free subsystems S^i only in some finite regions X_i, which can be determined by

$$\Delta x^i(0) \in X_i \,.\text{AND}.\, \Delta u_i = -k_i^T \Delta y^i \rightarrow \|\Delta x^i(t)\| < M_i e^{-\Pi_i t} \|\Delta x^i(0)\|, \forall t \in T \tag{2.6.10}$$

where the M_i are positive constants, while for $\Delta x^i(0) \notin X_i$ the above stability condition is not satisfied. Thus, the subsystems are locally exponentially stable in finite regions $X_i \subset R^{n_i}$.

It should be noticed that the local gains in (2.5.9) can be synthesized by minimizing the local standard quadratic criterion. If it is possible to minimize this criterion by output feedback only (under the supposition that the local subsystems are observable and controllable), then the synthesized local control is suboptimal compared with local control minimization of the same criterion using complete feedback. However, we shall not consider the problem of the suboptimality of the output feedback gains. We limit ourselves to the case described when local gains in (2.6.9) are chosen by pole placement methods.

(Lb) The second possibile local control synthesis, which shall briefly be considered here, is by minimizing the local quadratic criterion associated with the subsystem S^i when the complete subsystem state vector x^i is measurable (i.e. if $k_i = n_i$) or the subsystem state is reconstructed by the observer (see appendix 2.C.).

Analogous to the considerations in para. 2.5. the control synthesis based on minimizing the standard quadratic criterion will be discussed. In order to achieve decoupling of the control synthesis with respect to the chosen subsystems, the criterion will be considered in the form

$$J(\Delta x(0), \Delta u) = \int_0^{\infty} e^{2\Pi t}(\Delta x^T Q \Delta x + \Delta u \underline{R} \Delta u)dt, \qquad (2.6.11)$$

where $Q \varepsilon R^{N\times N}$, $Q = \mathrm{diag}(Q_i)$, $\underline{R} \varepsilon R^{n\times n}$, $\underline{R} = \mathrm{diag}(r_i)$, $Q_i \varepsilon R^{n_i \times n_i}$, $\forall i \varepsilon I$, the Q_i are symmetric non-negative definite matrices, $r_i \varepsilon R^1$, $r_i > 0$, $\forall i \varepsilon I$, $\Pi \geq 0$.

It should be noticed that, although the system S is observed in the finite time interval T, the criterion (2.6.11) is defined for an infinite time interval. Since the subsystems models S^i are time-invariant they are valid for $t\varepsilon(\tau, \infty)$, while $H^o(t, \Delta q) = H^o(\tau, \Delta q)$ and $h^o(t, \Delta q) = h^o(\tau, \Delta q)$, $\forall t \varepsilon (\tau, \infty)$. The matrices Q and $\underline{R}$, and the number Π should be chosen so that $x^o(t)$, $u^o(t)$ should be tracked as required in the finite time interval T, if the decoupled model of the system is considered.

Since decentralized control structure should be applied, we would not directly synthesize the control minimizing (2.6.11), as we did in para. 2.5. but the control will be chosen in the form (2.6.6).

When synthesizing the local control the system S will be considered as a set of subsystems S^i (2.6.8) free of coupling ΔP_i^* (assuming that $\bar{h}_i$ and $\bar{H}_{ii}$ are chosen).

If we minimize the criterion (2.6.11) with the constraint (2.6.8) the local control synthesis is reduced to the minimization of the local criterion:

$$J_i(\Delta x^i(0), \Delta u_i^L(t)) = \int_0^\infty e^{2\Pi t}[\Delta x^{iT} Q_i \Delta x^i + \Delta u_i^L r_i \Delta u_i^L]dt, \ \forall i \varepsilon I \quad (2.6.12)$$

for each subsystem S^i (2.6.8) separately, and $J = \sum_{i=1}^{n} J_i$.

Under the assumption that the pair $(\bar{A}^i, \bar{b}^i)$ is completely controllable and the input constraint $N(t, \Delta u_i^L)$ is not taken into account, the optimal control minimizing (2.6.12) with the constraint (2.6.8) is in the form [25]

$$\Delta u_i^L(t) = -r_i^{-1}\bar{b}^{iT}K_i\Delta x^i(t), \quad (2.6.13)$$

where $K_i \varepsilon R^{n_i \times n_i}$ is a symmetric positive definite matrix and the solution of an algebraic Riccati equation. Under the assumption that $Q_i = \bar{C}^i\bar{C}^{iT}$, where $\bar{C}^i \varepsilon R^{n_i \times n_i}$, and the pair $(\bar{A}^i, \bar{C}^i)$ is completely observable, the subsystem (2.6.8) with closed feedback loop is exponentially stable around $x^{io}(t)$, $u^{io}(t)$ with stability degree Π. The optimal cost of the criterion (2.6.12) is

$$J_i^*(\Delta x^i(0)) = \Delta x^{iT}(0)K_i\Delta x^i(0), \quad \forall i \varepsilon I \quad (2.6.14)$$

for subsystem S^i (2.6.8).

If the amplitude constraint on the input $N(t, \Delta u_i^L)$ is taken into account, it can be shown that the exponential stability of the subsystem (2.6.8) can be guaranteed for the initial conditions belonging to the finite region of the subsystem state space, i.e. for $\Delta x^i(0) \varepsilon X_i$, it follows that J_i^* is given by (2.6.14) and $||\Delta x^i(t)|| \leq ||\Delta x^i(0)||e^{-\Pi t}$, $\forall t \varepsilon T$. For $\Delta x(0) \notin X_i$ these conditions need not be satisfied.

The local control (2.6.9) or (2.6.13) does not take into account the dynamics of the whole system, i.e., of the mechanism S^M. If we now consider the response of the complete system S when the control (2.6.13) is applied, two problems arise (as we said in para. 2.2.):

(a) The problem of the global system stability.

(b) The problem of the "suboptimality" of the control (2.6.13) with respect to the criterion (2.6.11).

2.6.2. Stability

We want to determine whether or not the global system S is practically stable around $x^o(t)$, $\forall t \varepsilon T$ in the sense of the definition in para. 2.2. However, we shall first determine whether or not the system is asymprotically stable around $x^o(t)$, $\forall t \varepsilon T$, and then check to see that the conditions of the practical stability of the global system are satisfied. This approach might be conservative but it is very simple and appropriate for the class of systems considered. It can be applied because for all $t \varepsilon T$ the point $\Delta x(t) = 0$ is the equilibrium point for the model S.

The general approach investigating the stability of large-scale systems by stability analysis of subsystems and examination of their interconnections has been applied to both Liapunov stability and for input-output stability analysis [32]. Since stability in finite regions of the state space is being considered, we shall use the composite system method for Liapunov stability analysis. In order to estimate the finite region in which the composite system S is asymptotically stable we shall apply Weissenberger´s method [33].

Thus, one should estimate the finite region in the state space R^N in which global system S is asymptotically stable. If we neglect coupling, the system would be exponentially stable in the region given by

$$X = X_1 \times X_2 \times \dots \times X_n . \tag{2.6.15}$$

In order to estimate the region of asymptotic stability of the system S, the characteristics of the subsystems should be expressed by non-negative continuous scalar functions v_i (subsystems Liapunov functions). Since the subsystems are exponentially stable, there exist functions v_i such that

$$\eta_{i1}||\Delta x^i|| \leq v_i(\Delta x^i) \leq \eta_{i2}||\Delta x^i|| \tag{2.6.16}$$

$$-\eta_{i3}||\Delta x^i|| \leq \dot{v}_i(\Delta x^i) \leq -\eta_{i4}||\Delta x^i|| \qquad \forall t \varepsilon T, \quad \forall i \varepsilon I, \tag{2.6.17}$$

where $\eta_{ik}>0$ are scalars for $k = 1, 2, 3, 4$, $v_i \geq 0$ and $v_i : R^{n_i} \rightarrow R^1$. Here, $\dot{v}_i$ is the derivative of v_i along the trajectory of the uncoupled subsystem S^i.

The choice of functions v_i depends on the subsystem characteristics and there are a number of effective procedures for determining Liapunov functions for low-order subsystems. It has been shown that the "optimal" analysis of aggregate stability of the system S (the less "conservative" analysis) is obtained if we find the subsystems Liapunov functions which estimate as well as possible the degree Π_i of the exponential stability of uncoupled subsystems, i.e., if we choose the functions v_i such that [31]

$$\dot{v}_i(\Delta x^i) = (\text{grad } v_i)^T \Delta\dot{x}^i \leq -\eta_{i4}\eta_{i2} v_i \leq -\Pi_i v_i, \quad \forall i \varepsilon I \tag{2.6.18}$$

Eq. (2.6.18) means that the function v_i exactly estimates the degree of exponential stability of the subsystem S^i. In the case (Lb), when local control is given by (2.6.13), Liapunov functions are chosen in the form (analogous to (2.5.12))

$$v_i = (\Delta x^{iT} K_i \Delta x^i)^{1/2}, \qquad \forall i \varepsilon I. \tag{2.6.19}$$

Since the control (2.6.13) is synthesized by minimizing (2.6.12), it follows that for the subsystem (2.6.8) with closed feedback loop,

$$\dot{v}_{i(\text{along } (2.6.8),\ (2.6.13))} \leq -\Pi v_i - \frac{1}{2} \frac{\lambda_m(W_i)}{\lambda_M(K_i)} v_i, \ \forall i \varepsilon I \tag{2.6.20}$$

where $W_i = Q_i + K_i \bar{b}^i r_i^{-1} \bar{b}^{iT} K_i$.

Due to the constraint on the input amplitude, conditions (2.6.18) or (2.6.20) are satisfied only for $x^i(0) \varepsilon X_i$. The finite region X_i can be estimated by the Liapunov function

$$\tilde{X}_i = \{\Delta x^i : v_i(\Delta x^i) \leq v_{io} . \text{AND}. \Delta x^i \varepsilon X_i\}, \forall i \varepsilon I, \ \tilde{X}_i \subseteq X_i, \tag{2.6.21}$$

where $v_{io} > 0$ should be chosen so that $\tilde{X}_i$ is the best possible estimation of X_i (the X_i are the regions of exponential stability of the subsystems). The region

$$\tilde{X}(0) = \tilde{X}_1 \times \tilde{X}_2 \times \ldots \times \tilde{X}_n, \quad \tilde{X}(0) \subset R^N \tag{2.6.22}$$

should be the best possible estimation of X (the region of asymptotic stability of the system S with no interconnections).

Since $\lim_{\Delta x \to 0} \Delta P_i \to 0$, $\forall i \varepsilon I$, the numbers $\xi_{ij} < +\infty$ can be determined and shown to satisfy the inequalities [34]

$$(\text{grad } v_i)^T \Delta P_i(t,\Delta x) \leq \sum_{j=1}^{n} \xi_{ij} v_j, \quad \forall (t,\Delta x)\varepsilon T\times\tilde{X}(0) \text{ and } \forall i\varepsilon I \tag{2.6.23}$$

where $\xi_{ij} \geq 0$ for $i \neq j$, $\forall i$, $j\varepsilon I$.

Now, we want to determine whether or not the composite system S is asymptotically stable in the region $\tilde{X}(0)$. According to [33], a sufficient condition for asymptotic stability is that

$$Gv_o < 0, \tag{2.6.24}$$

where $v_o \varepsilon R^n$, $v_o = (v_{1o}, v_{2o},\ldots,v_{no})^T$, $G\varepsilon R^{n\times n}$, and the elements of G are given (in the case of local control (2.6.13)) by

$$G_{ij} = -(\Pi + \frac{\lambda_m(W_i)}{2\lambda_M(K_i)})\delta_{ij} + \xi_{ij}, \tag{2.6.25}$$

where δ_{ij} is the Kronecker delta. If the condition (2.6.24) is fulfilled, the region $\tilde{X}(0)$ is an estimate of the region of global system asymptotic stability.

If condition (2.6.24) is satisfied, then the region $\tilde{X}(t)$ in the state-space which contains the state of the system during the tracking of $x^o(t)$ can be estimated by:

$$\max_{i\varepsilon I}(v_i(\Delta x^i(t))/v_{io}) \leq \max_{i\varepsilon I}(v_i(\Delta x(0))/v_{io})\exp(-\eta t), \tag{2.6.26}$$

where $\eta>0$ is calculated from

$$\eta = \min_{i\varepsilon I}(-v_{io}^{-1} \sum_{j=1}^{n} G_{ij}v_{jo}) \tag{2.6.27}$$

We can now check the practical stability of the system S. The condition (2.6.26) gives the estimation of the shrinkage of the bounds of the region $\tilde{X}(t)$ containing the solution of the system S. If

$$\bar{X}^I \subseteq \tilde{X}(0) \text{ and } \tilde{X}(t) \subseteq \bar{X}^t(t) \ \forall t\varepsilon T, \ \tilde{X}(t) \subseteq \bar{X}^F(t), \ \forall t\varepsilon T_s, \tag{2.6.28}$$

then the system S is practically stable around $x^o(t)$ with only local control. If the conditions (2.6.28) are not satisfied, we cannot guarantee that the chosen local control can stabilize the global system in desired manner. If we neglect the possible conservatism of this method of stability analysis, we can conclude that the destabilizing effect of the coupling among subsystems might be too strong. In order to overcome this, two approaches are possible. The first is to choose "stronger" local control by increasing the prescribed stability degree Π, or to redefine in an appropriate way the weighting matrix Q_i and the numbers r_i, $\forall i \varepsilon I$. By doing this we shall increase the absolute values of the negative diagonal elements of the matrix G and, because of an increase in the local feedback gains, the finite regions X_i in which the subsystems are exponentially stable, are decreased. The second method is to introduce the global control Δu_i^G as a function of the coupling among subsystems and to minimize the destabilizing effect of coupling.

It is possible to stabilize the system in a number of ways (as we explained in para. 2.2.), with the same decentralized information structure or with the additional global feedback loops. We shall not consider other possible methods of control synthesis.

Since $\lim\limits_{\Delta x \to 0} \Delta P_i^* \to 0$, $\forall i \varepsilon I$, let us assume that the coupling belongs to the class of functions which can be estimated by

$$|\Delta P_i^*| \leq \sum_{j=1}^{n} \sum_{k=1}^{n} \xi_{ijk}^{(1)} ||\Delta x^k|| \; ||\Delta x^j|| + \sum_{j=1}^{n} \xi_{ij}^{(2)} ||\Delta x^j|| \tag{2.6.29}$$

$\forall (t, \Delta x(t)) \varepsilon T \times \tilde{X}$. In this case the numbers ξ_{ij} in (2.6.23) are given by

$$\xi_{ij} = \frac{1}{2} \frac{\lambda_M(K_i)}{\lambda_m^{1/2}(K_i)} \bar{\bar{f}}^i \frac{\xi_{ij}^{(2)} + \sum\limits_{k=1}^{n} \xi_{ijk}^{(1)} \max\limits_{\tilde{X}_k} ||\Delta x^k||}{\lambda_m^{1/2}(K_j)}, \tag{2.6.30}$$

where $||\bar{f}^i|| = \bar{\bar{f}}^i$. This form of coupling will be considered shortly when different forms of global control will be discussed.

2.6.3. Suboptimality

If the conditions for the practical stability of the system are satisfied, one should investigate the suboptimality of the decentralized control (2.6.13) when it is applied to the coupled system S, i.e. one should determine the increase in the cost of the criterion (2.6.11) when the decentralized control (2.6.13) is applied to the coupled system S. It is obvious that when the control (2.6.13) is applied to S, the cost of the criterion (2.6.11) will be different from $J^*=\sum_i J_i^*$ (given by (2.6.14)). It can be shown that the maximum cost of the criterion (2.6.11) with the control (2.6.13) applied to the over-all system S can be estimated by [16, 27]

$$J(\Delta x(0)) \leq \left(1+\frac{2\xi}{\min\limits_{i\varepsilon I}\left(\frac{\lambda_m(W_i)}{2\lambda_M(K_i)} - \sum\limits_{j=1}^{n}\xi_{ji}\right)}\right)J^*(\Delta x(0)),\ \Delta x(0)\varepsilon\tilde{X}(0), \qquad (2.6.31)$$

where $\xi = \sum\limits_{i=1}^{n}\sum\limits_{j=1}^{n}\xi_{ij}$. The proof of (2.6.31) is given in appendix 2.A.

Here, it is assumed that condition (2.6.24) is satisfied. Obviously, expression (2.6.31) represents an upper bound of the criterion cost when the decentralized control is applied to the coupled system. This means that we assumed that the coupling increases the cost of the criterion compared to the cost when the system has no coupling $J>J^*$. However, the coupling could decrease the cost of the criterion (2.6.11), i.e., it could be that $J<J^*$. Still, in control synthesis our aim is to decrease the maximum possible cost of the criterion, i.e., we want to ensure that the chosen control law is not "too suboptimal" for any point in the observed region in the state space $X^t(t)$. The suboptimality of the chosen control can be changed in two ways: 1. by introducing global control to reduce the influence of coupling or 2. by various distributions of the system model between the part associated with subsystems and the part associated with coupling. This involves choosing $\bar{h}_i$ and $\bar{H}_{ii}$. Obviously, by examining the practical stability of the global system and the suboptimality we can determine the local control synthesis which will satisfy conditions of practical stability (assuming that the system can be practically stabilized with the chosen control structure). With the chosen local control the conditions for practical stability are checked and the suboptimality is estimated. If the results are unsatisfactory a new local control is chosen until the desired conditions are achieved. One thus iterates the procedure for "optimal" choice of the control with chosen decentralized structure (see para. 2.8).

2.6.4. Global control synthesis

In para. 2.6.1. the local control at the stage of perturbed dynamics is synthesized in two forms (2.6.9) and (2.6.13). In this paragraph we shall introduce global control to ensure the practical stability of the global system S. The global control can be introduced in both above mentioned cases of local control. However, in order to synthesize the global control more precisely, and in order to analyze more precisely the effect of the global control on the suboptimality of the control applied, we shall introduce global control only in the case when the local control is chosen by minimizing the local criterion (2.6.12). We shall illustrate global control when the local control has the form (La) by particular examples in ch. 3. In order to change the suboptimality of the chosen decentralized control we apply a global control which will take into account the coupling ΔP among subsystems, i.e., the dynamics of the mechanism S^M. Thus

$$\Delta u_i^G(t) = \phi_i(\Delta x^i, \Delta \bar{P}_i^*(t, x)), \quad \forall i \in I \tag{2.6.32}$$

where $\phi_i : R^{n_i} \times R^1 \to R^1$. If we want to synthesize the global control law by minimizing the criterion (2.6.11), we will encounter considerable numerical difficulties and the realization of such a control law might be difficult and expensive. Since our aim is to minimize the maximum possible cost of the criterion, we shall examine a particular control law which has to minimize the destabilizing effect of the coupling ΔP_i on global system stability and on the suboptimality of the decentralized control. The global control has to compensate the destabilizing effect of coupling and to improve the tracking of $x^o(t)$. The global system stability and the suboptimality of the chosen control are estimated by the numbers ξ_{ij}. Thus, we choose our global control according to the condition that the numbers $\xi_{ij}^* < +\infty$ satisfying the inequality

$$(\text{grad} v_i)^T \bar{f}^i \Delta P_i^* + (\text{grad} v_i)^T \bar{b}^i \bar{N}(t, \Delta x^i, \Delta u_i^G) \leq \sum_{j=1}^{n} \xi_{ij}^* v_j, \quad \forall i \in I \tag{2.6.33}$$

$\forall (t, x) \in T \times \tilde{X}(t)$, where $\xi_{ij}^* \geq 0$, for $i \neq j$, also satisfy $\xi_{ij}^* \leq \xi_{ij}$, $\forall i, j \in I$. Starting from condition (2.6.33) we choose the following particular form of global control [14]

$$\Delta u_i^G(t) = -K_i^G [(\text{grad} v_i)^T \bar{b}^i]^{-1} (\text{grad} v_i)^T \bar{f}^i \Delta \bar{P}_i^*, \quad \forall i \in I \tag{2.6.34}$$

assuming that $(\text{grad} v_i)^T \bar{f}^i \Delta P_i^* > 0$ and $(\text{grad} v_i)^T \bar{b}^i \neq 0$. We shall also suppose that numbers $\bar{\xi}_{ij}^* < +\infty$ can be found which satisfy the inequality

$$|[(\text{grad} v_i)^T \bar{b}^i]^{-1} (\text{grad} v_i)^T \bar{f}^i \Delta P_i^*| \leq \sum_{j=1}^{n} \bar{\xi}_{ij}^* v_j / \varepsilon_i, \quad \forall i \in I \tag{2.6.35}$$

for $(t, \Delta x) \in T \times \tilde{X}(t)$, where $\bar{\xi}_{ij}^* \geq 0$, $\forall i, j \in I$. The condition (2.6.35) can be satisfied assuming that

$$\lim_{(\text{grad} v_i)^T \bar{b}^i \to 0} [(\text{grad} v_i)^T \bar{b}^i]^{-1} (\text{grad} v_i)^T \bar{f}^i \Delta P_i^* \to M_i \quad \text{for } \Delta x^j \neq 0, \; j = 1, 2, \ldots, i-1, i+1, \ldots, n, \; M_i < \infty$$

The inequality (2.6.35) can also be treated as follows. The control (2.6.34) is applied in the region $X^{\varepsilon} \subseteq \tilde{X}(0)$ in which $|(\text{grad} v_i)^T \bar{b}^i| \geq \varepsilon_i$, for $\forall i \varepsilon I$, $\varepsilon_i > 0$, (i.e. $X^{\varepsilon i} \subseteq \tilde{X}_i$ and $X^{\varepsilon} = X^{\varepsilon 1} \times X^{\varepsilon 2} \times \ldots \times X^{\varepsilon n}$). If $\Delta x \notin X$ and $\Delta x \varepsilon \tilde{X}(0)$ the control (2.6.34) reaches its amplitude constraint so the inequalities (2.6.33) and (2.6.35) still hold. Here, $\bar{N}(t, \Delta x^i, \Delta u_i^G)$ denotes the amplitude constraint on the global control (a constraint of saturation type). In order to apply global control, the finite regions of the local subsystems exponential stability should be determined, taking into account the input amplitude constraint $\bar{u}_m^i$, which is below the exact input amplitude maximum u_m^i. This means that a portion of the input amplitude is aimed at the global control signal.

$$|\bar{N}(t, \Delta x^i, \Delta u_i^G)| \leq u_m^i - u^{io}(t) - \Delta u_i^L(\Delta x^i) \qquad (2.6.36)$$

In expression (2.6.34), $\Delta\bar{P}_i^*: T \times R^N \to R^1$ denotes the value which should determine the character of the global control and which should be the function of the coupling among subsystems. The choice of $\Delta\bar{P}_i^*$ should ensure that $\xi_{ij}^* \leq \xi_{ij}$, $\forall i,j \varepsilon I$, i.e. should compensate for the effect of coupling on the system stability.

The stability of the system S, when the global control is introduced, can be examined by the procedure described in para. 2.6.2., when the ξ_{ij}^* are substituted for ξ_{ij}. Since $\Delta\bar{P}_i^*$ is chosen so that $\xi_{ij}^* \leq \xi_{ij}$, the asymptotic stability of the global system $\bar{S}$ can be ensured. Furthermore, the appropriate choice of global control can ensure the practical stability of the system S around $x^o(t)$. However, the system stability can be improved by different local gains. In order to compare the stabilization of the system by global control and stabilization using local control, we shall examine the suboptimality of the control (2.6.13), (2.6.34). It can be shown (see appendix 2.A) that the maximum cost of the criterion (2.6.11) when controls (2.6.13) and (2.6.34) are applied to the composite system S, can be estimated by

$$J(\Delta x(0)) \leq \left(1 + \frac{2(\xi^* + \bar{\xi}^*) + \sum_{i=1}^{n} r_i K_i^{G2} \max_{j \varepsilon I} (\bar{\xi}_{ij}^*)^2 / \varepsilon_i^2}{\min_{i \varepsilon I} \left(\frac{\lambda_m(W_i)}{2\lambda_M(K_i)} - \sum_{j=1}^{n} \xi_{ji}^*\right)}\right) J^*(\Delta x(0)), \qquad (2.6.37)$$

where $\Delta x(0) \varepsilon \tilde{X}(0)$, $\xi^* = \sum_i \sum_j \xi_{ij}^*$ and $\bar{\xi}^* = \sum_i K_i^G \sum_j \bar{\xi}_{ij}^*$ and the numbers $\bar{\xi}_{ij}^*$ satisfy (2.6.35).

By comparing (2.6.31) and (2.6.37), one can determine the conditions under which the introduction of global control is "less suboptimal" than the application of only decentralized control (2.6.13). Furthermore, from (2.6.37) the global gains $K_i^G \varepsilon R^1$, $\forall i \varepsilon I$, can be determined so that the estimation of the suboptimality of the chosen control is minimal. When doing this, it should be remembered that ξ_{ij}^* are also functions of K_i^G.

We now consider several different realizations of the global control $\Delta\bar{P}^*$.

(Ga) Since the coupling ΔP_i among subsystems S^i (d.o.f.) in the mechanical large-scale system is represented by generalized forces which can be measured directly, it is possible to introduce force feedback as the global control. In this case, $\Delta P_i = \Delta\bar{P}_i^*$ (supposing that $\bar{h}_i = 0$, $\bar{H}_{ii} = 0$, $\forall i \varepsilon I$) i.e. the global control is represented by nonlinear feedback with respect to total coupling.

Under the assumption that the global control is below its amplitude constraint (2.6.36) and $X^\varepsilon = \tilde{X}(t)$, $\forall t \varepsilon T$, (2.6.33) transforms to

$$(\text{grad} v_i)^T \bar{f}^i \Delta P_i (1-K_i^G) \leq \sum_{j=1}^{n} \xi_{ij} (1-K_i^G) v_j \quad \forall (t, \Delta x) \varepsilon T \times \tilde{X}(t) \qquad (2.6.38)$$

i.e., it follows that $\xi_{ij}^* = (1-K_i^G)\xi_{ij}$, $\forall i,j \varepsilon I$. Since $\Delta\bar{P}_i^* = \Delta P_i$, it follows that $\bar{\xi}_{ij}^* = \xi_{ij}$, $\forall i,j \varepsilon I$. Thus, (2.6.37) becomes

$$J(\Delta x(0)) \leq \left[1 + \frac{2\xi + \sum_{i=1}^{n} r_i K_i^{G2} \max_{j \varepsilon I} \xi_{ij}^2 / \varepsilon_i^2}{\min_{i \varepsilon I} \left(\frac{\lambda_m(W_i)}{2\lambda_M(K_i)} - \sum_{j=i}^{n} \xi_{ji} (1-K_j^G) \right)}\right] J^*(\Delta x(0)) \qquad (2.6.39)$$

where $\Delta x(0) \varepsilon \tilde{X}(0)$. One can determine the global gains K_i^G which minimize the estimation of the suboptimality (2.6.39) of the control (2.6.13), (2.6.34). In the special case when $K_1^G = K_2^G = \ldots = K_n^F = K^G$ by minimizing (2.6.39) we get

$$K^G = (B/C - 1)\left(\sqrt{1 + \frac{2\xi}{A} \frac{C^2}{(B-C)^2}} - 1\right), \qquad (2.6.40)$$

where $A = \sum_{i=1}^{n} r_i \max_{j \varepsilon I} \xi_{ij}^2 / \varepsilon_i^2$, $B = \min_{i \varepsilon I} \frac{\lambda_m(W_i)}{2\lambda_M(K_i)}$, $C = \max_{i \varepsilon I} \sum_{j=1}^{n} \xi_{ji}$.

If (2.6.39) is compared with (2.6.31) it can be seen that global control (2.6.34) reduces the suboptimality of the applied control only if the global gain is within the bounds

$$0 \leq K^G \leq \frac{2\xi C}{A(B-C)}. \tag{2.6.41}$$

For the values of K^G which satisfy the inequalities (2.6.41) the global control reduces the estimation of the suboptimality of the local control. In the case when $2\xi C \geq A(B-C)$, the choice $K^G = 1$ (which means $\xi^*_{ij} = 0$, $\forall i,j \in I$, complete decoupling of the system S into the set of the subsystems S^i), is valid, since only in this case the decoupled system is "less suboptimal" than when local control is applied to the coupled system. It should be mentioned that due to (2.6.33), $K^G > 1$ is not valid, i.e. the supposition in (2.6.38) that $\xi^*_{ij} = (1-K^G_i)\xi_{ij}$ holds only if $0 \leq K^G_i \leq 1$.

Thus, we can decide when the application of global control in the form of the force feedback is useful for the reduction of the suboptimality of the chosen local control. The introduction of such global control increases the number of feedback loops for n only, and the nonlinearity of the i-th global control depends only on the state coordinates x^i of the subsystem S^i. Thus, all good features of the decentralized control are preserved (the simplicity of the structure, the control modularity, realibility and so on). However, it is necessary to introduce force transducers which represent the additional technical problems but which also increases the global control price [35].

(Gb) The second method of realizing global control is by on-line calculation of the coupling among subsystems. The coupling among subsystems is given by (2.6.1) and thus, $\Delta\bar{P}^*_i$ in (2.6.34) represents the calculated forces. However, since ΔP_i is a function of all state coordinates x and since the calculation of ΔP_i might be very complex in some cases, this on-line calculation would destroy all the advantages of the decentralized control. On the other hand, the real-time calculation of ΔP_i requires the use of a powerful digital computer, thus increasing the price of the control system.

In order to simplify the realization of the global control, it is possible to use approximate relations for on-line calculation of the coupling among the subsystems. However, the approximate expressions for the coupling should be chosen for each concrete system. We shall examine the case when the coupling is estimated by (2.6.29). Starting from (2.6.29), two different approximate models for coupling will be considered.

(Gb.1) The function $\Delta\bar{P}_i^*$ is chosen in the form which is linear in state coordinates. The calculation of the global control (2.6.34) thus reduces to linear terms in the estimation of the coupling (2.6.29). Let us suppose that $\Delta\bar{P}_i^*$ is in the form

$$\Delta\bar{P}_i^* = \sum_{j=1}^{n} \xi_{ij}^{(2)} ||\Delta x^j|| \tag{2.6.42}$$

When the global control (2.6.34) with (2.6.42) is applied, it follows from (2.6.33) that

$$\xi_{ij}^* \leq \frac{1}{2} \frac{\lambda_M(K_i)}{\lambda_m^{1/2}(K_i)} \bar{\bar{f}}^i \frac{[\xi_{ij}^{(2)}(1-K_i^G) + \sum_{k=1}^{n} \xi_{ijk}^{(1)} \max_{\tilde{X}_k} ||\Delta x^k||]}{\lambda_m^{1/2}(K_j)} \tag{2.6.43}$$

and from (2.6.35),

$$\bar{\xi}_{ij}^* = \frac{1}{2} \frac{\lambda_M(K_i)}{\lambda_m^{1/2}(K_i)} \bar{\bar{f}}^i \frac{\xi_{ij}^{(2)}}{\lambda_m^{1/2}(K_j)} \tag{2.6.44}$$

Here, it is again assumed that global control is below its amplitude constraint (2.6.36). Introducing (2.6.43), (2.6.44) and (2.6.30) in (3.6.37) we obtain the estimation of the suboptimality of the decentralized control (2.6.13) together with global control (2.6.34), (2.6.42).

$$J(\Delta x(0)) \leq \left(1 + \frac{2\xi + A^1 K^{G2}}{B^1 + K^G C^1}\right) J^*(\Delta x(0)), \tag{2.6.45}$$

where $A^1 = \sum_{i=1}^{n} \left[\frac{r_i}{4} \frac{\lambda_M^2(K_i)}{\lambda_m(K_i)} \max_{j\varepsilon I} \frac{(\xi_{ij}^{(2)})^2}{\lambda_m(K_j)} \bar{\bar{f}}^{i2}/\varepsilon_i^2\right]$,

$$B^1 = \min_{i\varepsilon I} \left(\frac{\lambda_m(W_i)}{2\lambda_M(K_i)} - \sum_{j=1}^{n} \bar{\bar{f}}^j \frac{\sum_{k=1}^{n} \xi_{jik}^{(1)} \max_{\tilde{X}_k} ||\Delta x^k|| + \xi_{ji}^{(2)}}{2\lambda_m^{1/2}(K_i)} \cdot \frac{\lambda_M(K_j)}{\lambda_m^{1/2}(K_j)}\right),$$

$$C^1 = \frac{1}{2} \max_{i\varepsilon I} \sum_{j=1}^{n} \frac{\lambda_M(K_j)}{\lambda_m^{1/2}(K_j)} \bar{\bar{f}}^j \frac{\xi_{ji}^{(2)}}{\lambda_m^{1/2}(K_i)},$$

The expression (2.6.45) is obtained under the assumption that $K_1^G = K_2^G = \ldots = K_n^G = K^G$. Expression (2.6.45) can be analyzed in the same way as (2.6.39). Expressions (2.6.40), (2.6.41) can also be used in this case when A^1, B^1 and C^1 are substituted for A, B and C, respectively.

Let us consider the case when the global control uses $\Delta\bar{P}_i^*$ which compensates not only the linear terms in the estimation (2.6.39), but also second-order terms. However, we still want to use the linear form of $\Delta\bar{P}_i^*$ so we choose $\Delta\bar{P}_i^*$ in the form

$$\Delta\bar{P}_i^* = \sum_{j=1}^{n} \Big(\sum_{k=1}^{n} \xi_{ijk}^{(1)} \max_{\tilde{X}_k^1} ||\Delta x^k|| + \xi_{ij}^{(2)} \Big) ||\Delta x^j|| \qquad (2.6.46)$$

where $\tilde{X}_k^1 \varepsilon R^{n_k}$, $k\varepsilon I$, are the finite regions in the subsystems state spaces, in which the coupling is observed for the global control synthesis. Let $\max_{\tilde{X}_k^1} ||\Delta x^k||$ be denoted by ξ_k^1.

Let us look at the estimation of the maximum criterion cost in the region $\tilde{X}(0) = \tilde{X}_1 \times \tilde{X}_2 \times \ldots \times \tilde{X}_n$, where $\tilde{X}_k \subseteq \tilde{X}_k^1$, $\forall k\varepsilon I$, when the global control (2.6.34), (2.6.40) is applied.

$$J(\Delta x(0)) \leq \Big(1 + \frac{2\xi + A^2 K^{G2}}{\min\limits_{i\varepsilon I} \frac{\lambda_m(W_i)}{2\lambda_M(K_i)} - B^2 + K^G C^2}\Big) J^*(\Delta x(0)) \qquad (2.6.47)$$

for $\Delta x(0)\varepsilon\tilde{X}(0)$, where

$$A^2 = \sum_{i=1}^{n} \Big[r_i \bar{\bar{f}}^{i2} \frac{\lambda_M^2(K_i)}{4\lambda_m(K_i)} \max_{j\varepsilon I} \frac{1}{\lambda_m(K_j)} \Big(\sum_{k=1}^{n} \xi_{ijk}^{(1)} \xi_k^1 + \xi_{ij}^{(2)} \Big)^2 / \varepsilon_i^2 \Big],$$

$$B^2 = \max_{i\varepsilon I} \frac{1}{2} \sum_{j=1}^{n} \bar{\bar{f}}^j \frac{\sum\limits_{k=1}^{n} \xi_{jik}^{(1)} \xi_k + \xi_{ji}^{(2)}}{\lambda_m^{1/2}(K_i)} \frac{\lambda_M(K_j)}{\lambda_m^{1/2}(K_j)},$$

$$C^2 = \max_{i\varepsilon I} \frac{1}{2} \frac{1}{\lambda_m^{1/2}(K_i)} \sum_{j=1}^{n} \bar{\bar{f}}^j \frac{\sum\limits_{k=1}^{n} \xi_{jik}^{(1)} \xi_k^1 + \xi_{ji}^{(2)}}{\lambda_m^{1/2}(K_j)} \lambda_M(K_j),$$

and $\xi_k = \max_{\tilde{X}^k} ||\Delta x^k||$. Here, it is assumed that $K_1^G = K_2^G = \ldots = K_n^G = K^G$. The suboptimality of this case can be further analyzed as before.

(Gb.2) In the second case, the total approximate model of the coupling

(2.6.29) is used for on-line calculation of $\Delta\bar{P}_i^*$, i.e. the global control should reduce both the linear and quadratic terms in the estimation of the coupling (2.6.29). Thus, $\Delta\bar{P}_i^*$ is taken in the form

$$\Delta\bar{P}_i^* = \sum_{j=1}^{n} \sum_{k=1}^{n} \xi_{ijk}^{(1)} ||\Delta x^k|| \; ||\Delta x^j|| + \sum_{j=1}^{n} \xi_{ij}^{(2)} ||\Delta x^j|| \qquad (2.6.48)$$

In this case, the estimation of the maximum criterion cost in the region $\tilde{X}(0)$ (when $K_1^G = K_2^G = \ldots = K_2^G = K_n^G$) has the form (2.6.48) but ξ_k in A^2 and C^2 should be substituted for ξ_k^1.

If $\tilde{X}_k \subseteq \tilde{X}_k^1$ implies $\xi_k^1 \geq \xi_k$, $\forall k \varepsilon I$, then it is obvious that the estimation of the suboptimality of the control with (2.6.48) is less than the estimation of the suboptimality when the global control with (2.6.46) is applied. This means that a more precise estimation of the coupling is used for the on-line calculation of the global control, the control becomes "less suboptimal". However, the more precise calculation of the coupling makes the control more expencive than does the linear approximation of the coupling. Only in the case when $\xi_k = \xi_k^1$, $\forall k \varepsilon I$ (i.e. when $\tilde{X}_k^1 = \tilde{X}_k$, $\forall k \varepsilon I$) the linear approximation of the coupling makes the same estimation of the suboptimality of the global control as the quadratic approximation of the mechanism´s dynamics. This shows the importance of the precise estimation of the finite regions in which the system state can exist during tracking and the importance of appropriate knowledge of dynamical models during the control synthesis. Thus, when the dynamics of the system is well known and the estimation of the regions where the state can exist is precise, the global gains can be determined so that the control applied is not "too suboptimal".

This procedure makes possible a relatively simple control synthesis because it permits the use of experience and knowledge about the particular system. That is, when the optimal regulator is synthesized it is difficult to anticipate what will be the consequence of the particular choice of weighting matrices, so the application of knowledge of the system characteristics is not possible. In the above control synthesis it is possible at every stage to anticipate the consequences of control choice. The choice of the local control is very simple. When the global control is introduced the characteristics of coupling among subsystems can be precisely taken into account. The control synthesis can be performed in several simple steps. The above stability analysis and suboptimality estimation enable one to quite precisely determine whether or not the desired conditions are satisfied. Thus, they make it possible to synthesize a satisfactory control in several iterations (see para. 2.8).

2.6.5. "Distribution of the model" between local subsystems and coupling

It is obvious that the relation between the degree of suboptimality of only the local control and the degree of suboptimality when the global control is introduced, depends on the choice of parameters $\bar{H}_{ii}$ and $\bar{h}_{ii}$ as defined in (2.6.3) and (2.6.4). With a different choice of $\bar{H}_{ii}$ and $\bar{h}_{ii}$ we can vary the distribution of the model between the part which is associated with local subsystems (2.6.8) and the part which is associated with the coupling among subsystems. Obviously, by increasing $\bar{H}_{ii}$ and $\bar{h}_i$ the larger part of the model is associated with the models of the local subsystems and the coupling ΔP_i^* is reduced (this holds only if $\bar{H}_{ii} \leq \min H_{ii}^o(t, \Delta x)$ and if $|\bar{h}_i| \leq \min|h_i^o(t, \Delta x)|$, where the minimum is taken over the set product $(t, \Delta x)\varepsilon T\times\tilde{X}(t))$.

Since $\bar{H}_{ii}$ represents the factor of inertia, by increasing $\bar{H}_{ii}$ we increase the inertia of actuators, so the local gains $-r_i^{-1}\bar{b}^{iT}K_i$ will be higher. On the other hand, by increasing $\bar{H}_{ii}$ the inertia of the coupling is reduced. However, variation of the suboptimality of control (2.6.13) when $\bar{H}_{ii}$ is varied should be examined acc. to (2.6.31), taking into account the fact that the performance index cost in the decentralized case can be estimated by $J^* \leq \sum_i \lambda_M(K_i)||\Delta x^i(0)||^2$ and that by increasing $\bar{H}_{ii}$, $\lambda_M(K_i)$ is also increased.

The parameter $\bar{h}_i$ is concerned with those terms in the coupling ΔP_i^* which are linear in state coordinates of the i-th d.o.f. With mechanical systems such terms are the moments due to gravitational forces. By increasing $\bar{h}_i$ the models of subsystems become "more unstable" (when considered without control), but the coupling is reduced. In appendix 2.B the variation of suboptimality of control (2.6.6), (2.6.13) or (2.6.9) is analyzed in the special case of first-order subsystems. This analysis depends on the choice of elements $\bar{h}_i$.

It should be noticed that by introducing $\bar{H}_{ii}$ and $\bar{h}_i$ in the subsystem models we design local control by taking into account estimations of the inertia and gravitational moments of the mechanical part of the system. If the global control (2.6.34) is introduced by applying ΔP_i^*

in the form (Ga) or (Gb) (while some terms in $H^o_{ii}(t, \Delta q)$ or $h^o(t, \Delta q, \Delta\dot{q})$ are computed either exactly or using approximate expressions), the inertia and gravity factors are also taken into account in the control synthesis. However, in this case, control is synthesized on the basis of exact values (or of better estimation) of inertia or gravity moments. In this case, nonlinear feedback has to be used as global control. When we introduce $\bar{H}_{ii}$ and $\bar{h}_i$ and synthesize local control only, the linear feedback should be realized. By comparing suboptimality (2.6.31) when $\bar{H}_{ii}\neq 0$ and when $\bar{H}_{ii}=0$ (2.6.36) but when global control computes the inertia of the mechanism, we can estimate which of the control laws is "less suboptimal". This analysis will be illustrated by an example in para. 3.3.3.

2.6.6. Decoupled control of manipulators based on asymptotic regulator properties*)

Para. 2.5. described the control synthesis $u(t)$, $\forall t \varepsilon T$, based on the optimal regulator. Another approach to control synthesis is based on decoupled control, which was described in paras. 2.6.1 - 2.6.5.

In this paragraph we solve the problem of decoupling manipulation systems by using asymptotic regulator properties. As we said in para. 2.5., such an approach results from one of the basic problems with the synthesis of the optimal regulator, namely, how to choose the weighting matrices of the quadratic performance index in order to achieve the desired closed-loop system properties (such as the imposed stability conditions and compensation of coupling). This problem has not been solved for the general case. Thus, when designing such regulators, the optimization procedure has to be repeated several times for various values of weighting matrices, until the desired performances is achieved.

However, a significant reduction in the number of these iterations in regulator design can be achieved by using its asymptotic properties when weighting coefficients by the control are reduced. The basic idea is not to apply a singular criterion, when $Q \geq 0$ and $\underline{R} = 0$, since this leads to bang-bang control, but to implement model properties of asymptotic regulators when the matrix $\underline{R}$ is positive definite and its ele-

---

*) This paragraph was written by Mr N.Kirćanski, M.Sc.

ments are significantly smaller than the elements of the matrix Q.

The theory of asymptotic regulators has recently been developed. In [36] an atempt has been made to implement the theoretical methods for determining the weighting matrices Q and $\underline{R}$ on the basis of required asymptotic eigenvalues and eigenvectors of the system.

In this paragraph the above results are used for control synthesis of manipulation systems.

We consider the linearized time-invariant model of the manipulation system (2.6.6) and the standard quadratic performance index (2.6.7), the weighting matrices of which are defined by $Q = \tilde{H}^T\tilde{H} = Q^T > 0$, $\underline{R} = \rho R^o = \underline{R}^T > 0$, with ρ being a positive real constant and $H \varepsilon R^{n\times N}$.

We assume further that (1) rank $(\tilde{H}\bar{\tilde{B}}^o) = n$ and (2) the zeros of det $(\tilde{H}(\sigma I_N - \tilde{A}^o)^{-1}\bar{\tilde{B}}^o)$ are distinct, have negative real parts and do not belong to the spectrum of $\tilde{A}^o$. Then the optimally controled system has the following model properties:

A. As $\rho \to 0$, there are N-n eigenvalues of the form

$$\sigma_i(\rho) \to \sigma_i^o, \quad |\sigma_i^o| < \infty, \qquad i=1,\dots,N-n$$

with associated eigenvectors

$$\tilde{x}_i(\rho) \to (\sigma_i^o I_N - \tilde{A}^o)^{-1}\bar{\tilde{B}}^o \nu_i^o, \tag{2.6.49}$$

where σ_i^o and ν_i^o are defined by

$$\tilde{H}(\sigma_i^o I_N - \tilde{A}^o)^{-1}\bar{\tilde{B}}^o \nu_i^o = 0. \tag{2.6.50}$$

B. As $\rho \to 0$ there are n eigenvalues of the form

$$\sigma_i(\rho) \to \sigma_i^\infty/\sqrt{\rho}, \quad \tilde{x}_i(\rho) \to \bar{\tilde{B}}^o \nu_i^\infty, \quad |\sigma_i^\infty| < \infty, \tag{2.6.51}$$

where σ_i^∞ and ν_i^∞ are defined by

$$R^o = \tilde{N}^{-T} S^{-2} \tilde{N}^{-1}, \quad \tilde{W}^T\tilde{W} = (\tilde{H}_o\bar{\tilde{B}}^o)^{-T}\tilde{N}^{-T}\tilde{N}^{-1}(\tilde{H}_o\bar{\tilde{B}}^o)^{-1}, \tag{2.6.52}$$

where $\tilde{N} = [\nu_1^\infty, \dots, \nu_n^\infty]$ and $S = \mathrm{diag}\,(\sigma_1^\infty, \dots, \sigma_n^\infty)$. The ν_i^∞'s are real and negative. $\tilde{W}$ is a nonsingular n×n weighting matrix defined by $\tilde{H} = \tilde{W}\tilde{H}_o$, where $\tilde{H}_o$ has the form $\tilde{H}_o = [\tilde{H}_{11} \mid I_n]$.

On the basis of these properties a procedure has been developed for determining weighting matrices of the manipulation system (2.5.6), with (a) asymptotic eigenvalues (modes) of the system and (b) asymptotic eigenvectors of the system all being given in advance. The conditions (a) define the positional and velocity modes of the manipulator, and the conditions (b) define the distribution of these modes to subsystems, which determines the coupling among manipulator subsystems [37].

Using these properties one obtains the complete, unique characterization of quadratic weights.

Let us apply these properties to control synthesis of a strongly coupled manipulation system with actuators which can be described by second-order subsystems, i.e. let us assume $n_i=2$. The system matrices $\bar{\tilde{A}}^o$ $\bar{\tilde{B}}^o$ may be represented in the form:

$$\bar{\tilde{A}}^o = \begin{bmatrix} 0 & I_n \\ \hline A_{21} & A_{22} \end{bmatrix}, \qquad \bar{\tilde{B}}^o = \begin{bmatrix} 0 \\ \hline B_{21} \end{bmatrix}, \tag{2.6.53}$$

where $A_{21} \varepsilon R^{n\times n}$, $A_{22} \varepsilon R^{n\times n}$, $B_{21} = \text{diag}(b_1,\ldots,b_n)$ and $b_i \varepsilon R^1$ by rearranging states in the following manner:*)

$$\Delta x = [\Delta q^1 \Delta q^2 \ \ldots \ \Delta q^n \Delta\dot{q}^1 \Delta\dot{q}^2 \ \ldots \ \Delta\dot{q}^n]^T,$$

where $\Delta x^i = [\Delta q^i \ \Delta\dot{q}^i]$ is the state of the i-th actuator S^i.

Using the relation (2.6.49) in the form $(\sigma_i^o I_N - \bar{\tilde{A}}^o) x_i^o = \bar{\tilde{B}}^o \nu_i^o$ and dividing the eigenvector x_i^o into two subvectors $x_i^o = [x_i^{o1T} \mid x_i^{o2T}]^T$, with $x_i^{o1} \varepsilon R^n$ and $x_i^{o2} \varepsilon R^n$, it follows that

$$\sigma_i^o x_i^{o1} - x_i^{o2} = 0, \ -A_{21} x_i^{o1} + (\sigma_i^o I_n - A_{22}) x_i^{o2} = B_{21} \nu_i^o \tag{2.6.54}$$

---

*) Here, we consider the special case when matrix $\bar{\tilde{B}}^o$ is diagonal in its lower half.

It is obvious from (2.6.54) that (a) eigenvectors associated with asymptotically finite modes of the manipulation system (2.5.6) may be divided into two linearly dependant n×1 subvectors, where the coefficients of proportionality are corresponding eigenvalues.

If we substitute $B_{21} = \text{diag}\,(b_1,\ldots,b_n)$ into (2.6.54) it follows that

$$\nu_i^o = \text{diag}(b_1^{-1},\ldots,b_n^{-1})\,[-A_{21} + \sigma_i^o(\sigma_i^o I_n - A_{22})]\,x_i^{o1} \tag{2.6.55}$$

The latter expression implies that for any subvector of the eigenvector x_i^o it is possible to evaluate the ν_i^o satisfying the relation (2.6.49). Hence, we can compute the normalized matrix $\tilde{H}_o$ using (2.6.50). As $\tilde{W}$ is a nonsingular matrix and $\tilde{H} = \tilde{W}\tilde{H}_o$, it follows from (2.6.50) that

$$\tilde{H}_o(\sigma_i^o I - \tilde{\bar{A}}^o)^{-1}\bar{B}^o \nu_i^o = 0 \qquad \forall i\varepsilon I \tag{2.6.56}$$

Equation (2.6.56) contains n×n linearly independant scalar equations each with n×n elements from the matrix $\tilde{H}_o$. Thus, all elements of the matrix $\tilde{H}_o$ are uniquely defined. The procedure may be further simplified by substituting (2.6.49) into the latter equation:

$$[\tilde{H}_{11} \mid I_n]\begin{bmatrix} x_i^{o1} \\ \hline \sigma_i^o \; x_i^{o1} \end{bmatrix} = \tilde{H}_{11} x_i^{o1} + \sigma_i^o x_i^{o1} = 0, \tag{2.6.57}$$

where the subvector x_i^{o1} is arbitrary. From (2.6.57) for i=1,...,n we obtain $\tilde{H}_{11}$ as follows

$$\tilde{H}_{11} = -[x_1^{o1},\ldots,x_n^{o1}]^{-1}[\sigma_1^o x_1^{o1},\ldots,\sigma_n^o x_n^{o1}] \tag{2.6.58}$$

The expression (2.6.58) shows how to evaluate the normalized matrix $\tilde{H}_o = [\tilde{H}_{11} \mid I_n]$ when the desired finite asymptotic eigenvalues σ_i^o and associated eigenvectors x_i^{o1}, $i\varepsilon\{1,\ldots,n\}$ of the manipulation system (2.6.53) are given. This makes it possibile to evaluate the weighting matrix Q when arbitrary asymptotic eigenvectors and their distribution among subsystems are given. This is a key feature of asymptotic regulators which enables the designer to achieve the decoupled control of the observed system. For example, if one assumes that $x_i^{o1} = [0\ldots0 \overset{(i)}{1} 0\ldots0]^T$ the mode $x_i(t) = e^{\sigma_i t} x_i^o$ is distributed over only the i-th degree of freedom of the manipulator, and asymptotic decoupling of the system is achieved.

Let us now consider the second part of the theorem concerning the asymptotic infinite modes.

From (2.6.51), it follows that $x_j^{\infty 1}$ and $x_j^{\infty 2} = B_{21}\nu_j^{\infty}$. It follows from this that $x_j^{\infty 2} = \mathrm{diag}(b_1,\ldots,b_n)\nu_j^{\infty} = [b_1\nu_j^{\infty i},\ldots,b_n\nu_j^{\infty n}]^T$. To achieve asymptotic decoupling i.e. $x_j^{\infty 2} = [0\ldots0\ x_{jd}^{\infty}\ 0\ldots0]^T$, it can be assumed that the vector ν_j^{∞} is $\nu_j^{\infty} = [0\ldots0\ \nu_{jd}^{\infty}\ 0\ldots0]^T$, where $\nu_{jd}^{\infty} = x_{jd}^{\infty}/b_j$.

Now, matrix $\tilde{N}$ from (2.6.52) becomes $\tilde{N} = [\nu_1^{\infty}\ldots\nu_n^{\infty}] = \mathrm{diag}(\nu_{1d}^{\infty}\ldots\nu_{nd}^{\infty})$. Matrix R^O is obtained in the following form,

$$R^O = \tilde{N}^{-T}S^{-2}\tilde{N}^{-1} = \mathrm{diag}[(\nu_{1d}^{\infty}\sigma_1^{\infty})^{-2}\ldots(\nu_{nd}^{\infty}\sigma_n^{\infty})^{-2}].$$

Using the expression (2.6.52), the following is obtained for $\tilde{W}$

$$\tilde{W}^T\tilde{W} = \mathrm{diag}[(b_1\nu_{1d}^{\infty})^{-2}\ \ldots\ (b_n\nu_{nd}^{\infty})^{-2}] \tag{2.6.59}$$

since $\tilde{H}_O\bar{\tilde{B}}^O = [\tilde{H}_{11} \mid I_n] \left[\begin{array}{c} 0 \\ \hline \mathrm{diag}(b_1\ldots b_n) \end{array}\right] = \mathrm{diag}(b_1\ldots b_n)$.

From (2.6.59) we obtain

$$\tilde{W} = \mathrm{diag}[(b_1\nu_{1d}^{\infty})^{-1}\ \ldots\ (b_n\nu_{nd}^{\infty})^{-1}]. \tag{2.6.60}$$

In a special case, when $\nu_{id}^{\infty} = b_i^{-1}$, i.e. $x_{id}^{\infty} = 1$, it follows that

$$R^O = \mathrm{diag}[(\frac{\sigma_1^{\infty}}{b_1})^{-2}\ \ldots\ (\frac{\sigma_n^{\infty}}{b_n})^{-2}],\ \tilde{W} = I_n \tag{2.6.61}$$

The above procedure enables one to determine the weighting matrices of the quadratic criterion using asymptotic eigenvalues and eigenvectors of the manipulation system defined in advance.

It has been proved that there are N-n dominant (finite) system poles, which can be defined in advance. In each corresponding eigenvector n coordinates can also be defined in advance. Thus, the matrix Q is specified according to the described procedure and is exact to within the magnitude of one multiplicative weighting matrix $\tilde{W}$.

It has also been proved that there exist n poles of the manipulation system which tend to infinity when $\rho \to 0$. By prescribing the ratios of these eigenvalues and vectors, the matrices $\tilde{W}$ and R^O can be completely determined. It follows that by prescribing the finite and infinite asymptotic modes, as well as the distribution of these modes over the state coordinates, the matrices Q and R^o can be fully determined.

The free parameter not specified by this procedure is the proportionality factor ρ. The value of this parameter can be determined by repeating for a few values of ρ the optimizational procedure used to solve the Riccatti equation

$$K\bar{\tilde{A}}^o + \bar{\tilde{A}}^{oT}K - \frac{1}{\rho} K\bar{\tilde{B}}^o {R^o}^{-1}\bar{\tilde{B}}^{oT}K + Q = 0, \quad D = -\frac{1}{\rho} {R^o}^{-1}\bar{\tilde{B}}^{oT}K. \qquad (2.6.62)$$

It is necessary to choose the greatest value of ρ, for which the system with closed feedback yields the model characteristics approximating in a satisfactory manner the desired asymptotic characteristics for $\rho \to 0$. Thus, the feedback gains are not too high and a satisfactory degree of the system decoupling is achieved.

In para. 3.2.4. the described procedure will be illustrated by an example of a particular manipulation system.

2.7. Discrete-Time Control Synthesis

The control of large-scale mechanical systems such as manipulation systems, involves numerous problems and the necessity of using a fast computer in order to achieve the sampling periods compatible with the system dynamics. The proposed centralized control (linear optimal regulator) and decentralized control (with or without introducing global control) can be realized by analogue techniques but this involves many technical problems. Since advanced technology has rapidly decreased the cost and size of microcomputers, it is convenient to realize the proposed control laws by microcomputer.

In order to realize the control of the manipulators by microcomputer the discrete-time form of the control should be synthesized. That is, the control is realized by programme instructions, while all computation recurs periodically. The input data for discrete-time control realizations are samples of state variables of the system and are periodically read by appropriate conversion modules. The output data from microcomputers are coordinates of the input vector. These are connected, using appropriate modules, to the inputs of actuators. The values of the control signals are constant during the sampling periods, i.e.

$$u(t) = u(kT_D) \text{ for } kT_D \leq t < (k+1)T_D, \quad k=0,1,2,... \tag{2.7.1}$$

where T_D is the sampling interval, i.e., $t_k = kT_D$ are time-equidistant sampling points.

Since the sampling operation appears twice between the object (system S - manipulator + actuators) and the control device (microcomputer), such a system belongs to a so-called sampled-data system. Thus, it is necessary to consider the model of the system S in a discrete-time domain, i.e., a sampled-data model of the system. Let us consider an n d.o.f. mechanical system S consisting of a mechanical part of the system S^M and n actuators S^i. The model of the system in a continous-time domain is given by (1.1.30) and (1.2.1). The model of the system can be considered in centralized form (2.2.1). Starting from (2.2.1) the model in a discrete-time domain can be obtained in the form:

$$S: x(k+1) = G_D(x(k)) + B_D(x(k))N(u(k)), \quad k=0,1,2,... \tag{2.7.2}$$

where $x(k) = x(kT_D) \varepsilon R^N$ is the state vector in the moment $t_k = kT_D$, $u(k) = u(kT_D)$ is the input on the interval $kT_D \leq t \leq (k+1)T_D$, $G_D: R^N \to R^N$, $B_D: R^N \to R^{N \times n}$, $N(u(k)) = (N(u^1(k)), N(u^2(k)), ..., N(u^n(k)))^T$.

We shall consider the synthesis of a linear optimal regulator in a discrete-time domain starting with a linearized model of the system (para. 2.7.1). We shall then present the decentralized control synthesis in a discrete-time domain (para. 2.7.2). In both cases we start with the perturbation model (supposing that the nominal control and trajectories are computed in advance and stored).

Since we suppose that some state coordinates are not measurable, a decentralized observer is synthesized. The local optimal regulator and observer in a discrete-time domain is synthesized for the subsystems chosen in para. 2.6. As we said in para. 2.6, the system performance with local regulators could be unsatisfactory due to a strong destabilizing effect of coupling. The same is true of local observers. In order to compensate for this effect of coupling the programmed and/or global control are introduced. In para. 2.7.2 we shall consider the case when programmed control is synthesized at the subsystems level but not on the basis of the centralized model of the system. Thus, the programmed nominal control ensures the tracking of the nominal trajectory only if decoupled subsystems are considered. Such a programmed control does not compensate the effect of the coupling. In some cases it is therefore necessary to introduce global control. Para. 2.7.2. presents the global control synthesis when the control is realized by on-line computation of the coupling among subsystems using an aproximate model.

2.7.1. Linear optimal regulator synthesis in a discrete-time domain

Let us assume that the nominal trajectory $x^o(0)$, $x^o(1),\ldots,x^o(L)$, where $\tau=LT_D$, and the corresponding programmed control $u^o(0)$, $u^o(1),\ldots$ $\ldots,u^o(L)$, synthesized on the basis of the centralized model of the system (2.7.2) are given. Let us consider the linearized model (2.5.1) of the deviation of the system from nominal trajectory and control.

Using the condition (2.7.1) and by integrating the model (2.5.1) over the interval $t\varepsilon\{kT_D, (k+1)T_D\}$, the following discrete system model is obtained

$$S:\Delta x(k+1)=\phi^o(k)\Delta x(k)+D^o(k)N(k, \Delta u(k)),\ \Delta x(0) \text{ is given,} \qquad (2.7.3)$$

where $\Delta x(k)=x(k)-x^o(k)\varepsilon R^N$ is the vector of the system state at the moment of sampling kT_D, $\Delta u(k)=u(k)-u^o(k)$ is the control vector on the interval $kT_D \le t < (k+1)T_D$, $\phi^o(k)\varepsilon R^{N\times N}$ and $D^o(k)\varepsilon R^{N\times n}$ are the matrices of the discrete model, which are obtained from the equations

$$\phi^o(k) = \int_{kT_D}^{(k+1)T_D} \tilde{A}^o(\tau)d\tau = \phi^o((k+1)T_D, kT_D),$$

$$D^o(k) = \int_{kT_D}^{(k+1)T_D} \phi^o((k+1)T_D, \tau)\tilde{B}^o(\tau)d\tau. \qquad (2.7.4)$$

$N(k, \Delta u(k))$ denotes the nonlinearity of the amplitude saturation type at the moment of sampling kT_D.

The system (2.7.3) belongs to the class of non-stationary discrete linear systems. From the point of view of practical application, it is often useful to deal with time-invariant manipulator models. Thus, instead of the linearized time-varying model (2.5.1) we can use the linear time-invariant model obtained by discreting the time-averaged model (2.5.6):

$$\Delta x(k+1) = \bar{\phi}^{o} x(k) + \bar{D}^{o} N(k, \Delta u(k)), \ \Delta x(0) \text{ is given} \tag{2.7.5}$$

where $\bar{\phi}^{o}$ and $\bar{D}^{o}$ are the matrices of the linear model and depend on the sampling period T_D:

$$\begin{aligned} \bar{\phi}^{o} &= \bar{\phi}^{o}(T_D) = \exp(\bar{\tilde{A}}^{o} T_D) \\ \bar{D}^{o} &= \bar{D}^{o}(T_D) = \int_0^{T_D} \exp(\bar{\tilde{A}}^{o}\, t)\bar{\tilde{B}}^{o} dt \end{aligned} \tag{2.7.6}$$

The control task in the discrete-time domain can be stated as follows. One determines control sequence $u(k)=u(kT_D)$, $k=1,\dots,L$ ensuring the practical stability of the system, i.e., for every initial condition $x(0)\varepsilon X^I$ the following should hold: $x(t)\varepsilon X^t$, $\forall t\varepsilon T$ and $x(t)\varepsilon X^F$, $\forall t\varepsilon T_s$. The regions X^I, X^t and X^F are defined in para. 2.2. Such a control sequence formed at the output of the discrete regulator ensures that after $L_s<L$ samples the manipulator approaches the nominal trajectory, $\tau_s = L_s T_D$.

The control synthesis can be performed on the basis of the quadratic criterion (2.5.2), where we assume that the matrices $Q \geq 0$ and $\underline{R} > 0$ are invariant and $\Pi = 0$. A criterion thus defined evidently takes into account the behaviour of the system not only at the moments of sampling, but also during the whole duration of the functional manipulator movement $t\varepsilon T$.

It has been shown [38] that the problem of minimizing the criterion (2.5.2) reduces to the equivalent problem in the discrete-time domain where, instead of the model (2.5.6), the sampled-data model (2.7.5) is considered, and instead of criterion (2.5.2) (for $\Pi = 0$), one considers the following sum

$$J(\Delta x(0), \Delta u(k), \tau) = \frac{1}{2}\sum_{k=0}^{L-1}[\Delta x^T(k)\hat{Q}\Delta x(k) + 2\Delta x^T(k)M\Delta u(k) + \Delta u^T(k)\hat{R}\Delta u(k)] + \frac{1}{2}\Delta x^T(L)Q_T\Delta x(L), \tag{2.7.7}$$

where the weighting matrices $\hat{Q}$, M and $\hat{R}$ are

$$\hat{Q} = \int_0^{T_D} e^{\bar{\tilde{A}}^{oT} t} Q e^{\bar{\tilde{A}}^{o} t} dt = \hat{Q}(T_D), \quad M = \int_0^{T_D} e^{\bar{\tilde{A}}^{oT} t} Q\underline{H}(t)\, dt = M(T_D) \tag{2.7.8}$$

$$\hat{R} = T_D\underline{R} + \int_0^{T_D} \underline{H}^T(t)Q\underline{H}(t)dt = \hat{R}(T_D), \text{ where } \underline{H}(t) = \int_0^t \exp(\bar{\tilde{A}}^o t)\bar{\tilde{B}}^o dt$$

It follows that the design of discrete regulators is reduced to determining the matrices of the model (2.7.5) and of the criterion (2.7.7). The problem is thus transferred from the continuous time domain to the corresponding discrete domain. The matrices in the discrete domain can be determined by numerically integrating (2.7.6) and (2.7.8).

When the matrices of the discrete-time model of the system are calculated the feedback matrix should be determined. The criterion (2.7.7) with the constraint (2.7.5) can be minimized using the maximum principle, dynamic programming etc. By using the optimization procedure one obtains the linear control law in the discrete-time domain $\Delta u(k) = -D(k)\Delta x(k)$, where the gain matrix of the discrete regulator is obtained in the form [39]

$$D(k) = (\hat{R} + \bar{D}^{oT}K(k)\bar{D}^o)(\bar{D}^{oT}K(k)\bar{\phi}^o + M^T), \tag{2.7.9}$$

where $K \varepsilon R^{N\times N}$ is the matrix obtained by solving the Riccati difference equation

$$K(k) = \hat{Q} + \bar{\phi}^{oT}K(k+1)\bar{\phi}^o + (\bar{D}^{oT}K(k+1)\bar{\phi}^o + M^T)^T \cdot$$
$$\cdot\ (\hat{R} + \bar{D}^{oT}K(k+1)\bar{D}^o)^{-1}(\bar{D}^{oT}K(k+1)\bar{\phi}^o + M) \tag{2.7.10}$$

Assuming that the time interval on which the system is observed $\tau >> T_D$, we can accept the solution of the corresponding Riccati algebraic equation (instead (2.7.10)).

The control is obtained in the form

$$\Delta u(t) = -\bar{D}\Delta x(k), \quad t\varepsilon(kT_D, (k+1)T_D), \tag{2.7.11}$$

where $k = 0, 1, \ldots, L-1$, $\bar{D} = (\hat{R} + \bar{D}^{oT}K\bar{D}^o)(\bar{D}^{oT}K\bar{\phi}^o + M^T)$.

Since in order to calculate the nominal control $u^o(t)$ it is necessary to use the complete model of the manipulation system (para. 2.3.), this part of the input vector cannot be, in general, synthesized in real time but it is synthesized off-line and stored in the microcomputer memory. One thus makes the nominal control discrete. The sampling rate should by high enough to ensure the satisfactory behaviour of the system under ideal conditions. Concerning the discrete regulator synthesis the matrices Q and R in performance index (2.5.7) can be chosen by (1) from experience and using simulation results, (2) using the asymptotic regulator properties (the decoupled control of manipulators in para. 2.6.6.), or in some other way. However, when the linear discrete regulator is synthesized, one more parameter should be chosen, namely, the sampling interval T_D. In para. 3.2.2. the performance of the particular manipulation system with control (2.7.11) for various values of sampling interval T_D will be examined.

2.7.2. Synthesis of decentralized regulator and observer in the discrete-time domain

The preceding paragraph presented the centralized linear regulator synthesis in a discrete-time domain. This paragraph briefly discusses the decentralized control synthesis in the discrete-time domain. In order to synthesize decentralized control the model of the system S is considered in the form (1.1.30), (1.2.1). The sampling period T_D is determined by the capabilities of the computer used, i.e., by the period necessary for control computation in one cycle. If the microcomputer is faster or if the chosen control law is easier to compute, the sampling period T_D is shorter. Assuming that the control is realized by a sufficiently fast computer system, one can suppose that the sampling period is sufficiently short compared with the system dynamics. We can then assume that the driving torques (forces) P_i in the mechanism joints vary slowly in comparison with T_D, so we can take it that $P_i(t) \approx P_i(kT_D)$ for $kT_D \leq t \leq (k+1)T_D$. Under these assumptions the models of the actuators (1.2.1) in the discrete-time domain can be written as [40]

$$S^i: x^i(k+1) = A_D^i x^i(k) + b_D^i N(u^i(k)) + f_D^i P_i(k), \quad \forall i \in I, \ k=0,1,2,\ldots$$

$$y^i(k) = C^i x^i(k), \tag{2.7.12}$$

where $A_D^i \in R^{n_i \times n_i}$, $A_D^i = \exp(A^i T_D)$, $b_D^i \in R^{n_i}$, $b_D^i = \int_0^{T_D} b^i e^{A^i \tau} d\tau$, $f_D^i \in R^{n_i}$, $f_D^i = \int_0^{T_D} f^i e^{A^i \tau} d\tau$.

According to (1.1.30) the driving torques are direct functions of state coordinates. The mathematical model of the mechanical part of the system in the discrete-time domain can therefore be written as the following set of algebraic equations

$$S^M : P(k) = F_D(x(k)), \quad k=0,1,2,3,... \tag{2.7.13}$$

where $F_D : R^N \to R^n$.

Let us consider the same control task. It should ensure that $\forall x(0) \varepsilon X^I$ implies $x(t) \varepsilon X^t$, $\forall t \varepsilon T$ and $x(t) \varepsilon X^F$, $\forall t \varepsilon T_s$. Let us also assume that the nominal trajectory is given: $x^o(0), x^o(1), ..., x^o(L)$. The nominal trajectory satisfies $x^o(0) \varepsilon X^I$, $x^o(k) \varepsilon X^t$, $\forall k \varepsilon L_t$, $x^o(k) \varepsilon X^F$ for $k=L_s, L_s+1, ..., L$, where $L_t = (k, k = 0,1,...,L)$. The tracking of the nominal trajectory should satisfy the conditions of practical stability and, in this case, the constraint on the control (2.7.1) should be taken into account.

As described in para. 2.2, various solutions of the control task can be used. One of them is the direct synthesis of the control u(k) on the basis of the centralized model of the system (2.7.2) which transfers the state from the region X^I to the region X^F within the settling time [41]. However, it can be shown that for mechanisms with many d.o.f. the model (2.7.2) is too complex for on-line implementation and requires the use of fast computers. This is unacceptable in practice from the point of view of price and system complexity. The second possibility was in the paragraph above, namely, linear regulator synthesis. Here, we shall consider the control law presented in para. 2.6. Decentralized control will be synthesized since it has significant adventages concerning the number of computer operations necessary for calculating the control u(k) during one sampling period. However, in order to compensate the effect of coupling it is convenient to introduce the nominal programmed control $u^o(k)$.

Let us assume that $u^o(k)$ can be determined so that

$$S: x^o(k+1) = G_D(x^o(k)) + B_D(x^o(k))u^o(k), \quad \forall k \varepsilon L_t \tag{2.7.14}$$

In other words, it should be assumed that the nominal trajectory $x^o(k)$, $\forall k \varepsilon L_t$, is such that $u^o(k)$ can be found to satisfy (2.7.14), $\forall k \varepsilon L_t$, i.e., let us assume that it is possible to determine $u^o(k)$, $\forall k \varepsilon L_t$ so that it transfers the state from $x^o(k)$ to $x^o(k+1)$. Let us now consider the model of the deviation of a system S from the nominal $x^o(k)$, $u^o(k)$.

The model of deviation from the nominal for the subsystem S^i has the form

$$S^i: \Delta x^i(k+1) = A_D^i \Delta x^i(k) + b_D^i N(k, \Delta u^i(k)) + f_D^i \Delta P_i(k), \quad \forall i \varepsilon I \qquad (2.7.15)$$

where $\Delta x^i(k) = x^i(k) - x^{io}(k)$, $\Delta u^i(k) = u^i(k) - u^{io}(k)$, $\Delta P_i(k) = P_i(k) - P_i^o(k)$.

Analogously, the model of the mechanical part of the system for deviation from the nominal has the form:

$$S^M : \Delta P(k) = \bar{F}_D(k, \Delta x(k)), \qquad (2.7.16)$$

where $\bar{F}_D: L_t \times R^N \to R^n$.

Since the subsystem output $y^i(k)$ is of lower order than n_i, it is possible to synthesize local control in the form of output feedback (2.6.9). The second possibility is local control synthesis by minimizing the local performance index and introducing an observer for state reconstruction.

As described in para. 2.6.1., we introduce the standard quadratic performance index which should be minimized by the control at the perturbed level. As in (2.6.11), in the discrete-time domain we introduce the criterion in the form (similar to (2.7.7))

$$J(\Delta x(0), \Delta u(k)) = \frac{1}{2} \sum_{k=0}^{L} [\Delta x^T(k) Q \Delta x(k) + \Delta u^T(k) \underline{R} \Delta u(k)], \qquad (2.7.17)$$

where $Q = \mathrm{diag}(Q_i)$, $Q \varepsilon R^{N \times N}$, $Q_i \varepsilon R^{n_i \times n_i}$ are the positive semidefinite matrices, $\underline{R} = \mathrm{diag}(r_i)$, $\underline{R} \varepsilon R^{n \times n}$, $r_i \varepsilon R^1$, $r_i \geq 0$. The relation between criterion (2.6.11) in the time continuous domain and criterion (2.7.17) in the discrete time domain can be described as in para. 2.7.1. An additional term should be introduced into criterion (2.7.17) but we shall ignore this in what follows.

We shall synthesize the control minimizing (2.7.17) on the basis of the approximate decoupled model of the system S.

$$S^i: \Delta x^i(k+1) = A_D^i \Delta x^i(k) + b_D^i N(k, \Delta u^i(k)) \qquad (2.7.18)$$

The subsystems (2.7.18) are coupled through $\Delta P_i(k)$ which is given by (2.7.16). The minimization of the performance index (2.7.17) under the constraint given by (2.7.18) reduces to n minimizations of local criteria:

$$J_i(\Delta x^i(0),\ \Delta u^i(k)) = \frac{1}{2}\sum_{k=0}^{L}\left[\Delta x^{iT}(k)Q_i\Delta x^i(k)+\Delta u^i(k)r_i\Delta u^i(k)\right] \quad (2.7.19)$$

under the constraints given by (2.7.18) for each i separately.

Assuming that subsystem (2.7.18) is controllable, the control minimizing (2.7.19) under the constraint (2.7.18) is obtained in the form [39]

$$\Delta u_i^L(k) = -r_i^{-1}b_D^{iT}(A_D^{iT})^{-1}(K_i(k)-Q_i)\Delta x^i(k), \quad \forall i\varepsilon I, \quad \forall k\varepsilon L_t \quad (2.7.20)$$

where $K_i \varepsilon R^{n_i \times n_i}$ is the solution of the Riccati difference equation. Instead the time-varying gains of the linear local regulator we can apply time-invariant gains (assuming that L is high enough):

$$\Delta u_i^L(k) = -r_i^{-1}b_D^{iT}(A_D^{iT})^{-1}(K_i(0) - Q_i)\Delta x^i(k), \quad \forall i\varepsilon I, \quad \forall k\varepsilon L_t \quad (2.7.21)$$

However, control (2.7.21) imposes feedback with respect to the complete state vector $\Delta x^i(k)$. Since only output $y^i(k)$ is measured, it is necessary to estimate the subsystem state.

As in the case of regulator synthesis, when synthesizing the observer we can start with the centralized model of the system (2.7.2) and synthesize the centralized observer. This should reconstruct the system state $x = (x^{1T}, x^{2T},\ldots,x^{nT})^T$ on the basis of the complete output $y = (y^{1T}, y^{2T},\ldots,y^{nT})^T$. However, the realization of such observers would be very complex, so we shall consider decentralized observer synthesis.

For each subsystem S^i (2.7.18) the local observer is introduced by reconstructing the subsystem state if the coupling is not taken into account. As in the case of local criteria, a local observer can be realized in various ways. Here, we adopt the Luenberger´s minimal observer [42] because of its convenience from the point of view of the computation which has to be realized in one sampling interval.

For the free subsystem (2.7.18) Luenberger´s minimal observer is introduced (assuming that the subsystem is completely observable):

$$z^i(k+1)=D^i z^i(k)+E^i\Delta y^i(k)+h^i\Delta u^i(k),\quad z^i(k)=\tilde{C}^i\Delta\hat{x}^i(k),\quad k=0,1,2,\ldots \tag{2.7.22}$$

where $D^i\varepsilon R^{\ell_i\times\ell_i}$, $E^i\varepsilon R^{\ell_i}$, $h^i\varepsilon R^{\ell_i}$, $\tilde{C}^i\varepsilon R^{\ell_i\times n_i}$ where $\ell_i=n_i-k_i$ is the order of the i-th observer, $z^i\varepsilon R^{\ell_i}$ is the state vector of the i-th observer. If the state estimation has to asymptotically approach $\Delta x^i(k)$, the following conditions must be satisfied (when coupling is not taken into account):

$$\tilde{C}^i A_D^i = D^i\tilde{C}^i + E^i\tilde{C}^i,\quad h^i = \tilde{C}^i b_D^i, \tag{2.7.23}$$

where the matrix D^i is chosen to be stable, i.e., its eigenvalues must satisfy $|\lambda(D^i)| \leq 1$. Once the matrix D^i is chosen, $\tilde{C}^i$ and h^i are determined from (2.7.23). The state vector $\Delta\hat{x}^i(k)\varepsilon R^{n_i}$ can be estimated as:

$$\Delta\hat{x}(k) = [C^{iT} \mid \tilde{C}^{iT}]^{-T}\{\Delta y^{iT}(k) \mid z^{iT}(k)\}^T \tag{2.7.24}$$

Satisfying the conditions (2.7.23) guarantees that if $\Delta\hat{x}^i(0)\neq\Delta x^i(0)$ (the initial condition guess error) it follows that $\Delta\hat{x}^i(k)\to\Delta x^i(k)$ for $k\to\infty$, if coupling is not taken into account. However, one should examine the effect of coupling on the state estimation by the decentralized observer (2.7.22).

Since the local observer (2.7.22) does not take into account the interconnections among subsystems ΔP_i or the estimations of the other subsystem states $\Delta\hat{x}^j(k)$, the asymptotic stability of observer (2.7.22) when applied to the coupled system S cannot be separately analyzed from the asymptotic stability analysis of the complete system. That is, if the local regulator (2.7.21), is realized with the estimated state $\Delta\hat{x}^i(k)$ instead of the state $\Delta x^i(k)$, then

$$\Delta u_i^L(k) = -r_i^{-1}b_D^{iT}(A_D^{iT})^{-1}(K_i(0) - Q_i)\Delta\hat{x}^i(k) \tag{2.7.25}$$

and the separation conditions of the estimator and regulator [43] are not satisfied. Thus, it is necessary to examine the asymptotic stability of the system and the observer together (see appendix 2.C.).

If the effect of coupling among subsystems it too strong, additional control should be introduced in order to compensate it. One way of doing this is to introduce global control as in para. 2.6.4. In the discrete-time domain global control can be introduced in the form corresponding to (2.6.37), namely [40],

$$\Delta u_i^G(k) = K_i^G(\Delta x^i(k))\Delta\bar{P}_i^*(k), \qquad (2.7.26)$$

where $K_i^G: R^{n_i} \to R^l$ is the global gain as a function of the subsystem state and $\Delta\bar{P}_i^*(k) \varepsilon R^l$ is the function of coupling ΔP_i acting on the subsystem S^i, as in para. 2.6.4. As stated in 2.6.4, $\Delta\bar{P}_i^*(x)$ may be realized by directly measuring the driving forces P_i or by on-line calculation of the coupling ΔP_i, using approximate expressions. The system performance is better since a more precise estimation of the dynamics of the mechanical part of the system is used for control. However, the implementation of a more precise estimation usually requires more computer computation. This results in a longer sampling period, i.e. the time taken to calculate the new values of the control vector is longer. This problem will be discussed in more detail in chap. 3. (para. 3.6.).

Since the nominal programmed control has been synthesized on the basis of the centralized model of the system, the effect of coupling at the stage of perturbed dynamics is significantly reduced so it is often not necessary to introduce global control. On the other hand, the nominal control has to be synthesized off-line since the complexity of the model requires the use of fast computers. Thus, we shall consider the case when the nominal programmed control is not synthesized on the global model of the system, but at the subsystems level. Let us consider the approximate model of the system in the form of decoupled subsystems.

$$S^i: x^i(k+1) = A_D^i x^i(k) + b_D^i N(u^i(k)), \quad \forall i \varepsilon I, \quad \forall k \varepsilon L_t \qquad (2.7.27)$$

$$y^i(k) = C^i x^i(k)$$

These subsystems are coupled through $P_i(k)$, given by (2.7.13) but at the first step of the control synthesis we shall synthesize the control for free subsystems. The control is synthesized in the form

$$u_{loc}^i(k) = \bar{u}^{io}(k) + \Delta u_i^L(k), \qquad \forall i \varepsilon I \qquad (2.7.28)$$

where $\bar{u}^{io}(k)$ is the programmed control realizing $x^{io}(k)$ if $x^i(0)=x^{io}(0)$ and no perturbation is acting on subsystem S^i, while $\Delta u_i^L(k)$ is the control ensuring the tracking of the nominal trajectory $x^{io}(k)$ if $x^i(0) \neq x^{io}(0)$. Let us suppose that the programmed control $\bar{u}^{io}(k)$ can be determined so that

$$x^{io}(k+1) = A_D^i x^{io}(k) + b_D^i \bar{u}^{io}(k), \quad \forall i \varepsilon I, \quad \forall k \varepsilon L_t \tag{2.7.29}$$

To determine the $\bar{u}^{io}(k)$ so that (2.7.29) holds is not always possible but we assume that the nominal trajectory is so chosen that $\bar{u}^{io}(k)$ can be determined.

When the local programmed control is synthesized, the control at the stage of perturbed dynamics is synthesized as above. One considers the models of the deviations of subsystems S^i from the nominal in the form (2.7.18), and the local controls are synthesized by minimizing the local criteria (2.7.19). The local controls are obtained in the form (2.7.21) or (2.7.25) if we assume that the subsystem state is estimated using the observer.

Obviously, the local control synthesis for tracking the nominal trajectory $x^{io}(k)$ need not be performed in the two stages of local programmed control $\bar{u}^{io}(k)$ satisfying (2.7.29) and the local regulator (2.7.25). It is possible to synthesize the control as the local servo-problem, i.e., the control minimizing the standard quadratic performance index in the form

$$J_i(x^{io}(k),\, u^i_{loc}(k),\, x^i(0)) = \frac{1}{2}\sum_{k=0}^{L-1}[\Delta x^{iT}(k) Q_i \Delta x^i(k) + $$

$$+ u^i_{loc}(k) r_i u^i_{loc}(k)] + \frac{1}{2}\,\Delta x^{iT}(L) Q_i^F \Delta x^i(L),$$

where $Q_i^F \varepsilon R^{n_i \times n_i}$ (the positive semi-definite matrix), should be synthesized under the constraint given by (2.7.27). The minimization of such criteria would yield the control which ensures the tracking of the nominal trajectory $x^{io}(k)$ and which would be locally optimal. This control would have the form (2.7.28) (i.e. it would consist of a programmed part and linear feedback by the subsystem state). However, the programmed control would not satisfy (2.7.29) (this is the so-called singular case when $r_i \to 0$) but it would be the solution of the corresponding difference equation. The on-line synthesis of such control would be much more complex than in the case of the programmed control $\bar{u}^{io}(k)$ which satisfies (2.7.29). Thus, we consider the singular case.

However, the coupling among subsystems (2.7.27) is given by $P_i(k)$ since the programmed control does not compensate the effect of the nominal coupling among subsystems. Thus, the effect of the coupling is much more significant than in the case when the nominal control is synthesized on the basis of the exact model of the system S. This means that the local control (2.7.28) and the local observer (2.7.22) do not always ensure a satisfactory tracking of $x^{o}(k)$. In this case, it may be necessary to introduce global control to compensate the effect of the coupling $P_i(k)$. Now, the global control is in the form

$$u_i^G(k) = K_i^G(x^i(k))P_i^*(k), \qquad (2.7.30)$$

where $P_i^*(k)\varepsilon R^1$ is the function of the total generalized force P_i acting on the i-th subsystem. This global control ensures the stability of the system with decentralized observer and decentralized regulator.

Controls (2.7.28) and (2.7.30) might be realized and the sampling period which should be achieved depends on the complexity of the global control calculation (using approximative relations). This also determines the complexity of the computer which must be used to achieve a sampling period compatible with the system dynamics.

2.8. An Interactive Computer-Oriented Algorithm for Control Synthesis

The control concept presented in this chapter has been used to develop an interactive computer-oriented algorithm for control synthesis of manipulators. This algorithm is a part of a general computer-oriented algorithm for design and control synthesis of manipulators [44]. It has been set on a computer and is called a simulator for the design of industrial manipulators. The concept of general simulator is based on procedures previously developed for automatic modelling of manipulator dynamics on digital computers (para. 1.1). Also fundamental to the realization of the simulator concept is an attempt to obtain, by using elaborate dynamical models, an optimal mechanical configuration of the manipulator and adequate dimensions of its link cross-sections with respect to the chosen criteria and constraints imposed [45]. In this paragraph attention will be focused on control aspect of the simulator concept. This is based on a sound knowledge of manipulation robots dynamics, since this book is concerned only with the problems of control synthesis.

Based on the approach in paras. 2.3, 2.6, 2.7. an algorithm for a digital computer has been constructed for a relatively simple control synthesis of arbitrary manipulation configurations and for various control tasks (concerning the lowest control levels).

This paragraph briefly describes the algorithm [46]. All phases of the algorithm are described without going into the theoretical foundations of the procedures. This has been presented in detail in previous paragraphs. All points in the algorithm where the participation of the user is required are noted in order to underline the interactive character of the algorithm. The application of the algorithm for control synthesis is illustrated by examples of industrial manipulators in ch. 3.

The flow-chart of the algorithm is presented in Fig. 2.1. Since the control synthesis is based on an exact mathematical model of the dynamics of the manipulation mechanism, the algorithm for automatic construction of the mathematical model of open kinematic chains on digital computers (described in ch. 1) is incorporated in the algorithm for control synthesis. By using input data this algorithm enables one to construct the mathematical model of a manipulation system with an arbitrary number of d.o.f. and an arbitrary configuration. The algorithm for control synthesis is therefore general and can be utilized for arbitrary manipulator configuration.

Since the algorithm is based on a two-stage control synthesis, several steps can be distinguished in the procedure and easily followed by the user.

1. First, the user inputs data on mechanism parameters and data on the chosen actuators. Here it is assumed that the models of actuators are linear and time-invariant, of arbitrary order, with the constraint of the amplitude saturation type in their inputs. It is easy to extend the algorithm to the general case when the actuator models are nonlinear. The choice of actuator remains with the user, i.e., using this algorithm the user can easily, examine various types of actuators when applied to the particular manipulator.

2. The user then sets the task to be performed by the manipulator; the most general task in manipulation robotics reduces to the transfer of an object along a prescribed trajectory in space, while the orientation of the object should change during the motion in some required manner (para. 3.3). Most manipulation tasks in practice can be

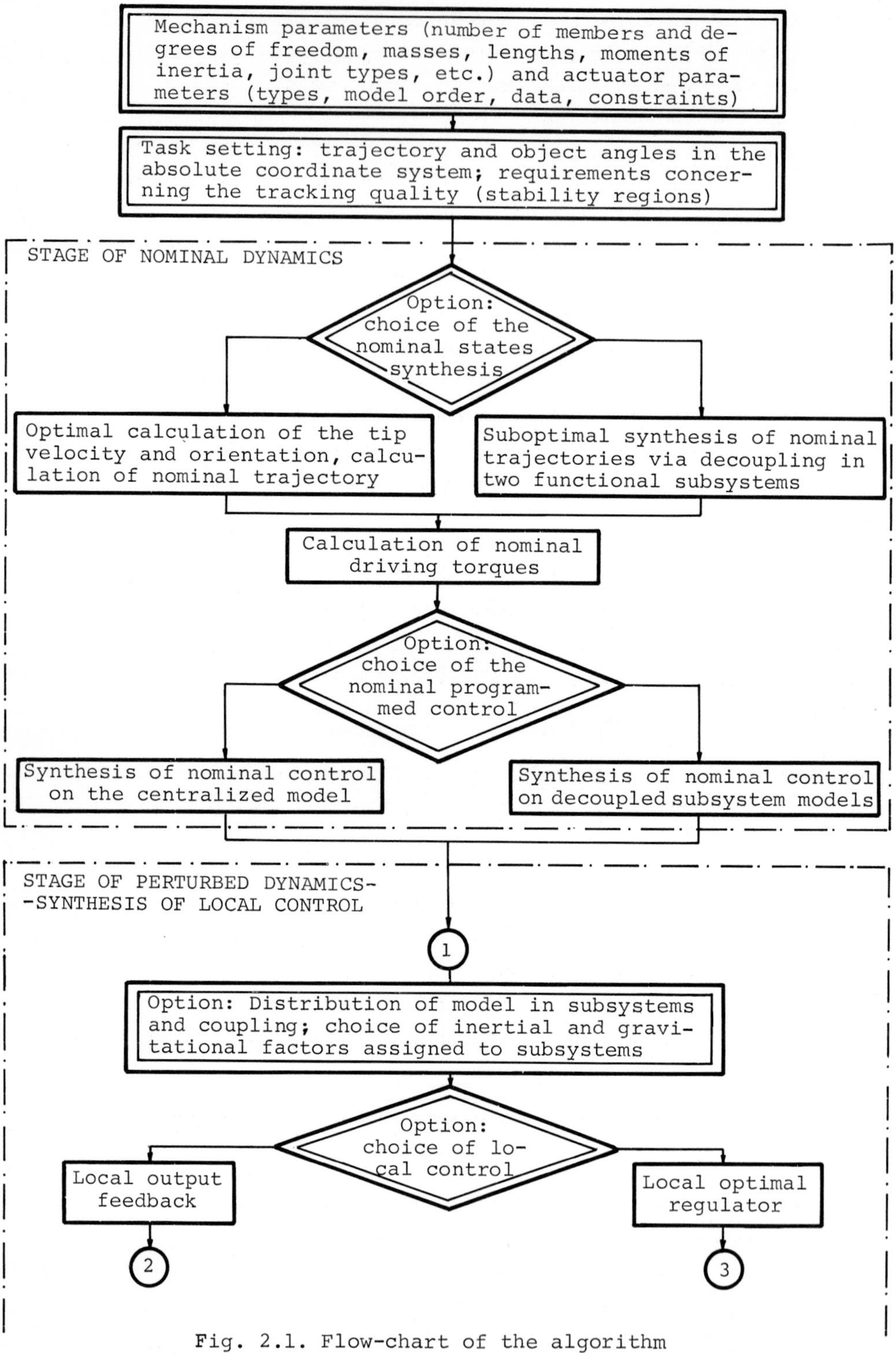

Fig. 2.1. Flow-chart of the algorithm

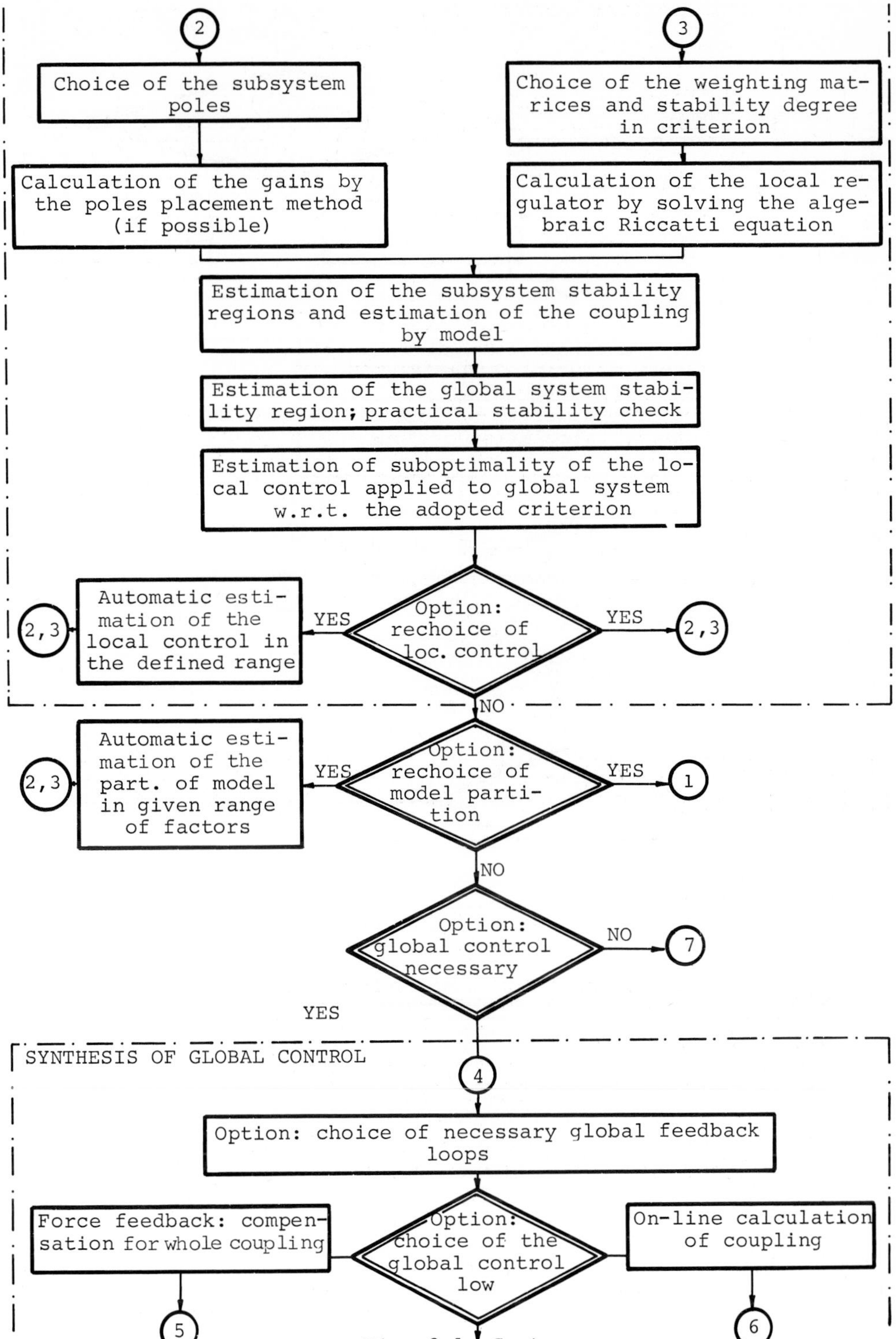
2
3
Choice of the subsystem poles
Choice of the weighting matrices and stability degree in criterion
Calculation of the gains by the poles placement method (if possible)
Calculation of the local regulator by solving the algebraic Riccati equation
Estimation of the subsystem stability regions and estimation of the coupling by model
Estimation of the global system stability region; practical stability check
Estimation of suboptimality of the local control applied to global system w.r.t. the adopted criterion
Automatic estimation of the local control in the defined range
2,3
YES
Option: rechoice of loc. control
YES
2,3
NO
Automatic estimation of the part. of model in given range of factors
2,3
YES
Option: rechoice of model partition
YES
1
NO
Option: global control necessary
NO
7
YES
SYNTHESIS OF GLOBAL CONTROL
4
Option: choice of necessary global feedback loops
Force feedback: compensation for whole coupling
Option: choice of the global control low
On-line calculation of coupling
5
6

Fig. 2.1. Cont.

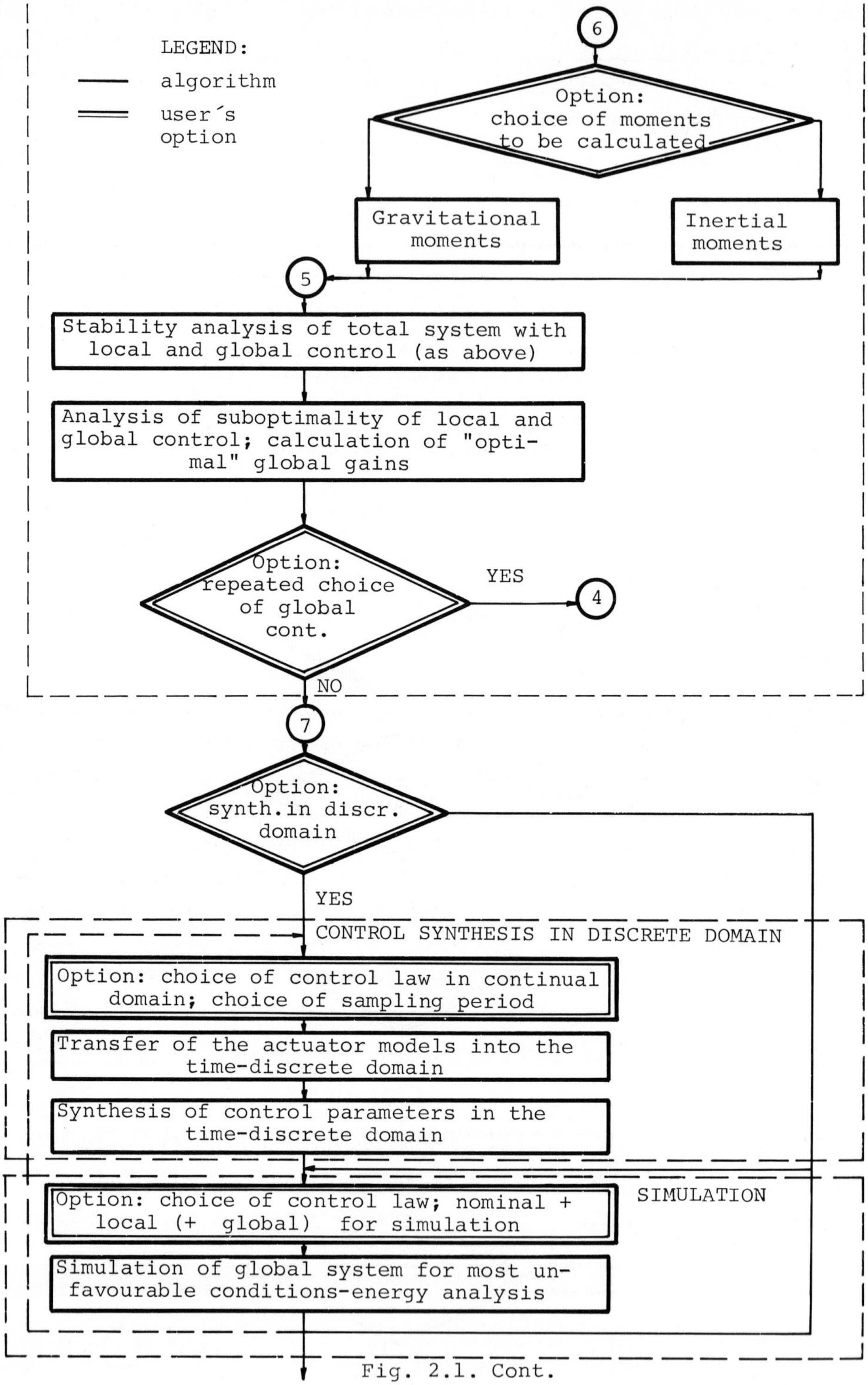
LEGEND:
algorithm
user´s option
6
Option: choice of moments to be calculated
Gravitational moments
Inertial moments
5
Stability analysis of total system with local and global control (as above)
Analysis of suboptimality of local and global control; calculation of "optimal" global gains
Option: repeated choice of global cont.
YES
4
NO
7
Option: synth.in discr. domain
YES
CONTROL SYNTHESIS IN DISCRETE DOMAIN
Option: choice of control law in continual domain; choice of sampling period
Transfer of the actuator models into the time-discrete domain
Synthesis of control parameters in the time-discrete domain
SIMULATION
Option: choice of control law; nominal + local (+ global) for simulation
Simulation of global system for most unfavourable conditions-energy analysis

Fig. 2.1. Cont.

reduced to this task. The user should prescribe the trajectory and the orientation of the object in space. He should also prescribe the finite regions in the state space X^I, X^t, X^F, and the time intervals τ and τ_s, i.e., the conditions of practical stability of the system should be defined.

3. *Synthesis of nominal dynamics.* The algorithm calculates the nominal trajectories of the manipulator coordinates and the nominal programmed control which realizes the nominal trajectories under ideal conditions for a chosen nominal initial condition. The user can choose various procedures for nominal control synthesis:

3.a. He can choose between the optimal synthesis based on the complete dynamic model (see paras. 2.3 and 3.2.1) and the suboptimal synthesis of nominal trajectories. In the second case the imposed task is solved by "decoupling" the manipulator into two functional subsystems, namely, the minimal configuration realizing the required trajectory of the object in space, and the gripper which should ensure the desired orientation of the object (see para. 3.3.1).

3.b. When the nominal motion is synthesized, the user can choose whether the programmed control should be synthesized on the basis of the complete centralized model of the system or on the models of the subsystems (actuators) neglecting the coupling among the degrees of freedom (para. 2.7.2). In the first case, the algorithm first has to calculate the nominal driving torques of the mechanism. The control synthesis in the second case is much simpler.

4. *Control synthesis at the stage of perturbed dynamics - local control synthesis.* When the nominal trajectory and control have been synthesized, the algorithm passes to the second stage of synthesis. The control is calculated which ensures the tracking of the nominal trajectory under the perturbed conditions. At this stage the approximate, decoupled model of the system is considered. The system is viewed as a set of subsystems associated with separate d.o.f. (i.e. actuators), and for each subsystem the local control is synthesized. The local control is synthesized in the following steps (para. 2.6).

4.a. The user choose the subsystem models by varying the factors of inertia and gravity of the mechanism which are associated with subsystem models (para. 2.6.5).

4.b. He can choose two different local control laws: by incomplete output feedback or by local square regulators (para. 2.6.1). In both cases, he can choose the parameters of the local control according to the requirements of the particular task (by choosing the poles of local subsystems or the weighting matrices in local criteria).

4.c. When the algorithm has calculated local gains by one of the two procedures, it checks whether the global system is stable when nominal and local control are applied. The stability is analyzed using the aggregation method for large-scale systems analysis described in para. 2.6.2. On the basis of the amplitude constraints upon the actuator inputs and on the basis of actuator characteristics, one estimates the finite regions in which the free subsystems (free of coupling) are asymptotically stable. One then estimates a finite region in which the global system is asymptotically stable. The practical stability of the global system can then be analyzed according to the finite regions and the settling time defined by the user. The algorithm can also analyze the suboptimality of the local control, when it is applied to the global system, with respect to the chosen criteria (para. 2.6.3). On the basis of these analyses the user can again choose the local control parameters and the models of the subsystems (or the choice of local control can be made iteratively by the algorithm itself). It should be mentioned that various methods for the synthesis of the local control, for low order subsystems, can also be incorporated in the algorithm.

5. *Control synthesis at the stage of perturbed dynamics - global control synthesis*. If the above analyses have shown that the satisfactory tracking cannot be achieved by local control only, the algorithm constructs the synthesis of global control which will compensate the destabilizing effect of coupling (para. 2.6.4). The global control synthesis is performed in the following steps.

5.a. The user can choose the global feedback structure, i.e., he can choose for which subsystem it is necessary to introduce global control and for which it is not.

5.b. He can choose the global control law. One way of doing this is to introduce the force-transducers in the mechanism joints and to apply force feedback to compensate the effect of coupling. The second is to calculate coupling on-line using approximate relations, in which case the algorithm enables the user to decide whether or not the inertia, gravity or other components of the driving torques (coupling) will be taken into account.

5.c. When the local and global control have been defined the algorithm examines the stability of the global system and the suboptimality of the chosen control and calculates the global gain for which the control is the least suboptimal (para. 2.6.4). Of course, the user can iteratively choose the structure and law of global control which is the most convenient from the point of view of system performance and the complexity of realizing the chosen control.

6. *Time-discrete control synthesis.* If the control synthesis is to be realized by micro-computers, the chosen control law should be synthesized in a time-discrete domain. The algorithm also can synthesize the control parameters in a time-discrete domain and the choice of sampling period can be analyzed (see para. 2.7).

7. *Simulation.* In order to verify the performance of the system with various control laws, the algorithm includes the block for simulating the system dynamics. The user can simulate the tracking of the calculated nominal trajectory with or without nominal control (centralized or decentralized), with various local and global control laws. On the basis of these simulations, the energy consumption can be analyzed together with the quality of tracking of the nominal with various control laws. This can help the user to decide which control law is the most convenient for the particular manipulation task.

It should be noted that this algorithm involves only the decentralized and global control synthesis at the stage of perturbed dynamics. However, the general simulator for the design of manipulators should also include some other options for control synthesis, e.g., a centralized optimal regulator (paras. 2.5 and 2.7.1), an asymptotic regulator (para. 2.6.6). Various forms of nonlinear and adaptive control also might be included.

2.9. Conclusion

This chapter considered in detail the problem of control synthesis for the class of large-scale mechanical systems (to which robots and manipulators belong). We described the synthesis of nominal programmed control and the linear centralized regulator. On the other hand, local control was synthesized on the basis of the decoupled model of the system and the global control was introduced. The procedures stated for synthesizing nominal control were described, together with the procedures for synthesizing local and global control.

By analyzing practical stability and estimating the suboptimality of the control, one can examine the system performance when the proposed control laws are applied. We have shown how local and global control can be chosen by iteration so as to ensure the practical stability of the system. However, if the nominal control and the centralized regulator or local and global control cannot satisfy all stability conditions (for example, if the tracking of nominal trajectory cannot be ensured for every point in the region of initial conditions X^I), then one can introduce the third stage in the control synthesis, the so-called stage of large perturbations [3, 9]. This stage represents preparing a number of nominal trajectories and nominal programmed controls realizing the same task but for various initial conditions. These nominals might be calculated in advance and memorized and the centralized regulator or the local and global control should realize them. The tracking of each memorized nominal trajectory is ensured in some finite region of initial conditions around the corresponding nominal initial condition. If the union of the regions of the initial conditions for all prescribed nominal trajectories completely covers the region X^I (defined when the task is set), then the prescription guarantees the practical stability of the system for each point in X^I. The control system should recognize which of the nominal trajectories memorized is the closest to the temporary state of the system. It should then be transferred to the tracking of the given nominal trajectory. The prescription of the nominal trajectories is necessary in the case when the regions X^I and X^t are not connected. At the stage of large perturbations the adaptation of the system to varying conditions can also be ensured if the nominal trajectories for the same control task is synthesized for various conditions (for example for various parameters of the system, various masses and dimensions of the object which is carried by the manipulator). When doing this, one should determine whether or not the unique choice of local and global control ensures the tracking of all prescribed nominal trajectories.

The practical stability analysis has been performed under the assumption that the perturbations are not acting on the system, except the perturbation of the initial conditions type in the initial moment of the movement. It is possible to apply the same procedure for the stability analysis in the case when the perturbations of the initial conditions type are acting during the process, for $t \varepsilon T$ (but which can transfer the state to finite region $\bar{X}^t(t)$, $\forall t \varepsilon T$, only). It should be noticed that, under the assumption that the nominal trajectory is tracked starting from the nominal state on the trajectory which is the closest to the perturbed state $\Delta x(t_1)$ for $t_1 \varepsilon T$, where t_1 is the instant when the perturbation has been acted [47], the analysis of the practical stability can be reduced to the procedure described in para. 2.6.2.

At last, the text about the general procedures for control synthesis has not included the possibilities to significantly simplify the control and the synthesis of various algorithms solving particular dynamic tasks, by introduction of the information on the forces (i.e. the force feedback). These problems will be discussed in ch. 3.

References

[1] Vukobratović, K.M., D.M.Stokić, "Postural Stability of Anthropomorphic Active Mechanisms", Mathematical Biosciences, Vol. 25, No 3/4, 1975.

[2] Stokić, M.D., M.K.Vukobratović, "Dynamic Control of Biped Posture", Mathematical Biosciences, Vol. 44, No. 2, 1979.

[3] Vukobratović, K.M., Legged Locomotion Robots and Anthropomorphic Mechanisms, Monograph, Institute "M.Pupin" (in English) Beograd, 1975, also published by Mir, Moscow, (in Russian) 1976.

[4] Reutemberg, J.N., Automatic Control, (in Russian), Nauka, Moscow, 1971.

[5] Sandell, N.R., P.Varaiya, M.Athans, "A Syrvey of Decentralized Control Methods for Large-Scale Systems", Proc. of the ERDA Conference on System Engineering for Power: Status and Prospects, 1976.

[6] Sandell, N.R., P.Varaiya, M.Athans, M.G.Safonov, "Survey of Decentralized Control Methods for Large-Scale Systems", IEEE Trans. on Automatic Control, AC-23, 108-128, 1978.

[7] Davison, E.J., "The Robust Decentralized Control of a General Servomechanism Problem", IEEE Tran. on Automatic Control AC-21,

14-24, 1976.

[8] Šiljak, D.D., Large-Scale Dynamic Systems: Stability and Structure, North-Holland, New York, 1978.

[9] Vukobratović, K.M., "How to Control the Artificial Anthropomorphic Systems". IEEE Trans. on Systems, Man and Cybernetics, SMC-3, 497-507, 1973.

[10] Vukobratović, K.M., D.M.Stokić, N.V.Gluhajić, D.S.Hristić, "One Method of Control for Large-Scale Humanoid Systems", Mathematical Biosciences, Vol. 36, No 3/4, 1977.

[11] Vukobratović K.M., D.M.Stokić, "Simplified Control Procedure of Strongly Coupled Complex Nonlinear Mechanical Systems", (in Russian), Avtomatika and telemechanika, No. 11, also in English, Autom. and Remote Control, Vol, 39, No. 11, 1978.

[12] Vukobratović, K.M., D.M.Stokić, "Contribution to the Decoupled Control of Large-Scale Mechanical Systems", Automatica, Jan., No. 1, 1980.

[13] Vukobratović, K.M., D.M.Stokić, D.S.Hristić, "A New Control Concept of Anthropomorphic Manipulators", Proc. of Second Conference of Remotely Manned Systems, Los Angeles, June, 1975.

[14] Vukobratović, K.M., D.M.Stokić, "Choice of Decoupled Control Law of Large Scale Mechanical Systems", Proc. of Second Symp. on Large-Scale Systems: Theory and Applications, Toulouse, 181-192, 1980, also in Large-Scale Systems-Theory and Application, June, 1981.

[15] Stokić, M.D., M.K.Vukobratović, "Decoupled Control of Non-Redundant Manipulators", (in Serbian), Proc. I Yugoslav Symp. on Industrial Robotics and Artif. Intell., Dubrovnik, 1979.

[16] Šiljak, D.D., M.Sundareshan, "A Multilevel Optimization of Large-Scale Dynamic Systems", IEEE Trans. on Automatic Control, AC-21, 79-84, 1978.

[17] Johnson, D.C., "Accomodation of External Disturbances in Linear Regulator and Servomechanism Problems",IEEE Trans. AC-16, 552--554, 1971.

[18] Athans, M., P.L. Falb, Optimal Control, McGraw-Hill, 1966.

[19] Vukobratović, K.M., Y.Stepanenko, "On the Stability of Anthropomorphic Systems", Mathematical Biosciences, Vol. 15, October, 1972.

[20] Vukobratović, K.M., D.M.Stokić, D.S.Hristić, "Dynamic Control of Anthropomorphic Manipulators", Proc. of IV International Symp. on Industrial Robots, Tokyo, 1974.

[21] Vukobratović, K.M., D.M.Stokić, "A Concept of Flight Dynamic Control", Proc. of VI IFAC Symp. on Automatic Control in Space, USSR, Erevan, 1974.

[22] Vukobratović, K.M., et al., "Analysis of Energy Demand Distribution within Anthropomorphic Systems", Trans. of the ASME, Journal of Dynamic Systems, Measurement and Control, Dec., 1973.

[23] Kahn, M.E., B.Roth, "The Near Minimum Time Control of Open Loop Articulated Kinematic Chains", Trans. of the ASME, Journal of Dynamic Systems, Measurement and Control, September, 164-172. 1971.

[24] Anderson, B.D., J.B.Moore, Linear Optimal Control. Prentice-Hall, Englewood Cliffs, New Jersey, 1971.

[25] Kalman, R.E., "Contribution to the Theory of Optimal Control". Buletin de la Sociedad Matematica Mexicana, 1960.

[26] Letov, A.M., "The Analytical Design of Control Systems", Automatica and Remote Control, Vol. 21, 389-393, 1960.

[27] Laub, A.J., I.F.N.Balley, "Suboptimality Bounds and Stability in the Control of Nonlinear Dynamic Systems", IEEE Trans. on Automatic Control, 396-399, 1976.

[28] Bailey, F.N., H.K.Ramapriyan, "Bounds on Suboptimality in the Control of Linear Dynamic Systems", IEEE Trans. on Automatic Control, October, 532-534, 1973.

[29] Medvedov, V.S., A.I.Maksimov, "Analytical Construction of Aircraft Spatial Motion Control Systems", Proc. of V IFAC Symp. on Automatic Control in Space, Genova, 1973.

[30] Aoki, M., "On Feedback Stabilizability of Decentralized Dynamic Systems", Automatica, Vol. 8, No. 2, 1972.

[31] Šiljak, D.D., "Multilevel Stabilization of Large-Scale Systems: A Spinning Flexible Spacecraft", Automatica, July, 1976.

[32] Araki, M., "Stability of Large-Scale Nonlinear Systems - Quadratic Order Theory of Composite System Method Using M-matrices", IEEE Trans. on Automatic Control, Vol. AC-23, No. 2, 129-142, 1978.

[33] Weissenberger, S., "Stability Regions of Large-Scale Systems", Automatica, Vol. 9, 653-663, 1973.

[34] Morari, M., G.Stephanopoulos, A.Routherford, "Finite Stability Region for Large-Scale Systems with Stable and Unstable Subsystems", Internat. Journal of Control, Vol. 26, No. 5, 805-815, 1977.

[35] Vukobratović, K.M., D.M.Stokić, "Significance of Force Feedback in Controlling Artificial Locomotion-Manipulation Systems", IEEE Trans. on Biomedical Engineering, December, 1980.

[36] Harvey, C.H., G.Stein, "Quadratic Weights for Asymptotic Regulator Properties", IEEE Trans. on Automatic Control, Vol. AC-23, 378-388, 1978.

[37] Vukobratović K.M., N.M.Kirćanski, "Decoupled Control of Industrial Manipulators via Asymptotic Regulators", (in Russian), Teknicheskaya kibernetika, No. 3, 1982.

[38] Levis, H.A., A.R.Schlueter, M.Athans, "On the Behaviour of Optimal Linear Sampled-Data Regulators", Int. J. of Control, Vol. 13, No. 2, 343-363, 1971.

[39] Cadzow, A.J., H.R.Martens, Discrete-Time and Computer Control Systems, Prentice-Hall, Englewood Cliffs, New Jersey, 1970.

[40] Stokić, M.D., M.K.Vukobratović, "Further Development of Dynamic Algorithms for Industrial Robots Control", International Conference on Systems Engineering, Coventry, England, 1980.

[41] Coiffet, P., J.N.Dumas, P.Molinier, J.Vertut, "Real Time Problems in Computer Control of Robots", Proc. VII ISIR, Japan, 145-152, 1977.

[42] Luenberger, D.G., "An Introduction to Observers", IEEE Trans. on Autom. Control, Vol. AC-16, No. 6, 1971.

[43] Chen, C.T., Introduction to Linear System Theory, Holt, Rinehart, and Winston, New York, 1970.

[44] Vukobratović, K.M., V.S.Cvetković, "Concept of Manipulation Robots Simulator", Prepr. of IV IFToMM Symp. on Theory and Practice of Robots and Manipulators, Warsaw, 1981.

[45] Vukobratović, K.M., V.Potkonjak, D.S.Hristić, "Contribution to the Computer Aided Design of Industrial Manipulators", Prepr. of IV IFToMM Symp. on Theory and Practice of Robots and Manipulators, Warsaw, 1981.

[46] Vukobratović, K.M., D.M.Stokić, M.V.Kirćanski, "A Procedure for Interactive Dynamic Control Synthesis of Manipulators", Prepr. of IV IFToMM Symp. on Theory and Practice of Robots and Manipulators, Warsaw, 1981.

[47] Sworder, D., "A Feedback Regulator for Following Reference Trajectory", IEEE Trans. on Automatic Control, Vol. AC-22, No. 6, 1977.

Appendix 2. A.
Suboptimality of Global Control

Starting from (2.6.11) and (2.6.12) with the control in the form (2.6.13), (2.6.34) we get:*)

$$J(x(0)) = \sum_{i=1}^{n} \int_{0}^{\infty} e^{2\Pi t}\{x^{iT}Q_i x^i + x^{iT}K_i b^i r_i^{-1} b^{iT} K_i x^i - 2b^{iT}K_i x^i u_i^G +$$

$$+ r_i K_i^{G2}[(\text{grad } v_i)^T b^i]^{-2}[(\text{grad } v_i)^T f^i \Delta P_i^*]^2\}dt \qquad (2.A.1)$$

Instead of the subsystem Liapunov function (2.6.12), we introduce

$$\hat{v}_i = e^{\Pi t} v_i = e^{\Pi t}\sqrt{x^{iT}K_i x^i}. \qquad (2.A.2)$$

Taking into account (2.6.20) in the region $\forall(t, x)\varepsilon T\times\tilde{X}(0)$,

$$\dot{\hat{v}}_{i(\text{along }(2.6.8,\ 13))} \leq -\frac{1}{2}\frac{1}{\hat{v}_i}(x^{iT}W_i x^i)e^{2\Pi t} + \sum_{j=1}^{n}\xi_{ij}^*\hat{v}_j \qquad (2.A.3)$$

Substituting (2.A.3) and (2.A.2) into (2.A.1) and taking into account the fact that $W_i = Q_i + K_i b^i r_i^{-1} b^{iT} K_i$, we get

$$J(x(0)) \leq -\int_{0}^{\infty} 2\left(\sum_{i=1}^{n}\hat{v}_i\dot{\hat{v}}_i - \hat{v}_i\sum_{j=1}^{n}\xi_{ij}^*\hat{v}_j\right)dt +$$

$$+\int_{0}^{\infty}\sum_{i=1}^{n}[2\hat{v}_i K_i^G \sum_{j=1}^{n}\bar{\xi}_{ij}^*\hat{v}_j + r_i K_i^{G2}\left(\sum_{j=1}^{n}\bar{\xi}_{ij}^*\hat{v}_j\right)^2/\varepsilon_i^2]dt \quad (2.A.4)$$

By introducing $\hat{v} = \sum_{i=1}^{n}\hat{v}_i$ into (2.A.4) we get following expression for the estimation of the maximum cost of the criterion

---

*) For the sake of simpler notation, in this appendix we use x^i, u_i^G instead Δx^i, Δu_i^G.

$$J(x(0)) \leq \sum_{i=1}^{n} (\hat{v}_i^2(0)) - \hat{v}_i^2(\infty)) + 2\sum_{i=1}^{n}\sum_{j=1}^{n} \xi_{ij}^* \int_0^\infty \hat{v}^2 dt +$$

$$+ 2\sum_{i=1}^{n}\sum_{j=1}^{n} K_i^G \bar{\xi}_{ij}^* \int_0^\infty \hat{v}^2 dt + \sum_{i=1}^{n} r_i K_1^{G2} \max_{j\varepsilon I} (\bar{\xi}_{ij}^*)^2/\varepsilon_i^2 \int_0^\infty \hat{v}^2 dt \tag{2.A.5}$$

On the basis of (2.A.3) we get

$$\dot{\hat{v}}_i \leq -\frac{1}{2}\frac{1}{\hat{v}_i}\frac{\lambda_m(W_i)}{\lambda_M(K_i)}\hat{v}_i^2 + \sum_{j=1}^{n} \xi_{ij}^* \hat{v}_j \tag{2.A.6}$$

Further,

$$\sum_{i=1}^{n} \dot{\hat{v}}_i \leq - \sum_{i=1}^{n} \hat{v}_i \min_{i\varepsilon I}\left(\frac{\lambda_m(W_i)}{2\lambda_M(K_i)} - \sum_{j=1}^{n} \xi_{ji}^*\right) \tag{2.A.7}$$

or $\hat{v}(t) \leq \hat{v}(0)e^{-\alpha t}$ if

$$\alpha = \min_{i\varepsilon I}\left(\frac{\lambda_m(W_i)}{2\lambda_M(K_i)} - \sum_{j=1}^{n} \xi_{ij}^*\right) > 0$$

Since $\hat{v}_i \geq 0$, $\forall i\varepsilon I$, it follows that

$$\sum_{i=1}^{n} \hat{v}_i^2(t) \leq \hat{v}^2(t) \leq 2 \sum_{i=1}^{n} \hat{v}_i^2(t).$$

From (2.A.7) it follows that $\hat{v}(\infty) = 0$ and $\sum_{i=1}^{n} v_i^2(\infty) = 0$, since $\alpha > 0$.

Introducing (2.A.7) in (2.A.5) we get:

$$J(x(0)) \leq \sum_{i=1}^{n} \hat{v}_i^2(0) + \frac{\xi^* + \bar{\xi}^*}{\alpha}\hat{v}^2(0) + \frac{\sum_{i=1}^{n} r_i K_i^{G2} \max_{j\varepsilon I} (\bar{\xi}_{ij}^*)^2/\varepsilon_i^2}{2\alpha}\hat{v}^2(0)$$

From the last equation one derives (2.6.37) for the estimation of the maximum cost of the criterion since $J^* = \sum_{i=1}^{n} \hat{v}_i^2(0)$. If $K_i^G = 0$, $\forall i\varepsilon I$, it follows that $\xi_{ij}^* = \xi_{ij}$ for $i,j\varepsilon I$, and $\bar{\xi}^* = 0$. Thus, the expression (2.6.37) becomes (2.6.31) for the estimation of the maximum criterion cost when only the decentralized control (2.6.13) is applied. It should be noted that in (2.6.37), one can use ξ instead of $\xi^* + \bar{\xi}^*$ but we keep this form in order to emphasize role of the global control.

Appendix 2.B.
Analysis of »Distribution of the Model« Between Subsystems and Coupling

We analyze the effect of the choice of the parameter $\bar{h}_i$, $\forall i \varepsilon I$ on the degree of suboptimality of the control in the forms (2.6.6), (2.6.13) and (2.6.34) in the special case when $n_i=1$, $\forall i \varepsilon I$, i.e., in the case of the first-order subsystems. We consider this special case because of its simplicity but the first-order subsystems can be regarded as aggregate models of higher-order subsystems, $n_i>1$, [1].

Let us consider the model of the system S around $x^o(t)$, $u^o(t)$ in the form

$$S^i : \dot{x}^i = a^i x^i + b^i N(u^i) + f^i P_i, \quad x^i(0) \text{ is given } \forall i \varepsilon I, \tag{2.B.1}$$

where, instead of Δx^i, Δu^i, ΔP_i, we use x^i, u^i, P_i for the sake of simplicity. Let us assume that a^i, b^i, $f^i \varepsilon R^1$, and $x^i \varepsilon R^1$, $u^i \varepsilon R^1$, $P_i \varepsilon R^1$, $\forall i \varepsilon I$. We also assume that the amplitude constraint on the control is given by $N(u^i)$, where $|u^i| \leq u^i_m$. The coupling P_i among subsystems is given in the form

$$P_i = H_i(t, x)\dot{x} + h_i(t, x), \qquad \forall i \varepsilon I \tag{2.B.2}$$

where $H_i : T \times R^n \to R^n$, $h_i : T \times R^n \to R^1$, $x = (x^1, x^2, \ldots, x^n)^T$ and $x \varepsilon R^n (N=n)$ with the condition $h_i(t, 0) = 0$.

Let us write the subsystems S^i in the form

$$S^i : \dot{x}^i = (a^i + \bar{h}_i f^i)x^i + b^i N(u^i) + f^i(P_i - \bar{h}_i x^i), \tag{2.B.3}$$

where $\bar{h}_i \varepsilon R^1$ is the arbitrary parameter which has to be chosen.

For free subsystems (considered without coupling),

$$S^i : \dot{x}^i = (a^i + \bar{h}_i f^i)x^i + b^i N(u^i). \tag{2.B.4}$$

Local control is chosen by minimizing local criteria. Thus,

$$J_i = \int_0^\infty e^{2\Pi t}(x^i q_i x^i + u^i r_i u^i)dt, \qquad \forall i\varepsilon I \tag{2.B.5}$$

Minimization of (2.B.5) results in control of the form

$$u_i^L = -r_i^{-1} b^i K_i^h x^i, \tag{2.B.6}$$

where K_i^h is the solution of the Riccati equation

$$2K_i^h(a^i+\Pi+\bar{h}_i f^i) = -q_i + K_i^h b^i r_i^{-1} b^i K_i^h. \tag{2.B.7}$$

That is,

$$K_i^h = \frac{(a_i+f^i\bar{h}_i+\Pi) + \sqrt{(a_i+f^i\bar{h}_i+\Pi)^2 + q_i b^{i2} r_i^{-1}}}{b^{i2} r_i^{-1}}. \tag{2.B.8}$$

The local subsystems (2.B.4) are exponentially stable in the region X_i, where

$$X_i = \{x^i : |x^i| \leq \frac{\bar{u}_m^i}{K_i^h b^i r_i^{-1}}\} \tag{2.B.9}$$

with stability degree Π. Let us assume that in the region $X = X_1 \times X_2 \times, \ldots \ldots, X_n$,

$$\sqrt{K_i^h}\ \mathrm{sgn}(x^i) f^i (P_i - \bar{h}_i x^i) \leq \sum_{\substack{j=1 \\ j\neq i}} \xi_{ij} \sqrt{K_j^h}|x^j| + (\xi_{ii} - f^i\bar{h}_i)\sqrt{K_i^h}|x^i|, \tag{2.B.10}$$

where $\xi_{ij} \geq 0$ for all $i, i\neq j$ and $j\varepsilon I$.

Let us introduce global control in the form

$$u_i^G = -\frac{K_i^G}{b^i} f^i P_i^*, \qquad i\varepsilon I, \tag{2.B.11}$$

where $K_i^G \varepsilon R^1$, $P_i^* : R^n \to R^1$. The criterion cost $J = \sum_i J_i$ when the control

$$u^i = u_i^L + u_i^G, \qquad i\varepsilon I \tag{2.B.12}$$

is applied becomes $J = \sum_{i=1}^{n} \int_0^\infty e^{2\Pi t}[x^i q_i x^i + r_i(u_i^L+u_i^G)^2]dt$

$$= \sum_{i}^{n} x^i(0)K_i^h\, x^i(0) + \int_0^\infty e^{2\Pi t}[2f^i(P_i-\bar{h}_i x^i)x^i K_i^h + r_i u_i^{G2}]dt \tag{2.B.13}$$

On account of (2.B.12), we get

$$\dot{x}^i = -\frac{W_i^h}{2K_i^h} x^i - \Pi x^i - K_i^G f^i P_i^* + f^i(P_i-\bar{h}_i x^i), \tag{2.B.14}$$

where $W_i^h = q_i + K_i^{h2} b^{i2} r_i^{-1}$. As in (2.B.10), let us write

$$\sqrt{K_i^h}\ \text{sgn}\ (x^i)f^i P_i^* \leq \sum_{j=1}^{n} \bar{\xi}_{ij}^* \sqrt{K_j^h}|x^j|, \tag{2.B.15}$$

$$\sqrt{K_i^h}\ \text{sgn}\ (x^i)f^i(P_i-\bar{h}_i x^i - K_i^G P_i^*) \leq \sum_{\substack{j=1\\ j\neq i}}^{n} \xi_{ij}^* \sqrt{K_j^h}|x^j| + (\xi_{ii}^* - f^i\bar{h}_i)\sqrt{K_i^h}|x_i| \tag{2.B.16}$$

As in appendix 2.A, it can be shown that (2.B.13) can be estimated by

$$J(x(0)) \leq (1 + \frac{2\sum_{i=1}^{n}(\sum_{j=1}^{n}\xi_{ij}-f^i\bar{h}_i) + \sum_{i=1}^{n} r_i K_i^{G2} \max_{j\varepsilon I} \bar{\xi}_{ij}^{*2}}{\min_{i\varepsilon I} (\frac{W_i^h}{2K_i^h} - \sum_{j=1}^{n}\xi_{ji}^* + \bar{h}_i f^i)}) \sum_{i=1}^{n} K_i^h x^{i2}(0) \tag{2.B.17}$$

in the region $\forall(t, x)\varepsilon T\times X$. In (2.B.17) all variables depending on the choice of $\bar{h}_i$ are denoted by the upper index h. It is possible to determine the minimum estimation of the maximum criterion cost (2.B.17) over all choices of $\bar{h}_i$. By differentiating (2.B.17) with respect to $\bar{h}_i$, $\forall i\varepsilon I$, we get a set of n second-order equations in $\bar{h}_i$, $i\varepsilon I$. We can therefore determine the $\bar{h}_i$ which minimizes the estimation of the criterion cost J for control (2.B.12). It is assumed that K_i^G has already been chosen*).

Let us consider (for the sake of simplicity) the special case when all free subsystems are equal: $a^i = a$, $b^i = b$, $f^i = f$, $\forall i\varepsilon I$, and $q^i = q = 0$, $r^i = r$, $K_i^h = K^h$, $\forall i\varepsilon I$. Then choose $\bar{h} = \bar{h}_i$ $\forall i\varepsilon I$. We obtain

$$J(x(0)) \leq (1 + \frac{2(\xi - nf\bar{h}) + B}{r^{-1}b^2K^h/2 - A + \bar{h}f})K^h \sum_{i=1}^{n} x^{i2}(0), \tag{2.B.18}$$

---

*) It is possible to find the minimum of (2.B.17) over $\bar{h}_i$ and K_i^G simultaneously. Then we obtain 2n second-order equations in $\bar{h}_i$ and K_i^G.

where $B = r \sum_{i=1}^{n} K_i^{G2} \max_{i\varepsilon I} \bar{\xi}_{ij}^{*2}$ and $A = \max_{i\varepsilon I} \sum_{j=1}^{n} \xi_{ji}^{*}$

The minimum of (2.B.18) over $\bar{h}$ is obtained from

$$\frac{\partial K^h}{\partial \bar{h}} + \frac{\left[-2nfK^h + (2\xi - 2nf\bar{h} + B)\frac{\partial K}{\partial \bar{h}}\right](r^{-1}b^2\frac{K^h}{2} - A + \bar{h}f)}{(r^{-1}b^2K^h/2 - A + \bar{h}f)^2} \tag{2.B.19}$$

$$- \frac{K^h(2\xi - 2n\bar{h}f + B)(\frac{r^{-1}b^2}{2}\frac{\partial K^h}{\partial \bar{h}} + f)}{(r^{-1}b^2K^h/2 - A + \bar{h}f)^2} = 0$$

Since K^h, based on (2.B.8), is given by

$K^h = 2(a+\Pi+f\bar{h})/b^2r^{-1}$, it follows that $\frac{\partial K^h}{\partial \bar{h}} = 2f/b^2r^{-1}$.

The solution of (2.B.19) for $\bar{h}$ is the "optimal" choice of $\bar{h}$, i.e. we get $\bar{h}$ for which the estimation of the maximum criterion cost is minimal. Thus,

$$\bar{h}_{opt} = \frac{A^1(n-1) + \sqrt{(n-1)\left[nA^{12} - 2(a+\Pi)(A^1n+B^1) + B^1A^1\right]}}{2f(1-n)} \tag{2.B.20}$$

where $A^1 = (a+\Pi) - A$, $B^1 = 2\xi + B$. Depending on the choice of K_i^G and the choice of P_i^* in (2.B.11) for the global control, $\bar{h}_{opt}$ takes various values. Putting $\bar{h}_{opt}$ in (2.B.18) we obtain the minimum estimation of the suboptimality. This might be examined for various choices of K_i^G and ΔP_i^*.

Reference

[1] Aoki, M., "Some Approximation Methods for Estimation and Control of Large-Scale Systems", IEEE Trans. on Automatic Control, Vol. AC-23, No 2, 1978.

Appendix 2. C.
Stability Analysis of System with Decentralized Regulator and Observer

In order to ensure sufficiently precise tracking of a required trajectory it is often desirable to realize local control as a feedback loop with respect to the complete state vector of the subsystem (see para. 2.6.1). However, it is not always convenient to measure the complete state vector. Thus, it might be necessary to introduce an observer to reconstruct the system state. A satisfactory state estimation might be achieved by a centralized and complete observer. However, such an observer may be complex and expensive to realize in the case of a large-scale multi-input multi-output system. Thus, it is convenient to synthesize a decentralized observer as in the decentralized control case. A low-order local observer is synthesized for each subsystem, while the interconnections are taken into account by the estimated states of other subsystems. Šiljak and Vukčević [1] have shown how the decentralized observer can be synthesized like the decentralized regulator. They have also shown that for the synthesized regulator and observer the separation property holds.

However, a synthesized "decentralized observer" requires on-line computation of the coupling acting upon each subsystem in terms of the estimated states and the outputs of other subsystem. When large-scale nonlinear systems are considered such a computation might be very complex, uneconomical for realization and unreliable. Thus, it is convenient to synthesize a decentralized observer which does not require calculation of the coupling among subsystems (as in para. 2.7.2). In general, such a set of local observers cannot be asymptotically stable when applied to interconnected system. However, if a regulator is introduced to stabilize the system, the whole system-observer-regulator may be stabilized. In this case, the separation property does not hold.

In this appendix the synthesis of a decentralized regulator and decentralized observer, with which the coupling among subsystems is not calculated, are considered in a time-continuous domain. The stability of the complete system, regulator and observer is considered and the region of the asymptotic stability of this ensemble is estimated using the method of Waissenberger [2] as in para. 2.6.2, where it is assumed that the local regulator and local observer stabilize subsystems in finite regions of the subsystems state-spaces. If the system performance, using only the local regulators and observers, is unsatisfactory, a global control is introduced as a feedback with respect to coupling, and information on coupling is introduced in the observer. It is possible, if necessary, to introduce global control in the form of force feedback [3], or the coupling can be computed on-line by different approximation models of the coupling [4]. This information about the coupling can also be used to improve the state reconstruction by the decentralized observer.

The mechanical system S (2.2.1) with n d.o.f. is considered. The same control task as stated in para. 2.2 is considered (the practical stability of the system).

Let suppose that the nominal trajectory $x^o(t)$, $\forall t \varepsilon T$ is given. The nominal programmed control $u^o(t)$, $\forall t \varepsilon T$ is synthesized first. This realizes $x^o(t)$ for $x(0) = x^o(0)$ under the assumption that no perturbations act upon the system. In order to ensure the tracking of $x^o(t)$, $\forall t \varepsilon T$ in a case when $x(0) \neq x^o(0)$, the model of deviation of the system from the nominal (2.6.1), (2.6.2) is considered. We suppose that only output Δy^i of each subsystem S^i is measurable (and is given by (2.6.7)).

As we mentioned, the control synthesis will be performed using the approximative decoupled model (2.6.8) of the system (by the model of the system we shall mean the model of deviation from the nominal). The control is to be synthesized in the form (2.6.6).

The local control $u_i^L(t)$ should stabilize the subsystem S^i (2.6.8) in such a way that for the decoupled model the tracking conditions are satisfied, i.e., that $\Delta x^i(0)) \varepsilon \bar{X}_i^I \to \Delta x^i(t) \varepsilon \bar{X}^t(t)$, $\forall t \varepsilon T$ and $\Delta x^i(t) \varepsilon \bar{X}_i^F(t)$, $\forall t \varepsilon T_s$, where it is assumed that $\bar{X}_i^I \subset R^{n_i}$, $\bar{X}_i^t(t) \subset R^{n_i}$, $\forall t \varepsilon T$, $\bar{X}_1^I \times \bar{X}_2^I \times \ldots \times \bar{X}_n^I = \bar{X}^I$, $\bar{X}_1^t(t) \times \bar{X}_2^t(t) \times \ldots \times \bar{X}_n^t(t) = \bar{X}^t(t)$, $t \varepsilon T$, $\bar{X}_i^F(t) \subset R^{n_i}$, $\bar{X}_1^F(t) \times \bar{X}_2^F(t) \times \ldots \times \bar{X}_n^F(t) = \bar{X}^F(t)$, $\forall t \varepsilon T_s$. Let us consider the case (Lb) when the local control is synthesized by minimizing the local standard quadratic criterion (2.6.12).

However, since subsystem outputs Δy^i are only measurable, it is necessary to reconstruct the subsystem state in order to apply the control (2.6.13) (if we do not want to use the output regulator). As with decentralized control synthesis, we shall observe the synthesis of the decentralized observer. For each subsystem S^i a local observer is introduced by means of which the state of the subsystem is reconstructed if the coupling from the rest of the system does not act. We shall adopt the Luenberger's minimal observer [5] in the following form (assuming that the subsystem S^i is completely observable).

$$\begin{aligned} \dot{z}^i &= D^i z^i + E^i \Delta y^i + h^i N(\Delta u^i), \\ z^i &= \tilde{C}^i \Delta \hat{x}^i, \end{aligned} \tag{2.C.1}$$

where $D^i \varepsilon R^{\ell_i \times \ell_i}$, $E^i \varepsilon R^{\ell_i}$, $h^i \varepsilon R^{\ell_i}$, $\tilde{C}^i \varepsilon R^{\ell_i \times n_i}$, $\ell_i = n_i - k_i$ is the observer order, $z^i \varepsilon R^{\ell_i}$ is the observer state vector. The observer (2.C.1) will provide the subsystem state reconstruction which will assymptotically approach Δx^i if the following conditions are satisfied

$$\tilde{C}^i A^i = D^i \tilde{C}^i + E^i C^i, \quad h^i = \tilde{C}^i b^i \tag{2.C.2}$$

The matrix D^i is chosen to be stable, i.e., its eigenvalues should satisfy the condition $\lambda(D^i) < -\alpha_i$, where α_i is the degree of the observer exponential stability. When D^i has been chosen, the matrices $\tilde{C}^i$, E^i and vector h^i are chosen to satisfy conditions (2.C.2). The estimated state vector is determined by Δy^i and the observer state z^i:

$$\Delta \hat{x}^i(t) = [C^{iT} \mid \tilde{C}^{iT}]^{-T} \{\Delta y^{iT}(t) \mid z^{iT}(t)\}^T \tag{2.C.3}$$

This choice of the observer ensures that the observer initial guess error $\Delta \hat{x}^i(t) \to \Delta x^i(t)$ for $t \to \infty$ for free subsystem S^i. However, since the subsystems S^i are interconnected by $\Delta P_i(t)$ it should be determined whether or not the decentralized observer (2.C.1) is asymptotically stable when applied to the global system S. The local regulator (2.6.13) is realized using the estimated subsystem state $\Delta \hat{x}^i$ (2.C.3) (instead of the real state $\Delta x^i(t)$):

$$\Delta u_i^L(t) = -r_i^{-1} b^{iT} K_i \Delta \hat{x}^i(t), \quad \forall i \varepsilon I \tag{2.C.4}$$

Since the local observer (2.C.1) does not take into account the coupling ΔP_i, or the estimated states of the rest of the subsystems $\Delta x^j(t)$, the asymptotic stability of (2.C.1), when applied to the coupled system S, cannot be analyzed separately from the asymptotic stability of the global system S, i.e., the separation property of the observer and regulator [6] does not hold. That is, the subsystems S^i with control (2.C.4) can be written in the form

$$S^i: \Delta\dot{x}^i = (A^i - b^i r_i^{-1} b^{iT} K_i)\Delta x^i + f^i \Delta P_i + b^i r_i^{-1} b^{iT} K_i w^i, \quad i\varepsilon I \quad (2.C.5)$$

where $w^i = \Delta x^i - \Delta\hat{x}^i$ and $w^i \varepsilon R^{n_i}$ is the deviation of the estimated state from the real state. (Expression (2.C.5) assumes that the input is beyond its amplitude constraint).

From (2.C.3) it follows that

$$w^i = \bar{C}^i \tilde{C}^i w^i = \bar{C}^i \bar{w}^i, \quad i\varepsilon I, \quad (2.C.6)$$

where $\bar{C}^i \varepsilon R^{n_i \times \ell_i}$, $\bar{w}^i = \tilde{C}^i w^i$, $\bar{w}^i \varepsilon R^{\ell_i}$. From (2.6.2), (2.C.1) and (2.C.2) we get

$$\dot{\bar{w}}^i = D^i \bar{w}^i + \tilde{C}^i f^i \Delta P_i, \quad i\varepsilon I. \quad (2.C.7)$$

It can be seen that the subsystems (2.C.5) and observers (2.C.7) are "coupled" through ΔP_i and w^i. It is therefore necessary to analyze the stability of the whole ensemble given by subsystems (2.C.5) and observers (2.C.7) (2n subsystems in all), taking into account their interconnections. The stability analysis will be made by using the decomposition- aggregate method for the asymptotic stability of a large-scale system [2], as in para. 2.6.2.

For this purpose we introduce the so-called Liapunov functions for subsystems S^i and for local observers. For subsystems S^i, Liapunov functions are introduced in the form (v_{2i-1} corresponding to S^i) (2.6.19)

$$v_{2i-1} = (\Delta x^{iT} K_i \Delta x^i)^{1/2}, \quad \forall i\varepsilon I \quad (2.C.8)$$

which satisfies (2.6.20):

$$\dot{v}_{2i-1} \text{ (along solution of (2.6.8), (2.6.13))} \le -\frac{\lambda_m(W_i)}{2\lambda_M(K_i)} v_{2i-1} - \Pi v_{2i-1}$$

$$\forall i\varepsilon I \quad (2.C.9)$$

The nonlinearity inequality (2.C.9) holds only for the initial conditions in the finite region $X_i \varepsilon R^{n_i}$. The regions X_i can be estimated by the Liapunov functions of the subsystems v_{2i-1}:

$$\tilde{X}_i = \{\Delta x^i : v_{2i-1}(\Delta x^i) \leq v_{2i-1,0}\}, \qquad i\varepsilon I, \tag{2.C.10}$$

where $v_{2i-1,0} > 0$ are numbers determining the regions $\tilde{X}_i C X_i$.

For the local observers free of coupling,

$$\dot{\bar{w}}^i = D^i \bar{w}^i. \tag{2.C.11}$$

we introduce Liapunov functions $v_{2i} \varepsilon R^1$, $v_{2i} > 0$ in the form

$$v_{2i} = (\bar{w}^i H_i \bar{w}^i)^{1/2}, \qquad \forall i\varepsilon I, \tag{2.C.12}$$

where the positive definite matrices $H_i \varepsilon R^{\ell_i \times \ell_i}$ are chosen so that

$$\bar{w}^{iT}(H^i D^i + D^i H^i)\bar{w}^i \leq -2\alpha_i v_{2i}^2, \qquad \forall i\varepsilon I,$$

i.e., the Liapunov functions (2.C.12) are chosen to be an exact estimation of the local observer stability degree (when the coupling is not taken into account), [7]. Thus,

$$\dot{v}_{2i} \text{ (along the solution (2.C.11))} \leq -\alpha_i v_{2i} \qquad \forall i\varepsilon I \tag{2.C.13}$$

Since the initial condition of the system S may belong only to the finite region $\bar{X}^I$, i.e., the initial state of the subsystem S^i may belong only to the finite region $\bar{X}_i^I$, it si obvious that an initial guess error of the observer $\bar{w}^i(0)$ may belong to a finite region in the observer (2.C.7) state-space. In other words we may suppose that $\bar{w}^i(0)\varepsilon\bar{X}_i C R^{\ell_i}$, where the region $\bar{X}_i$ is the projection of the region $\bar{X}_i^I$ from the space R^{n_i} onto the space R^{ℓ_i} (the mapping being made by the transformation C^{iT}). The regions $\bar{X}_i$ also can be estimated by the local observer Liapunov functions (2.C.12), so one obtains the regions

$$\tilde{\bar{X}}_i = \{\bar{w}^i : v_{2i}(\bar{w}^i) < v_{2i,0}\}, \qquad i\varepsilon I, \tag{2.C.14}$$

where $v_{2i,0} > 0$ are the members determining the regions $\tilde{\bar{X}}_i C \bar{X}_i$.

In order to examine the stability of the ensemble (2.C.5), (2.C.7) it is necessary to estimate the coupling among subsystems and observers in the region $\tilde{X} = \tilde{X}^1 \times \tilde{X}^2 \times \ldots \times \tilde{X}^n$ to which the system S state should belong. Since $\Delta P_i(t, 0) = 0$, $t \varepsilon T$, the numbers ξ_{ij} can be determined so that:

$$(\text{grad } v_{2i-1})^T f^i \Delta P_i \leq \sum_{j=1}^{n} \xi_{ij} v_{2j-1}, \quad \forall i \varepsilon I, \ \forall (t, \Delta x) \varepsilon T \times \tilde{X}, \tag{2.C.15}$$

where $\xi_{ij} \geq 0$, for $i \neq j$. For the same reasons, the numbers $\bar{\xi}_{ij}$ can be found so, that

$$(\text{grad } v_{2i})^T \tilde{C}^i f^i \Delta P_i \leq \sum_{j=1}^{n} \bar{\xi}_{ij} v_{2j-1}, \quad \forall i \varepsilon I, \ \forall (t, \Delta x) \varepsilon T \times \tilde{X} \tag{2.C.16}$$

According to the method of Weissenberger [2], the region of asymptotic stability of the whole system (system S with regulators (2.C.4) and observers (2.C.7) - in all 2n subsystems) can be estimated by

$$\bar{X} = \{\bar{x} : \max_{i \varepsilon I_1} (\frac{v_i}{p_i}) \leq \gamma_o\}, \ \bar{X} C R^{N_1}, \tag{2.C.17}$$

where $\gamma_o > 0$ is given by

$$\gamma_o = \min_{i \varepsilon I_1} (\frac{v_{io}}{p_i}), \ I_1 = \{i : i = 1, 2, \ldots, 2n\}, \tag{2.C.18}$$

and $\bar{x} = (\Delta x^T, \bar{w}^T)^T = (\Delta x^{1T}, \Delta x^{2T}, \ldots, \Delta x^{nT}, \bar{w}^{1T}, \bar{w}^{2T}, \ldots, \bar{w}^{nT})^T$, $\bar{x} \varepsilon R^{N_1}$ and $N_1 = N + \sum_{i=1}^{n} \ell_i$ is the order of the ensemble system. The region $\bar{X}$ (2.C.17) is the estimation of the region of asymptotic stability of the whole system if the following condition is satisfied

$$\bar{G} p < 0, \tag{2.C.19}$$

where $p = (p_1, p_2, \ldots, p_{2n})^T$, $p \varepsilon R^{2n}$, p_i are the arbitrary positive numbers and $G \varepsilon R^{2n \times 2n}$ is the matrix, the elements of which are given by

$$\begin{aligned}
\bar{G}_{2k-1,2j-1} &= -(\Pi + \frac{1}{2} \frac{\lambda_m(W_k)}{\lambda_M(K_k)}) \delta_{kj} + \xi_{kj} \\
\bar{G}_{2k-1,2j} &= \xi_k \delta_{kj} \\
\bar{G}_{2k,2j-1} &= \bar{\xi}_{kj} \\
\bar{G}_{2k,2j} &= -\alpha_k \delta_{kj}, \quad \forall k, \ j \varepsilon I
\end{aligned} \tag{2.C.20}$$

where δ_{kj} is the Kronecker delta and

$$\xi_k = \frac{1}{2}\,\frac{\lambda_M(K_k)}{\lambda_m^{1/2}(K_k)}\,\frac{\lambda_M^{1/2}(\bar{C}^{kT}K_k b^k r_k^{-1} b^{kT} b^k r_k^{-1} b^{kT} K_k \bar{C}^k)}{\lambda_m^{1/2}(H_k)}$$

Thus, by examining whether or not it is possible to determine the vector p so that it satisfies the condition (2.C.19), one can analyze the asymptotic stability of the system with decentralized regulator and observer. Using (2.C.17) we can estimate the region $\bar{X}$ in the space R^{N_1} (the state space of the whole system under consideration) in which the system is asymptotically stable. Furthermore, if

$$\bar{G}v_o \leq 0, \tag{2.C.21}$$

where $v_o = (v_{1,o}, v_{2,o}, \ldots, v_{2n,o})^T$, $v_o \varepsilon R^{2n}$, then $X = \tilde{X}^1 \times \tilde{X}^2 \times \ldots \times \tilde{X}^n \times \tilde{\bar{X}}^1 \times \tilde{\bar{X}}^2 \times \ldots \tilde{\bar{X}}^n$ is the estimation of the region of initial conditions for which the system is asymptotically stable, $X \subset R^{N_1}$. If the condition (2.C.19) or (2.C.21) is satisfied, then one can estimate the speed at which the whole system state approaches the origin of the state space (the speed at which the state approaches the nominal trajectory $x^o(t)$ and a zero error in the system state estimation). It has been shown [2] that

$$\max_{i\varepsilon I_1} \{\frac{v_i(t)}{p_i}\} \leq \max_{i\varepsilon I_1} \{\frac{v_i(0)}{p_i}\} \exp(-\eta t), \tag{2.C.22}$$

where $\eta>0$ is given by

$$\eta = \max_{i\varepsilon I_1} \{-p_i^{-1} \sum_{k=1}^{2n} \bar{G}_{ik} p_k\} \tag{2.C.23}$$

Thus, one can estimate the "shrinkage" of the region $\bar{X}(t)$ to which the state $\bar{x}$ of the whole system must belong. On the basis of this estimation, we can estimate the region $\bar{X}'(t)$ of the state-space of the system S, where $\bar{X}'(t) \subset R^N$, $\forall t\varepsilon T$, to which the state $\Delta x(t)$ must belong during the tracking. One can therefore determine whether on not the tracking conditions are satisfied:

$$\bar{X}^I \subseteq \bar{X}'(0), \quad \bar{X}'(t) \subseteq \bar{X}^t(t), \quad \forall t\varepsilon T, \quad \bar{X}'(t) \subseteq \bar{X}^F(t), \ \forall t\varepsilon T_S \tag{2.C.24}$$

If the conditions (2.C.19) or (2.C.21) and (2.C.24) are not satisfied, the asymptotic stability of the system cannot be guaranteed, i.e. it cannot be guaranteed that the tracking is satisfactory. One possibility is to alter the local regulator (by altering the weighting matrices Q_i, r_i and increasing the exponential stability degree Π) and the local observers (by increasing the exponential stability degree α_i). This increases the diagonal elements of the matrix $\bar{G}$ (by their absolute values), and the condition (2.C.19) is relaxed. One can thus ensure that the local regulators and observers offset the destabilizing effect of the coupling.

The second method of reducing the effect of coupling is by introducing global control. Global control should compensate the influence of coupling by decreasing the numbers ξ_{ij} and $\bar{\xi}_{ij}$ in (2.C.15) and (2.C.16). The global control is introduced in the form of feedback of information about coupling [3, 4]:

$$\Delta u_i^G(t) = -K_i^G(t, \Delta\hat{x}^i)\Delta P_i^*, \tag{2.C.25}$$

where $K_i^G : T\times R^{n_i}\to R^l$ is the global gain, which is a function of time and the reconstructed subsystem state, $\Delta P_i^* \varepsilon R^l$ is the information about the effect of coupling upon the i-th subsystem. ΔP_i^* can be obtained in various ways, as already described in para. 2.6.4.

(Ga) If we suppose that besides the subsystem outputs Δy^i, we can also measure the forces acting in each d.o.f., it is possible to obtain all the information about coupling, i.e., $\Delta P^* = \Delta P_i$. By force-feedbacks (2.C.25), we can therefore reduce the influence of coupling.

(Gb) It is possible to calculate on-line the coupling ΔP_i as a function of the reconstructed system state $\Delta\hat{x}$.

Since ΔP_i^* contains the information on the coupling ΔP_i, the use of both global control (2.C.25) and the local observer (2.C.1) can be modified in such a way that ΔP_i^* can be used.

$$\dot{z}^i = D^i z^i + E^i\Delta y^i + h^i N(\Delta u^i) + \tilde{C}^i f^i \Delta P_i^* \tag{2.C.26}$$

Thus, by introducing (2.C.25) the subsystems S^i (2.C.5) become

$$S^i: \Delta\dot{x}^i = (A^i - b^i r_i^{-1} b^{iT} K_i)\Delta x^i + f^i\Delta P_i - b^i K_i^G \Delta P_i^* +$$

$$+ b^i r_i^{-1} b^{iT} K_i \bar{C}^i \bar{w}^i \tag{2.C.27}$$

and the model of the deviation of the reconstructed subsystem state from the real state (2.C.7) becomes

$$\dot{\tilde{w}} = D^i \tilde{w}^i + \tilde{C}^i f^i (\Delta P_i - \Delta P_i^*). \tag{2.C.28}$$

In order to examine the stability of the system (2.C.27), (2.C.28) it is necessary to estimate by using Liapunov functions (2.C.8), (2.C.12) the coupling acting on the subsystems (2.C.27), (2.C.28). We shall suppose it is possible to determine the numbers satisfying the following inequelities.

$$(\text{grad } v_{2i-1})^T [f^i \Delta P_i - b^i K_i^G \Delta P_i^*(t, \Delta\hat{x})] \leq \sum_{j=1}^{n} (\xi_{ij}^* v_{2j-1} + \tilde{\xi}_{ij} v_{2j})$$

(2.C.29.1)

$$(\text{grad } v_{2i})^T [\tilde{C} f^i (\Delta P_i - \Delta P_i^*(t, \Delta\hat{x}))] \leq \sum_{j=1}^{n} (\bar{\xi}_{ij}^* v_{2j-1} + \tilde{\xi}_{ij}^* v_{2j})$$

(2.C.29.2)

$$\forall i \varepsilon I, \quad \forall (t, \Delta\bar{x}) \varepsilon T \times X, \quad \xi_{ij}^*, \ \bar{\xi}_{ij}^*, \ \tilde{\xi}_{ij}, \ \tilde{\xi}_{ij}^* \geq 0 \text{ for } i \neq j.$$

Obviously, these inequelities concern the case (Gb) when the coupling is calculated, using approximate models, as a function of the reconstructed state of the system $\Delta\hat{x} = (\Delta\hat{x}^{1T}, \Delta\hat{x}^{2T}, \ldots, \Delta\hat{x}^{nT})^T$. In case (Ga), when $\Delta P_i = \Delta P_i^*$, it is obvious that $\bar{\xi}_{ij}^* = \xi_{ij}^* = 0$, $\forall i,j \varepsilon I$. Now, the region of the asymptotic stability of (2.C.27), (2.C.28) can be estimated by applying the tests (2.C.19), (or (2.C.21)), and the region can be estimated by (2.C.17), (2.C.22), (2.C.23), where the elements of the matrix $\bar{G}$ are now given by

$$\begin{aligned}
\bar{G}_{2k-1,\ 2j-1} &= -\left(\frac{1}{2}\frac{\lambda_m(W_k)}{\lambda_M(K_k)} + \Pi\right)\delta_{kj} + \xi_{kj}^* \\
\bar{G}_{2k-1,\ 2j} &= \xi_k \delta_{kj} + \tilde{\xi}_{kj} \\
\bar{G}_{2k,\ 2j-1} &= \xi_{ij}^* \\
\bar{G}_{2k,\ 2j} &= -\alpha_k \delta_{kj} + \tilde{\xi}_{kj}^*, \quad \forall k,j \varepsilon I
\end{aligned} \tag{2.C.30}$$

The introduction of global control should make the numbers ξ_{ij}^* and $\bar{\xi}_{ij}^*$ smaller than ξ_{ij} and $\bar{\xi}_{ij}$ but at the same time the numbers $\tilde{\xi}_{kj}$ and $\tilde{\xi}_{ij}^*$ should not increase significantly. The global gains K_i^G should be determined so as to make matrix $\bar{G}$ as convenient as possible, and to obtain the widest possible region of the asymptotic stability of the system satisfying the condition (2.C.21). In doing so, it should be remembered that by introducing (2.C.25), (2.C.26) we depart from the decentralized structure of the control and the observer. Thus, only the necessary global feedback loops should be introduced to improve the tracking of the nominal trajectory. It should also be noted that the global control (2.C.25) is constrained by its amplitude:

$$|\Delta u_i^G| \leq u_m^i - \Delta u_i^L(\Delta x^i) \tag{2.C.31}$$

Thus, in (2.C.29.1) it should be remembered that the member Δu_i^G is under the constraint (2.C.31).

Now, the global system practical stability can be examined for various forms of global control and information about coupling introduced by local observers. The discrete-time version was discussed in para. 2.7.2 and a microcomputer realization will be presented in para. 3.6. An example of a decentralized regulator and observer synthesis together with stability analysis and global control synthesis will be presented in para. 3.3.3.

References

[1] Šiljak D.D., M.B.Vukčević, "On Decentralized Estimation", International Journal of Control, Vol. 27, No. 1, pp. 113-131, 1978.

[2] Weissenberger, S., "Stability Regions of Large-Scale Systems", Automatica, Vol. 9, 653-663, 1973.

[3] Vukobratović, K.M., D.M.Stokić, "Contribution to the Decoupled Control of Large-Scale Mechanical Systems", Automatica, Jan. No. 1

[4] Vukobratović, K.M., D.M.Stokić, "Choice of Decoupled Control Law of Large Scale Mechanical Systems", Proc. of Second Symp. on Large-Scale Systems Theory and Applications, Toulouse, 1980, also in Large-Scale Systems-Theory and Application, June 1981.

[5] Luenberger, D.G., "An Introduction to Observers", IEEE Trans. on Autom. Control, Vol. AC-16, No. 6, 1971.

[6] Chen, C.T., Introduction to Linear System Theory, Holt, Rinehart, and Winston, New York, 1970.

[7] Šiljak, D.D. "Multilevel Stabilization of Large-Scale Systems: A Spinning Flexible Spacecraft", Automatica, July, 1976.

Chapter 3
Control Synthesis for Typical Manipulation Tasks

3.1. Introduction

This chapter presents a few examples of manipulator control synthesis for some typical tasks frequently encountered in industrial practice. The control synthesis performed is based on the ideas of and procedures for the control synthesis presented in ch. 2. In this synthesis the manipulator mathematical model is used, the construction of which on a digital computer was presented in detail in ch. 1. We present the results of the control synthesis of a few concrete manipulation tasks.

The tasks encountered in industrial practice may be broken down into a few typical movements. As already stated in the Introduction, we shall consider in detail the elementary manipulation subtasks, the combination of which can perform the majority of the complex tasks appearing in industrial practice. For instance, the process of assembling mechanical subassemblies consists of several phases. The first task is to conduct the manipulator gripper to the working object, the second is grasping the object, the third is transferring the working object to the assembling site and the fourth is the assembly itself.

We shall consider the particular elementary tasks, in the case of manipulators without artificial intelligence properties but with autonomy, which are automated to perform full pre-defined tasks in the geometrical (kinematical) and control sense.

Typical manipulation tasks considered here can be divided into three groups [1, 2]:

I) transferring the manipulator tip (gripper and working object) along some prescribed trajectory in space;

II) transferring the working object along a given trajectory in space maintaining some given orientation of the object during the transfer;

III) the phase of assembly by the manipulator, which can be reduced to the insertion process of an object into a fixed hole.

Apart from that, a very significant task is the grasping of the object itself by means of the gripper. However, in the case when the object parameters (position and geometry) are precisely known in advance, control of this task is trivial. In the case, however, when these parameters of the object are not known in advance, this task falls within the scope of artificial intelligence, which is out of the scope of this book.

In this chapter the results of control synthesis for three types of industrial manipulators will be presented: an anthropomorphic type, a manipulator with linear degrees of freedom, and a semi-anthropomorphic manipulator*).

The anthropomorphic type manipulator UMS-1 (Universal Manipulation System No. 1) with six degrees of freedom is illustrated in Fig. 3.1. The second and third members are canes; the first three degrees of freedom are simple joints, while the gripper spherical joint can also be represented by three simple joints by introducing fictitious members of zero length and zero mass. Without trying to justify such a configuration choice, it can be concluded that such a configuration, compared with other possible mechanical schemes, poses more difficult problems in the process of control synthesis. In particular, the mathematical model of the dynamics of such a configuration is much more complex and more nonlinear than with some other manipulator versions (with linear degrees of freedom, and the like). In some ways, with this manipulator, the coupling among the individual degrees of freedom is better expressed. However, the procedures for the synthesis and the results of the control synthesis can be applied without difficulty to other manipulator types. Moreover, with the majority of other manipulator types the problems of synthesis are simplified so that the application of the general procedure for the control synthesis is made easier. This was precisely the reason for the choice of such a manipulator configuration for the application of the ideas about the synthesis of complex systems control (ch. 2). By considering the most difficult case, one obtains the basis for the application of the general

---

*) All three manipulation systems, the first Yugoslav industrial manipulators have been designed in the "Mihailo Pupin" Institute, Beograd, Yugoslavia.

postulates to much simpler and more suitable configurations. In addition, the surplus feedback loops are easily avoided (as will be illustrated by the example of the second manipulation system).

The anthropomorphic type manipulator (Fig. 3.1), which is considered in the control synthesis for all the tasks mentioned above possesses four members. The first three, with the degrees of freedom (q^1, q^2, q^3), form the so-called minimal manipulator configuration, while the fourth one, with its three degrees of freedom, represents the gripper. The seventh degree of freedom is the gripping itself but will not be considered in the simulation of dynamics, since, as we said earlier, the task of gripping the object with the gripper, will not be considered separately. We shall not discuss further details of manipulator design or the problems of the mechanical configuration realization of the robots considered.

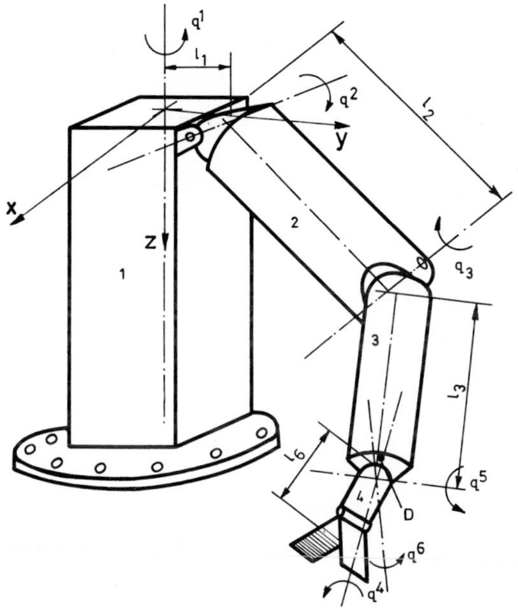

Fig. 3.1. Industrial anthropomorphic manipulator with six degrees of freedom (UMS-1)

In order to perform the synthesis of the control for such a manipulator and to check its behaviour, the mathematical model of the manipulator dynamics has been formed, and, based on the algorithms for forming the mathematical model of open chains on a digital computer (ch.

TABLE 3.1.

	i = 1	i = 2	i = 3	i = 4	i = 5	i = 6
ξ_i^1 (1 - cane, 0 - opposite)	0	1	1	0	0	0
m_i (kg)	6	2.7	2	0	0	2
ℓ_i (m)	0.11	0.334	0.35	0	0	0.15
ℓ_i^* (m)	0.055	0.157	0.175	0	0	0.075
J_{ix} (kgm^2)	-	-	-	0	0	0.0006
J_{iy} (kgm^2)	-	-	-	0	0	0.0008
J_{iz} (kgm^2)	0.0092	-	-	0	0	0.0008
J_{iS} (kgm^2)	-	0.000623	0.000623	-	-	-
J_{iN} (kgm^2)	-	0.001240	0.001240	-	-	-
$\vec{\tilde{e}}_i$	$(0,0,-1)^T$	$(0,1,0)^T$	$(0,0,-1)^T$	$(0,0,1)^T$	$(0,1,0)^T$	$(0.8,0.6,0)^T$
$\vec{\tilde{r}}_{i,i+1}$	$(-(\ell_1-\ell_1^*),0,0)^T$	$(-(\ell_2-\ell_2^*),0,0)^T$	$(-(\ell_3-\ell_3^*),0,0)^T$	$(0,0,0)^T$	$(0,0,0)^T$	-
$\vec{\tilde{r}}_{ii}$	$(\ell_1^*,0,0)^T$	$(\ell_2^*,0,0)^T$	$(\ell_3^*,0,0)^T$	$(0,0,0)^T$	$(0,0,0)^T$	$(\ell_6^*,0,0)^T$

1) the simulation of the manipulator dynamics has been performed. For the concrete manipulator the data needed for assembling the differenttial equations of the dynamics, according to the described algorithm are given in Table 3.1 (member lengths, positions of the mass centers, member masses, proper moments of inertia, rotation axes for the individual degrees of freedom, positions of the centers of rotation on the individual members, etc.). It is evident that for the concrete manipulator the number of pairs is n = 6, and a left coordinate system has been adopted. The initial projections of the vectors in the absolute coordinate system (Fig. 3.1) are $\vec{e}_1 = (0, 0, -1)^T$ and $\vec{r}_{o1} = (-1, 0, 0)^T$. Based on these data, the manipulator mathematical model has been assembled on a digital computer in the form (1.1.30).

It is supposed that D.C. servomotors are used as actuators (1.2.1), the mathematical models of which are given in the form of a system of linear differential equations with constant coefficients, of order $n_i = 3$. The coordinates of the state vector x^i are $x^i = (q^i, \dot{q}^i, i_R^i)^T$ where q^i is the angle, $\dot{q}^i$ is the angular velocity of the i-th degree of freedom, i_R^i - rotor current of the i-th servomotor, so that $k_i=2$ coordinates in the vector x^i coincide with the coordinates in the vector $q = (q^1, q^2, \ldots, q^6)^T$ and $\dot{q} = (\dot{q}^1, \dot{q}^2, \ldots, \dot{q}^6)^T$. The subsystem S^i outputs are given by

$$y^i = C^i x^i, \quad C^i = \begin{bmatrix} 1 & 0 & 0 \\ 0 & 1 & 0 \end{bmatrix}. \tag{3.1.1}$$

With the state vector $x^i(t)$, introduced in this way, the matrix A^i and the vectors b^i and f^i have the form

$$A^i = \begin{bmatrix} 0 & 1 & 0 \\ 0 & -F/J_R & C_M/J_R \\ 0 & -C_E/L_R & -r_R/L_R \end{bmatrix}, \quad b^i = (0, 0, \frac{1}{L_R})^T, \quad f^i = (0, -\frac{1}{J_R}, 0)^T, \tag{3.1.2}$$

where C_M, C_F are the proportionality coefficients of moment and electromotor force, generated by rotation and rotor current in a magnetic field, F is the viscous friction coefficient, L_R is the rotor inductance, r_R is the rotor resistance and J_R is the rotor moment of inertia. Here the indices i have been deleted although the motors need not be the same in all joints, since the energy demands are different. The physical input u^i for subsystem S^i is the rotor voltage and is subject to amplitude constraint. As a concrete case the motor from Globe Industries Division TRW Inc. type 102A200-8 was chosen, the parameters of which are given in Table 3.2.

As a second example of a manipulation system for which the control synthesis will be demonstrated, the manipulator UMS-2, illustrated in Fig. 3.2. will be presented. This manipulator also possesses in its basic (minimal) configuration three degrees of freedom (one rotation and two translations) and, in addition, has a gripper with three rotational degrees of freedom.

$$A^i = \begin{bmatrix} 0 & 1 & 0 \\ 0 & 0 & 2.83 \\ 0 & -703 & -100 \end{bmatrix} \quad b^i = \begin{bmatrix} 0 \\ 0 \\ 100 \end{bmatrix} \quad f^i = \begin{bmatrix} 0 \\ -.658 \\ 0 \end{bmatrix} \quad u_m^i = 27V, \ \forall i \varepsilon I$$

Table 3.2. Parameters of the actuator model (of the chosen D.C. motors) of UMS-1

The internal manipulator coordinates can be clearly seen in Fig. 3.2.: rotation around the vertical axis q^1, translation along the vertical axis q^2 and translation along the horizontal axis q^3, gripper rotation around the horizontal axis q^4, rotation around the axis orthogonal to the first gripper member in the horizontal plane q^5 and rotation around an axis, orthogonal to the previous two, q^6. This system is also powered by means of D.C. servomotors, the models of which are given by (3.1.2). The parameters of the manipulator are given in Table 3.3. and the motor data in Table 3.4.*).

This type of manipulator has been chosen as an opposite example to the UMS-1 in Fig. 3.1 because with the UMS-2 manipulator the coupling among the degrees of freedom is much less expressed, as is the system nonlinearity, so the control synthesis is much simpler. This example is intended to demonstrate the extent to which the control synthesis with simpler manipulator configurations is made easier. Hence, the model of this manipulation system has also been created on a digital computer and the simulation of its dynamics performed with various forms of synthesized control.

---

*) It is supposed that for the degrees of freedom of the minimal configuration the motors of type IG 2315-P20 are used, with different gear ratio and with correction of J_R by means of the corresponding constants of the linear kinematic pairs. Globe 102A200-8 motors are used for the gripper degrees of freedom.

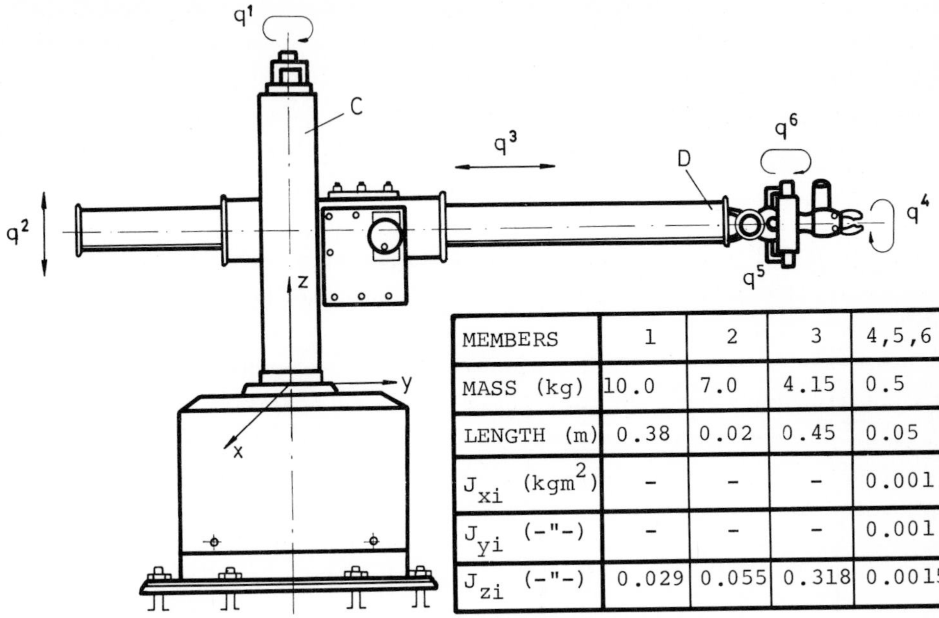

MEMBERS	1	2	3	4,5,6
MASS (kg)	10.0	7.0	4.15	0.5
LENGTH (m)	0.38	0.02	0.45	0.05
J_{xi} (kgm^2)	-	-	-	0.001
J_{yi} (-"-)	-	-	-	0.001
J_{zi} (-"-)	0.029	0.055	0.318	0.0015

Fig. 3.2. Industrial manipulator UMS-2

Table 3.3. Mechanism parameters

i	a^i_{22}	a^i_{23}	a^i_{32}	a^i_{33}	b^i_3	f^i_2	u_m
1	-2.	20.0	-1000.	-450.	2000.	-29.	12
2	-3.	0.18	-10^5	-450.	2000.	-0.004	12
3	-3.	0.13	-10^5	-450.	2000.	-0.0025	12
4,5,6	-1.	2.83	-700.	-100.	100.	-0.65	27

Table 3.4. Actuator parameters: $a^i_{22} = -F/J_R$, $a^i_{23} = C_M/J_R$, $a^i_{32} = -C_E/L_R$, $b^i_3 = -1/L_R$, $f^i_2 = -1/J_R$ (D.C. motors IG2315-P20 for three d.o.f. of minimal configuration and D.C. motors Globe 102A 200-8 for three d.o.f. of gripper are applied).

The synthesis will also be illustrated with a third example: the manipulation system UMS-3B, presented in Fig. 3.3. As in the two preceding cases, this manipulator possesses six degrees of freedom. Three of them form the minimal configuration, while the rest present the gripper. The individual degrees of freedom can be clearly seen in Fig. 3.3. The manipulator has five rotational and one linear degree of freedom. The manipulator works in the spherical coordinate system and is intended for spray-painting of objects of different form. The manipulator parameters (member lengths, masses, moments of inertia) are given in Table 3.5.

Unlike the UMS-1 and UMS-2, this manipulator is powered by means of hydraulic actuators. The hydraulic cylinders, controlled by hydraulic servovalves, are mostly modelled as being of 5-th order. However, since the servovalves used ("MOOG" series 77-200 and 77-100) possess a rather wide frequency band so that their dynamics does not influence the behaviour of the complete system, we adopt linear actuator models (1.2.1) of order $n_i = 3$, $\forall i \in I$. The hydraulic actuator models were adopted in the form

$$A^i = \begin{bmatrix} 0 & 1 & 0 \\ 0 & -\dfrac{B_c^i}{m_t^i} & \dfrac{S_c^i}{m_t^i} \\ 0 & -\dfrac{S_c^i 4\beta}{V_m^i} & -\dfrac{K_c^i 4\beta}{V_m^i} \end{bmatrix}, \quad b^i = \begin{bmatrix} 0 \\ 0 \\ \dfrac{K_q^i 4\beta}{V_m^i} \end{bmatrix}, \quad f^i = \begin{bmatrix} 0 \\ -\dfrac{1}{m_t^i} \\ 0 \end{bmatrix}, \quad (3.1.3)$$

where B_c^i is the viscous friction coefficient, S_c^i is the piston area, m_t^i is the piston mass (including the rod mass), β is the oil compressibility coefficient, V_m^i is the cylinder (working) volume, K_c^i is the gradient of the servovalve flow-pressure characteristic, K_q^i is the proportionality coefficient of the oil flow and servovalve coil current. Here is understood, that for the state vector of the hydraulic actuator $x^i = (\ell^i, \dot{\ell}^i, p^i)^T$, where ℓ^i is the piston travel, $\dot{\ell}^i$ is the piston velocity, p^i is the oil pressure in the cylinder (i.e. the pressure difference). The first three manipulator degrees of freedom are driven by means of hydraulic cylinders ("KNAPP"-Z.9.40/25) while the three d.o.f. of gripper are powered by wane actuators ("KNAPP" rotac D-10-250-1), having rotational motion. In this second case, the subsystem S^i state vector is taken as $x^i = (q^i, \dot{q}^i, p^i)^T$, where q^i, $\dot{q}^i$ are the corresponding angle and angular velocity, so that the actuator models differ from (3.1.3) but can be derived in complete analogy with (3.1.3). The actuator parameters are given in Table 3.6. The actuator

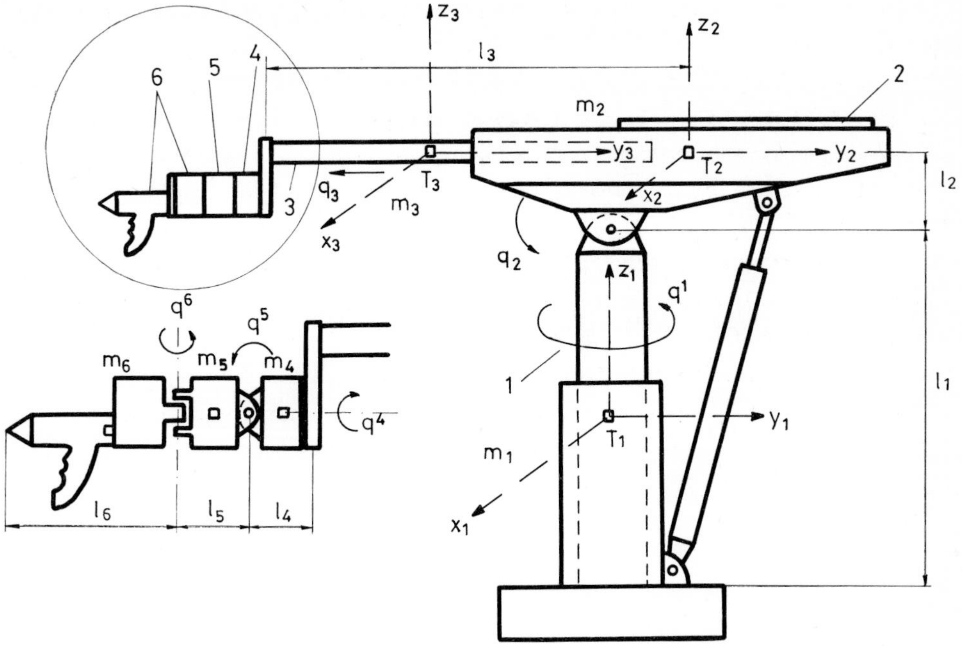

Fig. 3.3. Manipulation system UMS-3B

Members	1	2	3	4, 5	6
Mass m_i (kg)	-	27.8	22.34	2.33	3.33
Length ℓ_i (m)	1.2	0.142	1.14	0.14	0.26
J_{ix} (kgm^2)	0	2.98	1.21	0.004	.007
J_{iy} (kgm^2)	0	-	-	.004	.009
J_{iz} (kgm^2)	0.322	3.701	1.21	.004	.009

Table 3.5. Parameters of manipulator UMS-3B

i	a_{22}^i	a_{23}^i	a_{32}^i	a_{33}^i	b_3^i	f_2^i	u_m^i
1,2	-11.32	47.5	-1133.3	-66.18	74.9	-0.377	15. 30.
3	-9.77	41.04	-850.	-49.63	56.1	-0.325	30.
4,5,6	-166.	800	-300.	-80.	80.	-330	15.

Table 3.6. Actuator parameters for manipulator UMS-3B

inputs are presented by servovalve coil currents, which are constrained by amplitude.

As the first two d.o.f. are rotational and linear actuators are used to drive them, the connection between the actuator model state coordinates x^i and the state coordinates of the mechanical system part model (i.e. the corresponding angles and angular velocities) is not linear but, in the concrete case is given by $((g_1^i)^{-1}$ and $(g_2^i)^{-1}$ from (1.2.5)):

$$\ell^1 = 0.12\, q^1, \qquad \dot{\ell}^1 = 0.12\, \dot{q}^1$$

$$\ell^2 = (1.19^2 + 0.2^2 - 2.\cdot 0.2\cdot 1.19 \cos (q^2+1.4))^{1/2}, \qquad (3.1.4)$$

$$\dot{\ell}^2 = 2\cdot \dot{q}^2 \sin (q^2+1.4)\cdot 0.2\cdot 1.19/\ell^2$$

In addition, since the load of the linear actuators is created by forces and the generalized forces P_i of the rotational d.o.f. are torques, the vectors of the load distribution for these two d.o.f. are given by

$$f^1 = (0,\ 1/(m_t^1\ 0.12),\ 0)^T \quad f^2 = (0,\ 1/(m_t^2\ 0.2 \cos q^2),\ 0)^T \qquad (3.1.5)$$

The remaining four d.o.f. are powered by actuators, having the same type of motion as the corresponding d.o.f., so that each two actuator state vector coordinates x^i coincide with q^i, $\dot{q}^i$ for i = 3, 4, 5, 6.

Using relations (3.1.4) and (3.1.5), one can easily derive expression (1.2.6) for the generalized forces of the mechanical system part of the manipulator.

The reason for presenting this example of a manipulation system is that the manipulator in Fig. 3.3. is semi-anthropomorphic, so with respect to coupling among the d.o.f. it lies between the two previous examples. On the other hand, this example will serve to apply the described principles to systems with hydraulic actuators.

3.2. Task of Transferring the Manipulator TIP Along a Desired Trajectory

The simplest manipulation task considered will be to transfer the manipulator tip with the terminal device and the working object along a prescribed trajectory in space. In principle, the task can be defined as follows. Starting from its initial position the manipulator tip should be transferred to the given point without the trajectory of the tip between these two points being prescribed. In any case, the control could be synthesized by optimizing some criterion (2.2.2) [3, 4]. However, due to restrictions in the working space the movement of manipulator is strongly specified. Thus, some optimization procedures are impossible. Such case will be considered, together with the case when the velocity of the tip moving along a prescribed trajectory is optimized.

As any point in the working space (bounded by kinematic limitation of the observed manipulator) can be reached by only three d.o.f. of the manipulator, for the realization of such a task it is sufficient to consider the minimal configuration of the manipulator. Hence, we restrict ourselves in this paragraph to observing the first three d.o.f. $n = 3$, and we shall consider the other three d.o.f. (terminal device) to be fixed (q^4 = const, q^5 = const, q^6 = const). We shall suppose that all three d.o.f. are powered by D.C. motors. In this case, system S^M has three d.o.f. For the purpose of simplicity, we shall suppose that rotor inductivity can be neglected ($L_R \simeq 0$) and thus the actuator models reduce to the second-order linear system, $n_i = 2$, $\forall i \varepsilon I = \{i: i=1,2,3\}$. The actuator state vectors coincide with the subsystem outputs S^i, $x^i = y^i = (q^i, \dot{q}^i)^T$ and $i_R^i = (u^i - C_e \dot{q}^i)/r_R$.

Thus, instead of (3.1.2), we get

$$A^i = \begin{bmatrix} 0 & 1 \\ & \\ 0 & -C_M C_E / r_R J_R \end{bmatrix}, \quad b^i = (0,\ C_M / J_R r_R)^T, \quad f^i = (0,\ -1/J_R)^T \qquad (3.2.1)$$

The system model S (minimal configuration + actuators) is in the form (2.2.1), where the state vector is of the order $N = 6$ ($k_i = n_i$, $\forall i \varepsilon I$), $x = (q^1, \dot{q}^1, q^2, \dot{q}^2, q^3, \dot{q}^3)^T$. The system S output is given by $y = x$.

The problem of transferring the manipulator tip along its prescribed trajectory is defined as folows. Let us suppose that the initial manipulator state $x^o(0)$ and the end-point state $x^o(\tau_s)$ are given. We shall suppose that the trajectory of the manipulator tip between the states $x^o(0)$ and $x^o(\tau_s)$ is defined. We shall suppose that the possible deviations of the initial state from $x^o(0)$ are in a defined bounded region

$x(0)\varepsilon X^I$, where $x^o(0)\varepsilon X^I$. Let us suppose that the permissable deviations from the end-point state $x^o(\tau_s)$ are in the defined, bounded region X^F around $x^o(\tau_s)$. In this region the manipulator state must be found in $\tau_s<t<\tau$, where τ is the longest permissable time period for which the transfer of the manipulator is performed. Thus, $x^o(\tau_s)\varepsilon X^F$ and $x(t, x(0))\varepsilon X^F$, $\forall t\varepsilon T_s$ and $\forall x(0)\varepsilon X^I$. The allowable deviation of the manipulator state from the desired trajectory is also defined. Thus, the region X^t is defined, so that $x(t, x(0))\varepsilon X^t$, $\forall t\varepsilon T$, $\forall x(0)\varepsilon X^I$ and $x^o(t)\varepsilon X^t$. We shall suppose that no perturbations act on the system. Thus, the task of the manipulator tip transfer can be treated in the same way as the general control task defined in para. 2.2. As already stated, the region X^t might be very narrow since the manipulator tip motion is prescribed in advance. It should be noticed that if the manipulator tip motion from one defined point to another point is considered, usually the nominal initial state $x^o(0)$ and the nominal end-point state $x^o(\tau_s)$ are so defined that $\dot{q}^i(0) = 0$ and $\dot{q}^i(\tau_s) = 0$, $\forall i\varepsilon I$. However, it is not always necessary to introduce these constraints on the manipulator angular velocities.

3.2.1. Nominal dynamic synthesis

We consider the problem of synthesizing the nominal trajectory $x^o(t)$, $\forall t\varepsilon T$ and programmed inputs $u^o(t)$, $\forall t\varepsilon T$ for the given manipulator tip motion.

In this case, there is a unique correspondence between angles of the three d.o.f. and the tip coordinates (point D in Fig. 3.1. and 3.2.) in the Cartesian coordinates. The relation between coordinates of the point D(x, y, z) and the angle trajectories can be given in the form

$$\vec{S} = (x, y, z)^T = f(q) = f(q^1, q^2, q^3), \tag{3.2.2}$$

where $f:R^3\to R^3$, $f(q) = \{f_1(q), f_2(q), f_3(q)\}^T$.

In order to calculate the manipulator angles trajectories, when the point D trajectory in the space is given, let us consider small increments of the manipulator tip motion along a prescribed trajectory $\Delta\vec{S} = (\Delta x, \Delta y, \Delta z)^T$. Assuming that the increments are sufficiently small, it is possible to write the following relation between angles and Descartes coordinates [5, 6]

$$\left(\frac{\partial f_i}{\partial q^1}\right)\Delta q^1 + \left(\frac{\partial f_i}{\partial q^2}\right)\Delta q^2 + \left(\frac{\partial f_i}{\partial q^3}\right)\Delta q^3 = \Delta S_i, \quad i = 1,2,3 \tag{3.2.3}$$

$$A\Delta q = \Delta\vec{S}, \; \Delta q = (\Delta q^1, \Delta q^2, \Delta q^3)^T, \quad a_{ij} = \frac{\partial f_i}{\partial q^j}, \quad i,j = 1,2,3 \tag{3.2.4}$$

These a_{ij} are calculated for the previous point on the trajectory. Since

$$\Delta q = q(t_\ell) - q(t_{\ell-1}), \qquad \Delta\vec{S} = \vec{S}(t_\ell) - \vec{S}(t_{\ell-1}),$$

the elements of the matrix A are calculated for $q(t_{\ell-1})$ and, on the basis of (3.2.4), the angles in the next point on the tip D trajectory are calculated.

In principle, the algorithm for synthesizing the prescribed kinematics of the minimal configuration is stated as follows. The trajectory along which the point D should be moved is prescribed with a given velocity distribution $\dot{\vec{S}}^o(t)$, $\forall t \varepsilon T$, starting from the initial position of the manipulator (with angles $q^o(0)$ and coordinates $\vec{S}^o(0)$). For the point $q^o(0)$ the matrix A is calculated from (3.2.4) and the desired $\dot{\vec{S}}(0)$. Thus,

$$\dot{q}^o(0) = A^{-1}(0)\dot{\vec{S}}^o(0) \tag{3.2.5}$$

assuming that the matrix A is nonsingular.

Let us consider sufficiently short time intervals $\Delta t = t_\ell - t_{\ell-1}$, during which the values of matrix A elements are not changed significantly. For sufficiently small Δt we may assume that

$$q^o(t_\ell) \simeq q^o(t_{\ell-1}) + \dot{q}^o(t_{\ell-1})\Delta t \tag{3.2.6}$$

Thus, one calculates the angles for the next point on the desired trajectory of the manipulator tip. Using new $q^o(t_\ell)$, the matrix $A(t_\ell)$ is calculated and then the necessary angular velocity $\dot{q}^o(t_\ell)$ is calculated, etc. Thus, one obtains the nominal trajectories for all coordinates of the state vector $x^o(t)$, corresponding to the minimal configuration (three angles and three angular velocities are calculated).

When the nominal trajectories $x^o(t)$, $\forall t \varepsilon T$, are calculated on the basis of the model S^M (1.1.30) and (2.2.1) the nominal driving torques

$P^o(t)$, $\forall t \in T$ can be calculated. One also calculates the programmed control $u^o(t)$, $\forall t \in T$, defining the manipulator tip motion along the desired trajectory when there is no perturbation and $x(0) = x^o(0)$.

The motion of the manipulator for a specific control task is presented in Fig. 3.4. [7]. The manipulator tip should be moved from the point A, defined by $x^o(0) = (q^1(0), \dot{q}^1(0), \dots, \dot{q}^3(0))^T = (0.1, 0, -0.8, 0, 0.1, 0)^T$, to the point B, defined by $x^o(\tau) = (-0.4, 0, -0.9, 0, 1.9, 0)^T$ in the time interval $\tau_s = 1.8$ sec (this means that the manipulator has to reach the terminal position at a precisely defined time instant, $T_s = \{t: t=\tau=\tau_s\}$). The manipulator tip should move along a straight line between the points A and B. The tip acceleration should be constant and change its sign once during the movement. Namely, the manipulator tip acceleration should be:

$$a_{tip} = \begin{cases} a_{max} & \text{if } \quad ||\vec{s}^o(t) - \vec{s}^o(0)|| \leq 0.5 \text{ dist}, \\ -a_{max} & \text{if } \quad ||\vec{s}^o(t) - \vec{s}^o(0)|| > 0.5 \text{ dist}, \end{cases} \tag{3.2.7}$$

where $a_{max} = \frac{4\text{dist}}{\tau^2}$, $\text{dist} = ||\vec{s}^o(\tau) - \vec{s}^o(0)||$.

The results of the nominal dynamics synthesis for this particular motion are presented in Fig. 3.4 - 3.6. In Fig. 3.4. the motion is shown in three quarter projection in space (the gripper is assumed to be fixed with respect to the third member). In Fig. 3.5. the minimal trajectories are presented for all three manipulator angles and in Fig. 3.6. the corresponding driving torques are presented.

As the second example of nominal trajectory synthesis we shall consider the synthesis of nominal trajectories for the manipulator UMS-2 [8]. Let the basic configuration of the manipulator change proportionally during the functional movement from the initial position $q^o(0) = \{0, 0, 0\}$ to $q^o(\tau) = \{1{,}57[\text{rad}], 0{,}20[\text{m}], 0{,}30[\text{m}]\}$, $\tau = 2$ sec (Fig. 3.7). Such a prescription is possible due to the relatively simple kinematic structure of the basic manipulator configuration. The actuators are modelled by second-order systems (3.2.1). A velocity profile has been adopted such that, in the interval $t \in (0, \tau/2)$, the accelerations of the internal manipulator coordinates are constant and positive: $\ddot{q}^1 = 1{,}5$ rad/sec^2, $\ddot{q}^2 = 0{,}3$ m/sec^2, $\ddot{q}^3 = 0{,}2$ m/sec^2, and in the interval $t \in (\tau/2, \tau)$: $\ddot{q}^1 = -1{,}5$ rad/sec^2 $\ddot{q}^2 = -0{,}3$ m/sec^2, $\ddot{q}^3 = -0{,}2$ m/sec^2. The nominal trajectories, the torques (forces) and control inputs of the

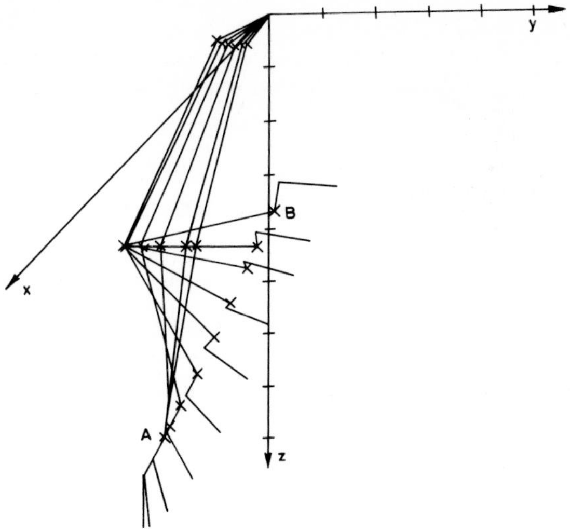

Fig. 3.4. Transfer of manipulator UMS-1 tip along a straight line (with fixed gripper)

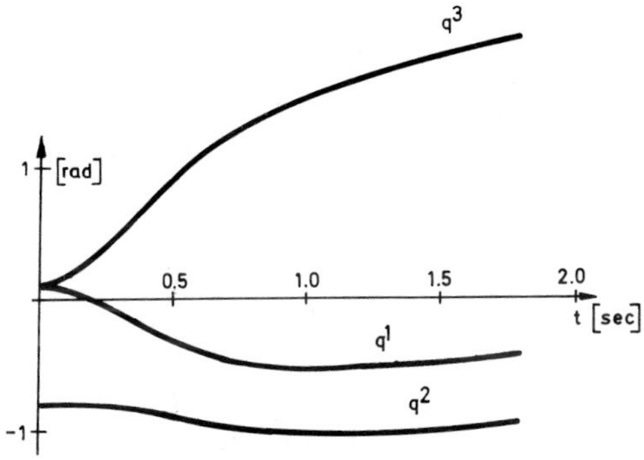

Fig. 3.5. Nominal trajectories of minimal configuration

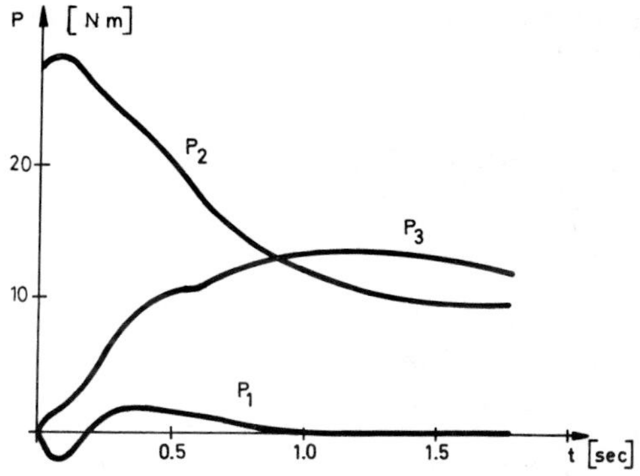

Fig. 3.6. Nominal driving torques of minimal configuration

basic system configuration are presented in Fig. 3.8.

In this example of the nominal trajectories for UMS-1 we assumed that the tip of the manipulator moves along a prescribed trajectory in space with a prescribed variation in the tip velocity along the trajectory. Now, we briefly consider the possibility of optimizing the tip velocity along the trajectory using principles of dynamic programming. Although we do not want to "optimally" synthesize nominal trajec-

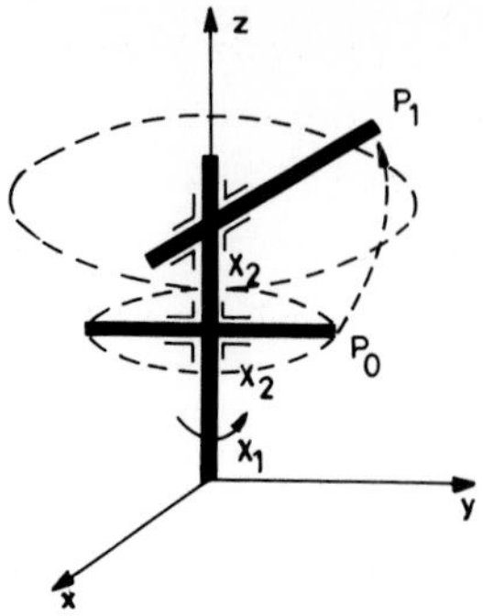

P_o - Initial tip position

P_1 - Terminal tip position

$q^1(\tau) - q^1(0) = 1.57[rad]$

$q^2(\tau) - q^2(0) = 0.20m$

$q^3(\tau) - q^3(0) = 0.30m$

Fig. 3.7. Functional movement of manipulator

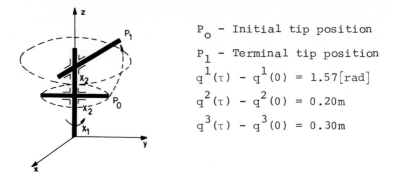

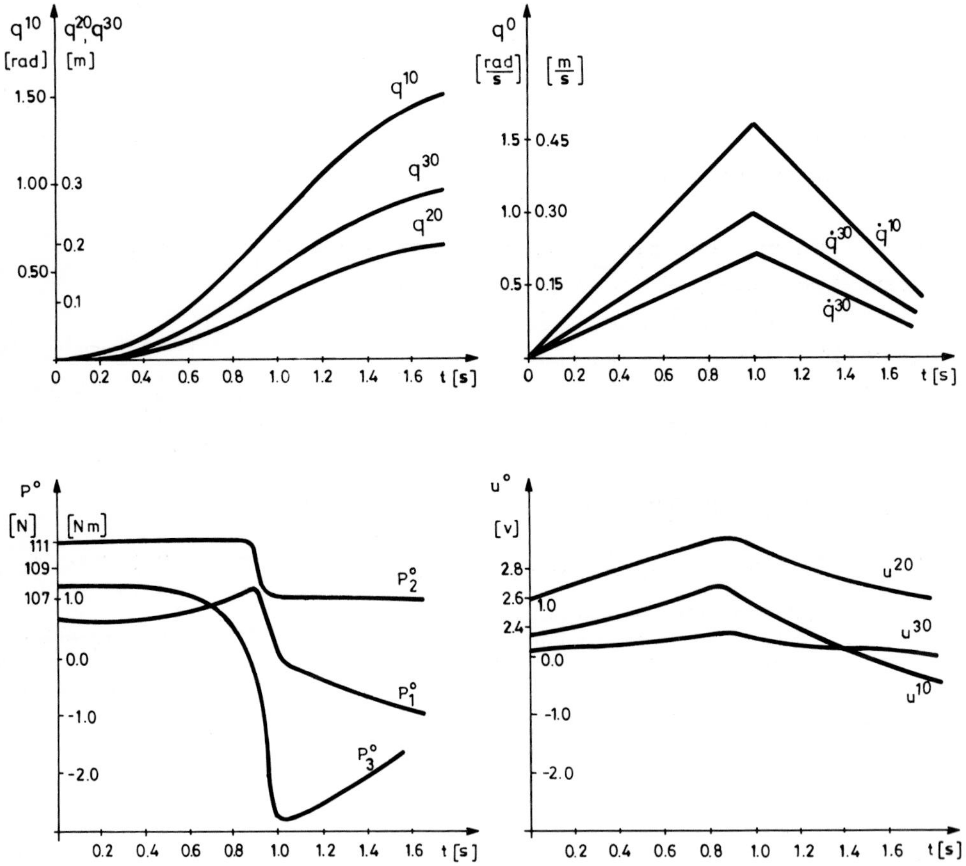

Fig. 3.8. Nominal dynamics for manipulator UMS-2

tories, assuming that the tip trajectory is determined by the nature of the task, we shall show how some particular parameters of the trajectory (e.g. the tip velocity) can be optimized*). Let us start from the general problem of optimizing the nominal trajectory (para. 2.3) and let us show how dynamic programming can be used to synthesize the optimal nominal trajectory.

Let us consider the manipulation system with n d.o.f., described by a continuous nonlinear, stationary system (2.2.1). Let us introduce the performance index in the form (2.2.2).

We assume that the state and inputs are constrained, i.e.

$$x(t) \varepsilon X^t \subset R^{2n}, \quad x(0) = x^o(0), \quad x(\tau_s) = x^o(\tau_s) \qquad (3.2.8)$$

The optimal control $u^o(t)$ conducting the system (2.2.1) along an optimal trajectory $x^o(0)$, minimizing the criterion (2.2.2) and satisfying (3.2.8), should be synthesized.

In order to apply dynamic programming it is necessary to consider a discrete-time model of the system:

$$\begin{aligned} x(k+1) &= x(k) + \Delta t\hat{A}(x(k)) + \Delta t\hat{B}(x(k))u(k) \\ x(k) &= x(k\Delta t), \quad u(k) = (k\Delta t) \end{aligned} \qquad (3.2.9)$$

$\Delta t = \frac{\tau_s}{L}$, where L is the number of discrete steps.

Criterion (2.2.2) in the discrete domain becomes

$$J_o = \sum_{k=0}^{L-1} L_1(x(k),\ u(k)) + g_1(x(L)), \qquad (3.2.10)$$

where $L_1(x(k),\ u(k)) = \Delta t L(x(k),\ u(k))$

$$g_1(x(L)) = g(x(\tau_s))$$

The task is to evaluate the control vectors $u^o(0),\ldots,u^o(L-1)$ and state vector $x^o(0),\ldots,x^o(L-1)$ which minimize the criterion J_o under the constraints (3.2.8).

---

*) This part of the paragraph was written by Mrs Manja Kirćanski, M.Sc.

Dynamic programming reduces this problem to a series of successive minimizations of the variable $u(k)$, $k = 0,\dots,L-1$, on the principle that each part of the optimal trajectory is an optimal trajectory.

In the first step of optimization we consider the last time-interval $[L-1, L]$. For various assumed values of the state vectors $x(L-1)\varepsilon X^t$ the control vectors $u(L-1)$ transferring the system (3.2.9) to the terminal state $x(L) = x^o(\tau_s)$ should be determined. On this interval, optimization is not performed since $x^o(\tau_s)$ is given, and the optimal cost of the criterion (denoted by S) is

$$S_{L-1}(x(L-1))=J_{L-1}(x(L-1),\ u(L-1))=L_1(x(L-1),\ u(L-1))+g(x(L)),$$

while the control is determined by

$$u(L-1)=\left[\hat{B}^T(x(L-1))\hat{B}(x(L-1))\right]^{-1}\hat{B}^T(x(L-1))\left[\frac{x(L)-x(L-1)}{\Delta t}-\hat{A}(x(L-1))\right].$$

In the (L-j)-th interval for each assumed value $x(L-j)\varepsilon X^t$ we determine the controls transferring the system to the states $x(L-j+1)$. We should choose those $u(L-j)$ which minimize the criterion over the whole interval $[L-j, L]$. For a particular pair $x(L-j)$, $x(L-j+1)$ the control is given by

$$u(L-j)=\left[\hat{B}^T(x(L-j))\hat{B}(x(L-j))\right]^{-1}\hat{B}^T(x(L-j))\left[\frac{x(L-j+1)-x(L-j)}{\Delta t}-\right.$$
$$\left.-\ \hat{A}(x(L-j))\right].$$

The criterion cost is given by

$$J^o_{L-j}(x(L-j),\ u(L-j)) = L_1(x(L-j),\ u(L-j)) + S_{L-j+1}(x(L-j+1)),$$

while the optimum cost of the criterion is

$$S_{L-j}(x(L-j)) = \min_{u(L-j)} J^o_{L-j}(x(L-j),\ u(L-j)).$$

In the last optimization step we assume that the system starts from $x(0) = x^o(0)$. For this state we calculate controls $u(0)$ which transfer the system to the state $x(1)$, and of these controls we choose an optimal $u^o(0)$ which yields the minimal cost of the criterion over the whole interval:

$$S_o = \min_{u(0)} J^o(x(0),\ u(0)).$$

The optimal trajectory is now defined backwards. The optimal control $u^o(0)$ corresponds to the optimal state $x^o(1)$. For this state the optimal control $u^o(1)$ has been determined, together with the optimal state $x^o(2)$, etc. We thus obtain the set of optimal control values $u^o(0)$, $u^o(1),\ldots,u^o(L-1)$, transferring the system along the optimal trajectory $x^o(0)$, $x^o(1),\ldots,x^o(L-1)$, $x^o(\tau_s)$.

This is a general procedure for optimal synthesis of manipulator nominal trajectory by means of dynamic programming. The criterion and type of the constraints can be arbitrary.

If the energy consumption criterion is determined and D.C. motors are used as actuators, the criterion (2.2.2) becomes

$$J = \int_0^{\tau_s} (u^T Q_1 u + x^T Q_2 u)\,dt, \tag{3.2.11}$$

where $Q_1 \varepsilon R^{n\times n}$ and $Q_2 \varepsilon R^{2n\times n}$. Matrix Q_1 is diagonal, $Q_1 = \mathrm{diag}(1/r_R^1,\ldots$ $\ldots,1/r_R^n)$, where r_R^i are the resistances of the D.C. motor rotor winding in the i-th joint of the manipulator. All elements of matrix Q_2 are zero, except for $q_{2(n+i),i} = -C_E^i/r_R^i$, $\forall i \varepsilon I$.

Corresponding to the criterion (3.2.11) in the discrete domain there are functions L_1 and g_1 such that

$$L_1(x(k),\, u(k)) = \Delta t(u^T(k)Q_1u(k)+x^T(k)Q_2u(k)), \tag{3.2.12}$$

$$g_1(x(L)) = 0$$

Let us consider in more detail the constraints on trajectories and how we can calculate the assumed values $x(L-j)\,\varepsilon X^t$, $j=0,\ldots,L$. We shall consider only nonredundant manipulators [9].

The dynamic programming algorithm allows for the introduction of all types of constraints encountered in practice. The constraints can be imposed on the external coordinates (which induces constraints on the state vector) or on the internal coordinates directly. In the domain of the external coordinates the region in the space through which the manipulator tip should move and the range of the gripper orientation change are usually defined. For each assumed tip position and gripper orientation it is possible to check whether any point of the manipulator is in contact with the obstacle. The state vector corresponding to contact points should be excluded from optimization procedure. The constraints on the internal coordinates include constraints on construction and constraints on maximum internal velocities. These constraints directly influence the set of possible state vectors X^t.

Now, let us describe how to calculate the state vectors $x(L-j)$ in the case of a manipulator with $n = 3$ d.o.f. In the $(L-j)$-th iteration step we assume a set of vectors in external coordinates, $\vec{S}^i(L-j)$, $i=1,\ldots,M$. For each $\vec{S}^i(L-j)$ we calculate corresponding vector in internal coordinates $q^i(L-j)$, according to (3.2.2). (This transformation is unique since we are considering a nonredundant manipulator). For each pair $q^i(L-j)$, $i=1,\ldots,M$ and $q^k(L-j+1)$, $k=1,\ldots,M$ we calculate the velocity in internal coordinates by

$$\dot{q}^m(L-j) = \frac{q^k(L-j+1) - q^i(L-j)}{\Delta t}, \tag{3.2.13}$$

where $i=1,\ldots,M$, $k=1,\ldots,M$, $m = (i-1)M + k$ and M is the number of assumed vectors of external coordinates.

Finally, the state vector is given by

$$x^m(L-j) = [q^{iT}(L-j), \dot{q}^{mT}(L-j)]^T, \quad m=1,\ldots,M^2. \tag{3.2.14}$$

Let us consider the problem of dimensionality. For a general nonredundant manipulator with n d.o.f. ($n \leq 6$) n parameters should by varied (n elements of the vector in external coordinates). For a manipulator with 6 d.o.f. the six parameters should be varied. This represents considerable dimensionality and in practice it is not possible to carry out the optimization. For a manipulator with 3 d.o.f. optimization can be carried out (assuming a sufficiently large computer memory).

We suggest an algorithm for synthesizing the optimal profile of the tip velocity, which requires variation of only one parameter. We consider the manipulator with $n = 3$ d.o.f. and optimize its movement along a prescribed trajectory in the external coordinates space. This trajectory is determined by the parameter equation

$$\vec{S} = \vec{S}(\lambda), \tag{3.2.15}$$

where λ is the scalar parameter, $\lambda \varepsilon [\lambda^o, \lambda^F]$, $\vec{S}(\lambda^o) = \vec{S}^o(0)$ and $\vec{S}(\lambda^F) = \vec{S}^o(\tau_s) = f(q^o(\tau_s))$. Functional dependance $\lambda = \lambda(t)$, $t \varepsilon [0, \tau_s]$ is not prescribed and it can be determined by optimization.

As an example of a trajectory (3.2.15) in the external coordinates space the linear change of external coordinates may be written (motion along the straight line beetween the terminal points) as

$$\vec{S}(t) = \vec{S}^{o}(0) + \lambda(t)(\vec{S}^{o}(\tau_s) - \vec{S}^{o}(0)) \tag{3.2.16}$$

where $\lambda(0) = \lambda^{o} = 0$, $\lambda(\tau_s) = \lambda^{F} = 1$. In this case, the rate of change of parameter λ is proportional to the velocities in the external coordinates, i.e.,

$$\dot{\vec{S}}(t) = \dot{\lambda}(t)(\vec{S}^{o}(\tau_s) - \vec{S}^{o}(0)).$$

Motion along a parabolic trajectory can be similarly determined.

At each step (L-j), j=0,...,L of the optimization procedure it is necessary to assume the set of parameters $\lambda^{i}(L-j)$, i=1,...,M and on the basis of (3.2.15) we calculate the corresponding vector in external coordinates $\vec{S}^{oi}(L-j)$. The corresponding set of state vectors $x^{m}(L-j)$, $m=1,\ldots,M^2$ is then determined. We can thus obtain the optimal profile of the manipulator velocity $\lambda^{o}(L-j)$, j=0,...,L and the corresponding optimal state vector trajectory $x^{o}(L-j)$ and control $u^{o}(L-j)$.

The algorithm for optimal velocity profile synthesis, based on a complete dynamic model of the manipulation system, has been illustrated by the manipulator UMS-1 (Fig. 3.1). Optimization has been performed on the energy consumption criterion (3.2.11) [9].

Matrices Q_1 and Q_2 are given by

$$Q_1 = \begin{bmatrix} 0,5 & 0 & 0 \\ 0 & 0,5 & 0 \\ 0 & 0 & 1 \end{bmatrix}, \qquad Q_2 = \begin{bmatrix} 0 & 0 & 0 \\ 0 & 0 & 0 \\ 0 & 0 & 0 \\ -3,55 & 0 & 0 \\ 0 & -3,55 & 0 \\ 0 & 0 & -1,35 \end{bmatrix}.$$

We have optimized the manipulator tip motion along a straight line and parabolic trajectory (between the two terminal points $x^{o}(0)$ and $x^{o}(\tau_s)$).

As an example of a straight line trajectory we considered the motion between the points $\vec{s}^o(0) = [0.49 \quad 0.36 \quad 0]^T[m]$ and $\vec{s}^o(\tau_s) = [0.44 \quad 0.47 \quad 0.1]^T[m]$ in the time period $\tau = 0.5$ s. The model was made discrete by using the sampling period $\Delta t = 0.02$ s. It was assumed that $\dot{\vec{s}}^o(0) = 0$ and $\dot{\vec{s}}^o(\tau_s) = 0$.

Fig. 3.9 shows the optimal change of the parameter $\lambda(t)$ and Fig. 3.10. shows the corresponding optimum profile of the velocity $\dot{\lambda}(t)$. From Fig. 3.10 it can be concluded that the better approximation of optimum change $\dot{\lambda}(t)$ can be obtained by incresing the density of the assumed values of the parameter λ at each optimization step or by reducing the time interval Δt.

Fig. 3.11 shows the optimal change in the external coordinates $\vec{s}^o(t)$ of the manipulator tip and corresponding velocity $\dot{\vec{s}}^o(t)$ when the initial velocity is zero. The internal coordinates and velocities, corresponding to this tip motion are presented in Fig. 3.12 (optimal trajectory of the state vector x). The maximum velocity and acceleration of the manipulator tip during this motion are 0.48 [m/s] and 5 [m/s^2]. Fig. 3.13 shows the optimal control signals. The wave character of the control signal is the consequence of the parameter λ and time t being discrete. By reducing the number of discrete steps the oscillations in the control are reduced. Fig. 3.14 shows the increas of energy consumption during the motion along the optimal trajectory.

The results of the optimal profile of velocity synthesis based on an exact dynamic model and energy consumption criterion lead to the following conclusions.

The most convenient motions from the point of view of energy consumption are when the tip velocity sharply increases in the first part of the trajectory and then increases move slowly up to some constant value. In the second half of the trajectory the braking is at first slower and then harder until the tip motion terminates. This velocity distribution is similar to a parabolic function.

We have also considered the optimization of the manipulator tip motion along the parabolic trajectory for motion between the points $\vec{s}^o(0) = [0.427 \quad 0.432 \quad 0.213]^T[m]$ and $\vec{s}(\tau_s) = [0.366 \quad 0.532 \quad 0.248]^T[m]$. The duration of motion is $\tau = 0.5$ s and the discrete intervals are $\Delta t = 0.02$s. Fig. 3.15 shows the optimal change of parameter λ by which ener-

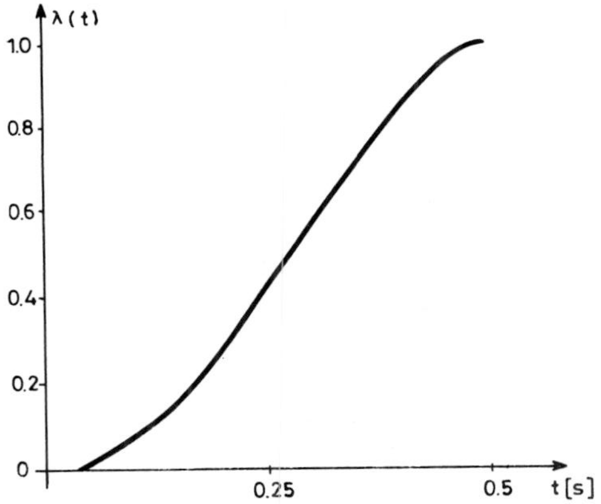

Fig. 3.9. Optimal change of parameter λ (for straight line tip trajectory)

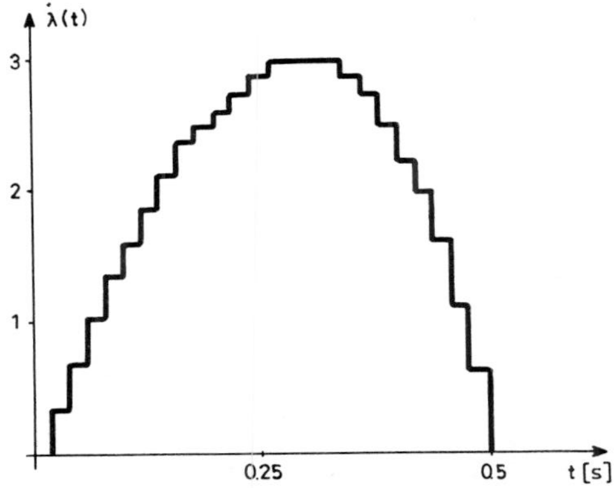

Fig. 3.10. Optimal profile $\dot{\lambda}(t)$

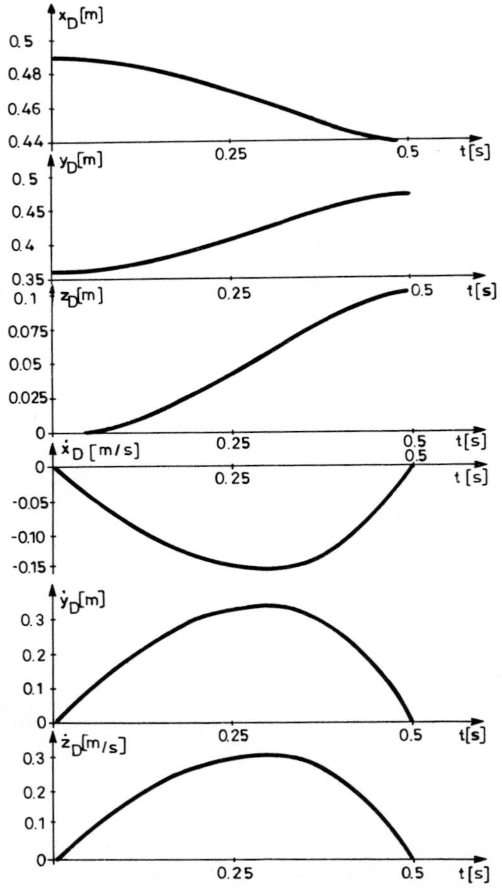

Fig. 3.11. Optimal change of external coordinates and their velocities

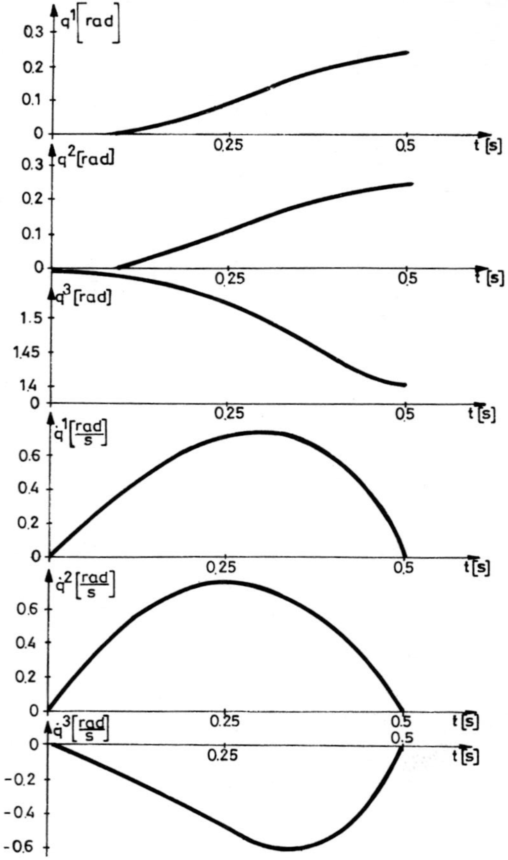

Fig. 3.12. Optimal change of state vector

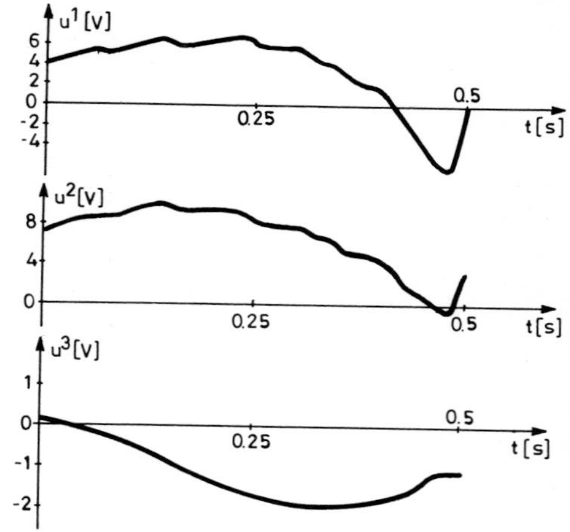

Fig. 3.13. Optimal nominal control

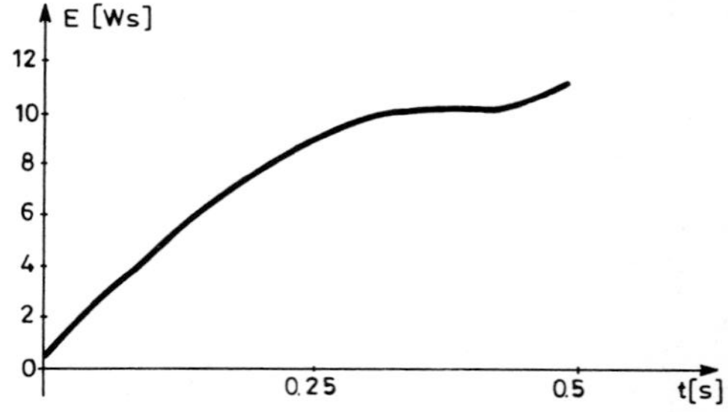

Fig. 3.14. Change of energy consumption

gy consumption is minimized. Fig. 3.16 shows the velocity profile $\dot{\lambda}(t)$. The optimal changes of external coordinates and velocities are presented in Fig. 3.17. Fig. 3.18 shows the optimal nominal trajectory of the state. The corresponding nominal control signals transferring the system along the optimal trajectory are presented in Fig. 3.19. The energy consumption is presented in Fig. 3.20.

The proposed algorithm for manipulator trajectory synthesis permits the evaluation of optimal trajectories using a complete dynamic model of the active mechanism. This procedure has the advantage of being very convenient for optimal synthesis when state vector trajectories and controls are subjected to various types of constraints (design constraints, presence of obstacles in the work-space, velocity and acceleration constraints, limited input signals for the actuators, etc.).

However, the deficiencies of the suggested procedure are in principle associated with the dynamic programming algorithm itself. They are: the dimensionality problem, relatively long optimization time and large computer memory requirements. On the other hand, this procedure is especially suitable for optimal velocity distribution synthesis for nonredundant manipulators. In this case the problem reduces to optimization with respect to one parameter only. The long optimization time and large memory capacities are, in the case of optimal velocity distribution synthesis, a consequence of the complexity of the nonlinear dynamic model and to a large extent depend on the desired optimization accuracy and the duration of the movement. On the other hand, dynamic programming is the only optimization technique which could be applied to such a complex, nonlinear and multi-constrained system.

The proposed procedure is assigned to optimal nominal trajectory synthesis in the off-line mode. This is not a great drawback for automatic industrial robots, considering that industrial tasks are very often completely defined in advance.

We have thus presented various ways of synthesizing nominal trajectories in the case of the simplest manipulation task: transfer of the manipulator tip between two terminal points in space.

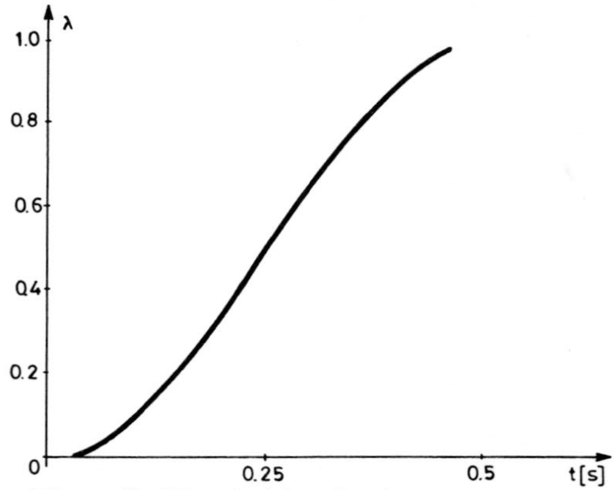

Fig. 3.15. Optimal change of parameter $\lambda(t)$ (for parabolic trajectory)

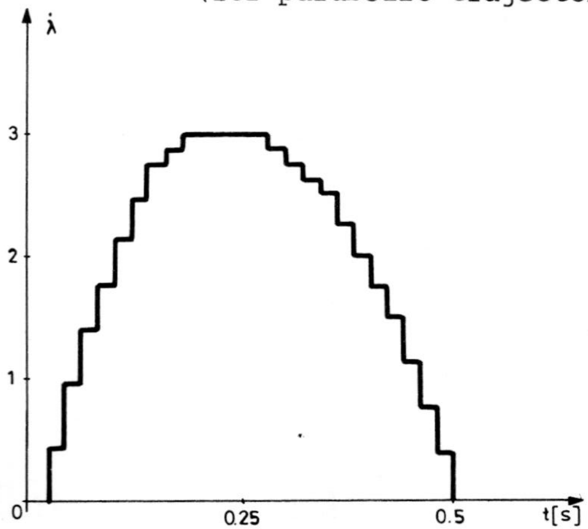

Fig. 3.16. Optimal profile of velocity $\dot{\lambda}(t)$ (for parabolic trajectory)

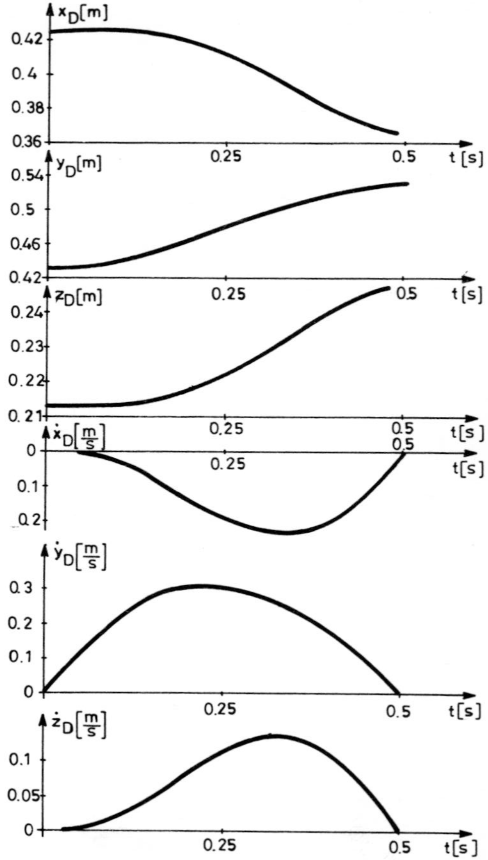

Fig. 3.17. Optimal change of vector in external coordinates and their velocities (for parabolic trajectory)

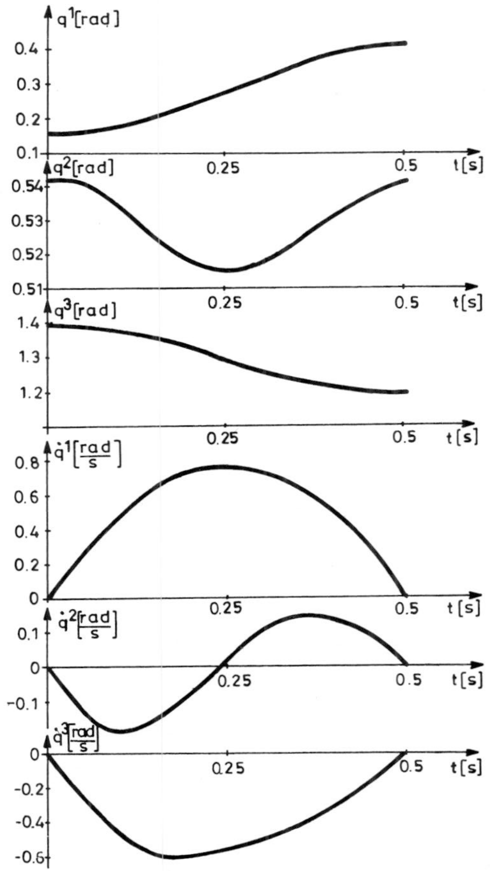

Fig. 3.18. Optimal change of state vector (for parabolic trajectory)

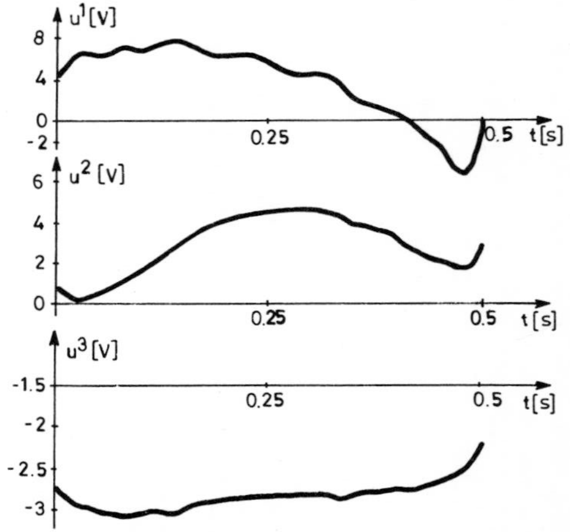

Fig. 3.19. Optimal nominal control (for parabolic trajectory)

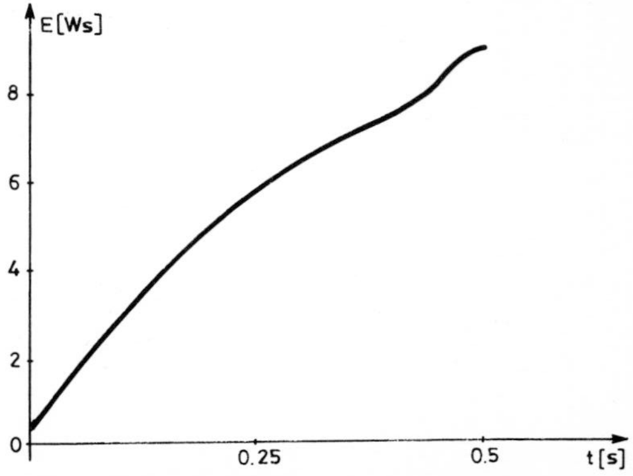

Fig. 3.20. Change of energy consumption (for parabolic trajectory)

3.2.2. Linear optimal regulator - control synthesis at the stage of perturbed dynamics

In the previous paragraph we presented the synthesis of the nominal trajectory $x^o(t)$ and programmed control $u^o(t)$, $\forall t \in T$. The control at the stage of perturbed dynamics should now be synthesized in order that $\forall \Delta x(0) \in \bar{X}^I \rightarrow \Delta x(t) \in \bar{X}^t$, $\forall t \in T$ and $\Delta x(t) \in \bar{X}^F$, $\forall t \in T_s$. We shall approach the problem of control synthesis at this stage by applying three different procedures described in ch. 2, namely, the use of a linear optimal regulator, decentralized control and the use of an asymptotic regulator. The results of discrete optimal regulator synthesis will also be presented.

First, we briefly describe the linear optimal regulator synthesis for two manipulation types with three d.o.f. powered by D.C. motors. The case of a time-invariant linear regulator given by (2.5.6) - (2.5.9) is considered. To this end we first linearize the model (2.4.1) of deviation around the nominal trajectory. As already stated in para. 2.5, the linearization can be carried out either analytically, by applying an algorithm for automatic linearization of the manipulator model on a digital computer (para. 1.3) or by identifying a linear, time-invariant model (matrices $\tilde{\bar{A}}^o$ and $\tilde{\bar{B}}^o$).

For the linear system given by the pair of matrices $\tilde{\bar{A}}^o$, $\tilde{\bar{B}}^o$ the feedback gain matrix (matrix D in Eq. (2.5.8)) should be determined.

As already stated, K is the solution of the Riccatti matrix equation (2.5.9). A numerical procedure for solving the set of algebraic quadratic equations (2.5.9), i.e. for determining matrix K, is based on the following well-known theorems.

Theorem 1. Each solution of the quadratic matrix equation (2.5.9) can be presented in the form

$$K = [z_1, z_2, \ldots, z_n][w_1, w_2, \ldots, w_n]^{-1}, \tag{3.2.17}$$

where $c_i \triangleq \frac{w_i}{z_i}$ are the eigenvectors of the matrix

$$E = \begin{bmatrix} \tilde{\bar{A}}^o + \Pi I & -\tilde{\bar{Q}} \\ -Q & -(\tilde{\bar{A}}^{oT} + \Pi I) \end{bmatrix}; \quad \tilde{\bar{Q}} = \tilde{\bar{B}}^o \underline{R}^{-1} \tilde{\bar{B}}^{oT} \tag{3.2.18}$$

Theorem 2. Let $c_i = \frac{w_i}{z_i}$, i=1,...,n be the eigenvectors corresponding to the eigenvalues σ_i, i=1,...,n, of matrix E with negative real parts. Then $K = [z_1,\ldots,z_n][w_1,\ldots,w_n]^{-1}$ is the positive definite solution of Eq. (2.5.9) and the σ_i, i=1,...,n are also the eigenvalues of the linear system with the closed feedback loop (2.5.10).

This procedure has been applied in optimal regulator synthesis with prescribed stability degree [10].

For the particular case of the minimal configuration of the manipulator UMS-1 the optimal regulator has been synthesized for various corresponding linear models (around the nominal of Figs. 3.4 - 3.6). The choice of the linearized model of the manipulator dynamics by an identification procedure has been analyzed [11]. By approximating dominant members of matrices $\tilde{\bar{A}}^o$ and $\tilde{\bar{B}}^o$ and by analyzing the optimal gains the following pair of matrices is chosen for the time-invariant linear model corresponding to the dynamic model of the minimal configuration of the manipulator UMS-1.

$$\tilde{\bar{A}}^o = \begin{bmatrix} 0 & 0 & 0 & 1 & 0 & 0 \\ 0 & 0 & 0 & 0 & 1 & 0 \\ 0 & 0 & 0 & 0 & 0 & 1 \\ 6.58 & 119.7 & 50 & -79.5 & -77.3 & 190.8 \\ 41.3 & 96 & 3.7 & -28.9 & -82.1 & 28.1 \\ 27.9 & -298 & -116.2 & 149.4 & 174.5 & -580.8 \end{bmatrix}$$

$$\tilde{\bar{B}}^o = \begin{bmatrix} 0 & 0 & 0 \\ 0 & 0 & 0 \\ 0 & 0 & 0 \\ 2.63 & 1.1 & -3.3 \\ 0.64 & 1.34 & -0.68 \\ -5.43 & -2.4 & 10 \end{bmatrix}$$

For this linear model the optimal regulator has been synthesized for various constant matrices Q and $\underline{R}$ and for various degrees of exponential stability Π. Gains D_{q1}, D_{q2}, D_{q3} (in feedback loops with respect to angles) for various values of matrix Q and for $\underline{R} = 0.1\ I$, $\Pi = 0$ are presented in Fig. 3.21, while in Fig. 3.22. these gains are presented as functions of $\underline{R}$ for Q = 1000 I, Π = 15. For Π = 0, Q = 1, $\underline{R}$ = I the distribution of gains in matrix D is better than in the case when Q = 1000 I, $\underline{R}$ = I or Q = 100 I, $\underline{R}$ = I but, for $\Pi \neq 0$, the gain matrix D_{Π} is almost insensitive to variation of the weighting matrices Q and $\underline{R}$. The elements of this matrix increase significantly by increasing the prescribed stability degree Π. It can be seen that the corresponding elements of matrix D_{20} are nearly equal when Q = I, $\underline{R}$ = I and when Q = 1000 I, $\underline{R}$ = 100 I. The same holds for D_{15} when Q = 1000 I and when Q = I, $\underline{R}$ = I. Assuming that Π = 15 is a sufficient degree of exponential stability and taking into account the fact the gains in matrix D_{15} are significantly less than the corresponding gains in matrix D_{20}, together with the conditions of the imposed control task for further analyzing, we adopt matrix D_{15} (Π = 15, Q = 1000 I, $\underline{R}$ = 10 I) as follows

$$D_{15} = \begin{bmatrix} 1147 & 75 & 148 & 45 & -10 & 14 \\ 285 & 1956 & 268 & -10 & 30 & -1.5 \\ 201 & -110 & 1873 & 21 & 0.5 & 10 \end{bmatrix}$$

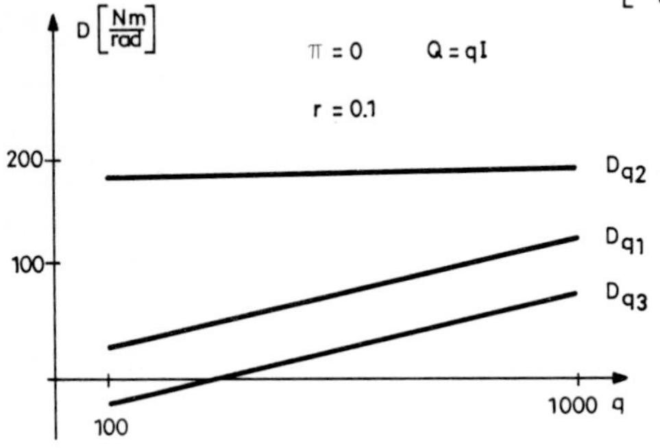

Fig. 3.21. Linear regulator gains as function of matrix Q (for UMS-1)

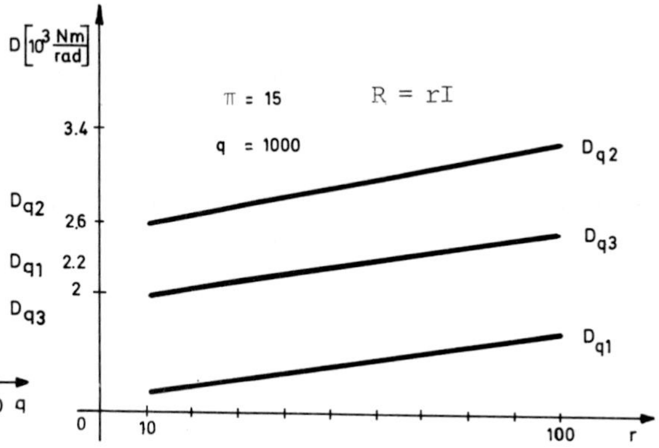

Fig. 3.22. Linear regulator gains as a function of matrix $\underline{R}$ (for UMS-1)

The linear regulator constructed in this way ensures the tracking of nominal trajectories of the nominal configuration of the manipulator UMS-1. It should now be determined whether or not the conditions of practical stability of system S are satisfied. For the control task in para. 3.2.1. (Figs. 3.4 - 3.6), let the regions of practical stability be defined as follows: $\bar{X}^I = \{\Delta x(0): |\Delta q^i(0)| \leq 0.2, \Delta\dot{q}^i = 0, \forall i \varepsilon I\}$, $\bar{X}^t(t) = \{\Delta x(t): ||\Delta x(t)|| \leq 0.6, \forall t\varepsilon(0, 0.3), ||\Delta x(t)|| \leq 0.01, \forall t\varepsilon(0.3, \tau)$, $\bar{X}^F = \{\Delta x(\tau): ||\Delta x(\tau)|| \leq 0.01, \tau_s = \tau = 2s\}$. In order to determine whether these regions of practical stability are satisfied by the linear regulator, the minimal configuration dynamics has been simulated with a linear regulator in closed feedback loop. The control has been taken in the form (2.5.8). In Fig. 3.23. the results of simulation are presented for the initial conditions on the boundary of the region X^I. The deviations of angles q^i, $\forall i\varepsilon I$ from the nominal trajectories $q^o(t)$, $\forall t\varepsilon T$ are presented. The driving torques developed by the optimal linear regulator when the tracking of the nominal trajectory as in Fig. 3.23 is realized, are presented in Fig. 3.24. According the developed driving torques D.C. motors can be chosen which produce the tracking of the nominal trajectory in the desired way. From Fig. 3.24 it is obvious that the conditions of practical stability are relatively severe and require high driving torques of the actuators.

It should be mentioned that we are considering the simulation of the tracking of the nominal trajectory with actuators in the form (3.2.1), where amplitude constraints on the inputs has not been taken into account In order to realize the desired tracking $\forall x(0)\varepsilon X^I$, an actuator which can develop sufficient driving torques should be chosen.

As the second example of optimal linear regulator synthesis we shall consider the manipulation system UMS-2. In this case, we shall present the synthesis of a time - varying linear regulator and the synthesis of a time - invariant linear optimal regulator*).

We shall consider only the basic (minimal) configuration of the manipulation system UMS-1 and the functional movement defined in para. 3.2.1 (Figs. 3.7 - 3.8). Applying the method developed for forming the linearized model of active mechanisms (para. 1.3), the matrices of the linear model are obtained in the form

$$A^o(t) = \left[a^o_{ij}\right]_{6\times 6} = \begin{bmatrix} 0 & I_3 \\ A^o_{21} & A^o_{22} \end{bmatrix}, \qquad B^o(t) = \begin{bmatrix} 0 \\ B^o_{21} \end{bmatrix}$$

---

*) This part of the paragraph was written by Mr Nenad Kirćanski, M.Sc.

where

$$A_{21}^{o} = \begin{bmatrix} 0 & 0 & a_{43}^{o} \\ 0 & 0 & 0 \\ 0 & 0 & a_{63}^{o} \end{bmatrix}, \quad A_{22} = \begin{bmatrix} a_{44}^{o} & 0 & a_{46}^{o} \\ 0 & a_{55}^{o} & 0 \\ a_{64}^{o} & 0 & a_{66}^{o} \end{bmatrix},$$

$$B_{21} = \mathrm{diag}(b_{41}^{o}\ b_{52}^{o}\ b_{63}^{o}).$$

The elements of matrices $a_{ij}^{o}(t)$ and $b_{ij}^{o}(t)$, $\forall t\varepsilon T$ are presented in Fig. 3.25. Analysis of the functions in Fig. 3.25 leads to the conclusion that the elements of matrix $\tilde{A}^{o}$, $a_{43}^{o}(t)$ and $a_{46}^{o}(t)$, $\forall t\varepsilon T$, change more than the other elements of matrices $\tilde{A}^{o}$ and $\tilde{B}^{o}$ during the movement. This is the result of the coupling between the first and third d.o.f. of the mechanism. Element $a_{43}^{o}(t)$ shows the influence of the third d.o.f. position on the dynamics of first d.o.f., and $a_{46}^{o}(t)$ represents the influence of the third d.o.f. velocity on the dynamics of the first d.o.f.

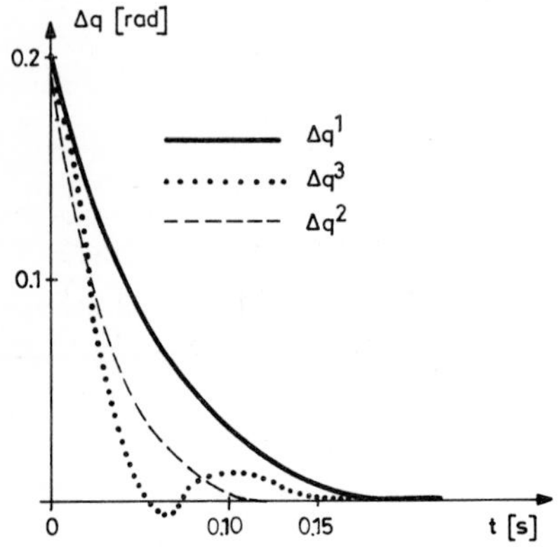

Fig. 3.23. Tracking of the nominal with linear regulator (for UMS-1)

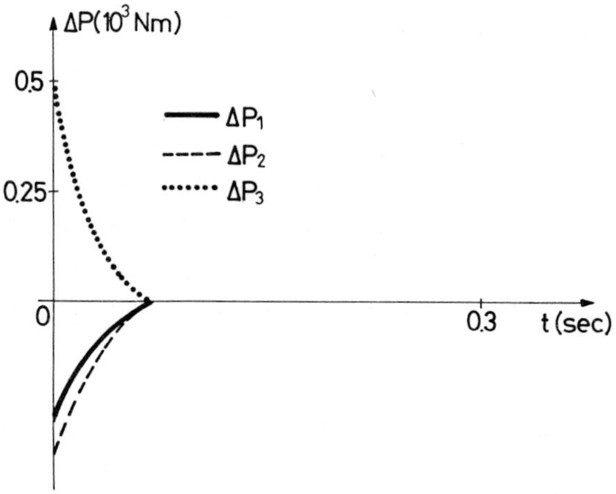

Fig. 3.24. Deviation of driving torques from the nominal for a system with linear regulator (for UMS-1)

According to the procedure for synthesizing the linear continuous time - varying regulator on the finite optimization interval, the regulator has been synthesized with the following weighting matrices: Q = diag(50. 50. 50. 1. 1. 1.), $\underline{R} = I$, $\Pi = 0$. The time-variable elements of the feedback matrix $D(t) = -\underline{R}^{-1}\tilde{B}^{oT}K(t)$ are presented in Fig. 3.26.

In order to examine the performance of the synthesized regulator when applied to the complete nonlinear model of the manipulator, the motion of the system with perturbations of the initial conditions has been simulated with initial conditions: $\Delta x(0) = \{-0.15$ [rad], -0.02 [m], -0.03 [m], -0.1 [rad/s], 0.025 [m/s], -0.04 [m/s]$\}$. The simulation results are presented in Fig. 3.27.

Although with such regulator the minimal cost of the performance index is obtained in the interval T for the defined weighting matrices Q and $\underline{R}$, implementation of time-variable gains (Fig. 3.26) is complex. For example, variable gains can be implemented by digital computer (in the feedback loop) in which data on the gains are stored in the computer memory. However, since the number of feedback loops in general is $N \times n$, it is obvious too large memory would be necessary.

From the previous analysis it follows (as already stated in para. 2.5) that, from the point of view of application, a regulator with time-invariant gains is much more convenient. In para. 2.5 we showed that such a regulator can be obtained using the model (2.5.6) and criterion (2.5.7).

In order to analyze in more detail the performance of the system with time - invariant regulators, we shall synthesize the regulators on the following regions (parts) of the considered functional movement of the manipulator UMS-2: the initial region X^{I}, the region of acceleration X^{II}, the transfer region X^{III} and the region of deacceleration X^{IV} (Fig. 3.28).

We present the characteristics of the regulators in the particular regions. For each regulator the poles of the closed-loop system will be determined.

Region X^{I}: In this region the matrix $\tilde{\bar{A}}^{o}$ is quasidiagonal, i.e., the system is noninteractive. We have chosen $\underline{R} = I$, $\Pi = 0$ and the weighting factors in matrix Q to be varied. For example, for Q = I the solution of the Riccati equation is given by

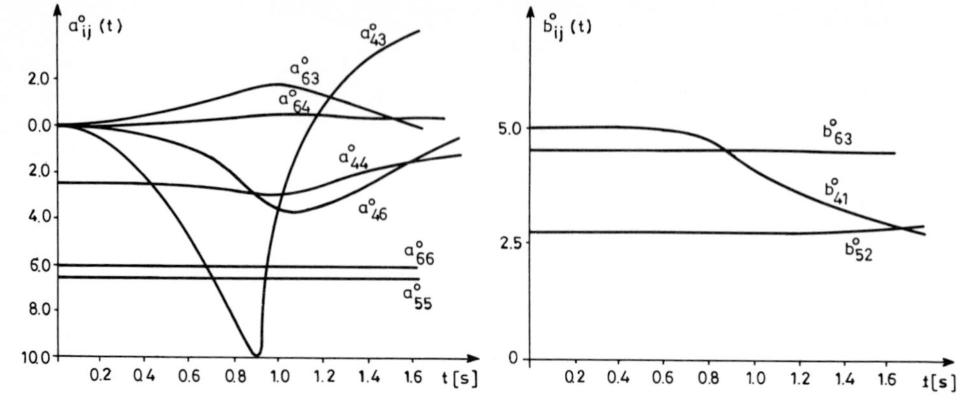

Fig. 3.25. Change of elements of linearized model matrices: $\tilde{A}^O(t)$ and $\tilde{B}^O(t)$ during functional movement (UMS-2)

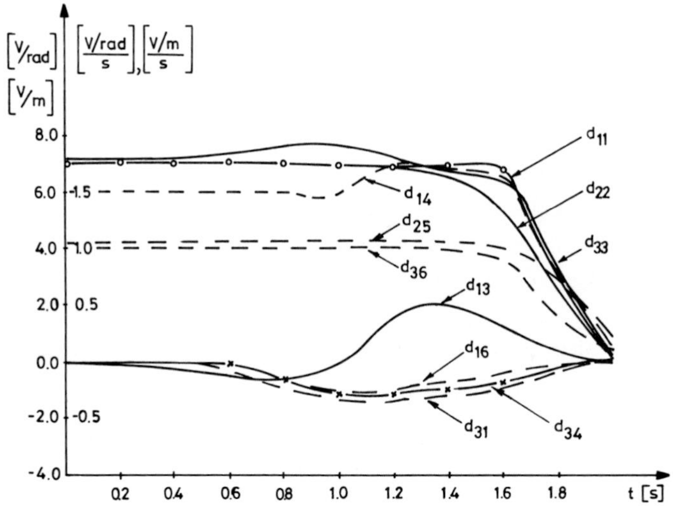

Fig. 3.26. Time - varying regulator feedback gains (UMS-2)

$$K = \begin{bmatrix} 1.29 & 0 & 0 & 0.20 & 0 & 0 \\ 0 & 2.53 & 0 & 0 & 0.33 & 0 \\ 0 & 0 & 1.82 & 0 & 0 & 0.22 \\ 0.20 & 0 & 0 & 1.55 & 0 & 0 \\ 0 & 0.33 & 0 & 0 & 1.15 & 0 \\ 0 & 0 & 0.22 & 0 & 0 & 0.1 \end{bmatrix}$$

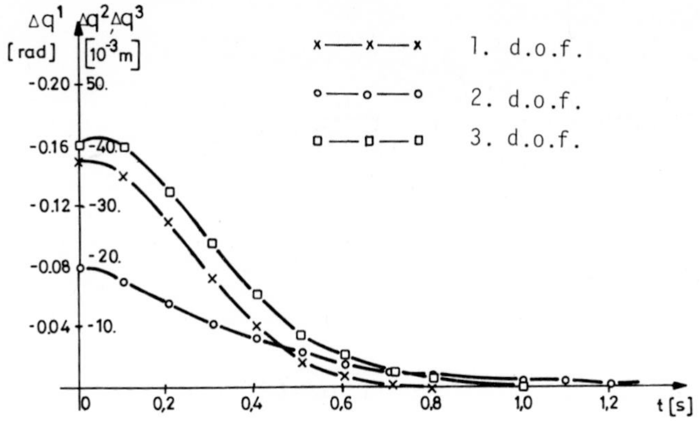

Fig. 3.27. Simulation of manipulation system with time - varying regulator (UMS-2)

The feedback matrix is also found to be

$$D = \underline{R}^{-1}(\tilde{\bar{B}}^{o})^{T}K = \begin{bmatrix} 2.24 & 0 & 0 & 1.97 & 0 & 0 \\ 0 & 2.24 & 0 & 0 & 1.17 & 0 \\ 0 & 0 & 2.24 & 0 & 0 & 1.44 \end{bmatrix}.$$

It follows that the system in the first region is noninteractive.

For the second example we choose Q = diag(10. 1. 1. 1. 1. 1.). In this case,

$$K = \begin{bmatrix} 5.0 & 0 & 0 & 0.63 & 0 & 0 \\ 0 & 2.53 & 0 & 0 & 0.33 & 0 \\ 0 & 0 & 1.8 & 0 & 0 & 0.22 \\ 0.63 & 0 & 0 & 0.215 & 0 & 0 \\ 0 & 0.33 & 0 & 0 & 0.114 & 0 \\ 0 & 0 & 0.22 & 0 & 0 & 0.1 \end{bmatrix}$$

$$D = \begin{bmatrix} 3.16 & 0 & 0 & 1.07 & 0 & 0 \\ 0 & 1.0 & 0 & 0 & 0.35 & 0 \\ 0 & 0 & 1.0 & 0 & 0 & 0.45 \end{bmatrix}$$

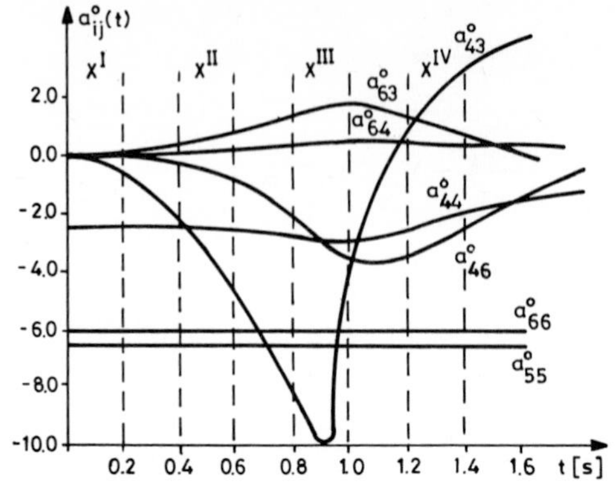

Fig. 3.28. Regions of functional movement (UMS-2)

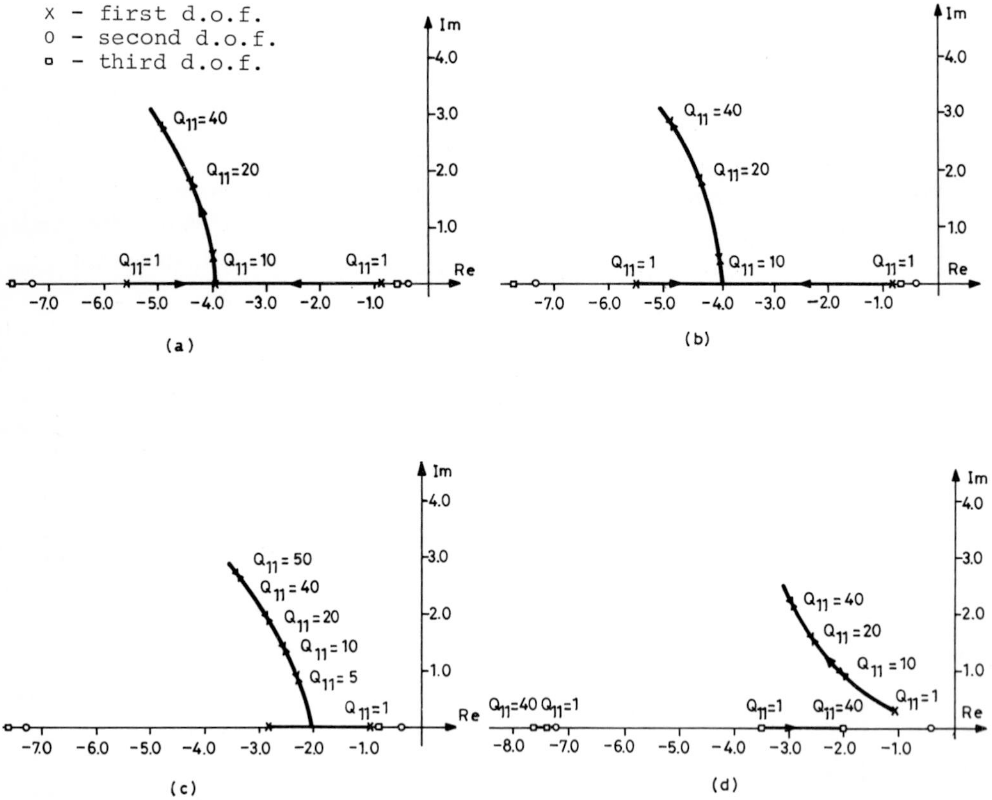

Fig. 3.29. Pole placement of the system as a function of Q_{11}

Such analysis has been carried out for various values of the elements of the weighting matrix Q. Results are summarized in Fig. 3.29a.

Region X^{II}: Since, in this case, non-diagonal elements appear in matrix $\tilde{\bar{A}}^o$, there is interaction between the first and third d.o.f. of the manipulator. Fig. 3.29b shows positions of the poles when the element Q_{11} of matrix Q is varied between 1 and 40. The other elements of Q are: $Q_{ii} = 1$, $i \neq 1$, $Q_{ij} = 0$ for $i \neq j$. It can be seen that the interaction between the first and third d.o.f. is very weak.

Region X^{III}: In this region there is strong interaction between the first and third d.o.f., as can be seen in Fig. 3.28 and Fig. 3.29c. By increasing the weighting factor Q_{11} we move the poles of the third d.o.f.

Region X^{IV}: Fig. 3.29d shows the displacement of the system poles in this region as the weighting factor Q_{11} is varied between 1 and 40. It can be seen that for a relatively small weighting factor $Q_{11} \simeq 1$ the poles of the first d.o.f. become complex.

In order to compare the feedback matrices D^k in these regions, the weighting matrices are chosen as Q = diag(50. 50. 50. 1. 1. 1.), $\underline{R} = I$. The feedback matrices $D^k = \underline{R}^{-1}(\tilde{\bar{B}}^{ok})^T K^k$ in these regions are

$$D^I = \begin{bmatrix} 7.071 & 0 & 0 & 1.510 & 0 & 0 \\ 0 & 7.071 & 0 & 0 & 1.040 & 0 \\ 0 & 0 & 7.071 & 0 & 0 & 1.080 \end{bmatrix}$$

$$D^{II} = \begin{bmatrix} 7.066 & 0 & -0.44 & 1.500 & 0 & -0.07 \\ 0 & 7.071 & 0 & 0 & 1.042 & 0 \\ -0.28 & 0 & 7.224 & 0 & 0 & 1.097 \end{bmatrix}$$

$$D^{III} = \begin{bmatrix} 4.88 & 0 & 0.192 & 1.263 & 0 & -0.3 \\ 0 & 7.07 & 0 & 0 & 1.04 & 0 \\ -5.11 & 0 & 7.3 & -1.32 & 0 & 1.455 \end{bmatrix}$$

$$D^{IV} = \begin{bmatrix} 7.07 & 0 & 0.83 & 2.03 & 0 & 0.035 \\ 0 & 7.07 & 0 & 0 & 1.04 & 0 \\ 0.06 & 0 & 7.13 & 0.06 & 0 & 1.085 \end{bmatrix}$$

The simulation results of the complete dynamic model of the manipulator UMS-2 (minimal configuration) are presented in Figs. 3.30a - 3.30b. By analyzing these results we can conclude:

(1) The system with the linear regulator D^I behaves like the system with a time - varying regulator (Fig. 3.27). This result is obvious since D^I has been synthesized in the first region of the movement.

(2) When the nominal trajectory is tracked by the D^{III} regulator the significant overshoot in the third state coordinate appears in the time interval $t \varepsilon (0,\ 0.5s)$, i.e., in the first region of the movement. The regulator has been synthesized in the third region in which the coupling between the first and third d.o.f. is the strongest. Thus the system performance in the first region $t \varepsilon (0,\ 0.5s)$ is unsatisfactory.

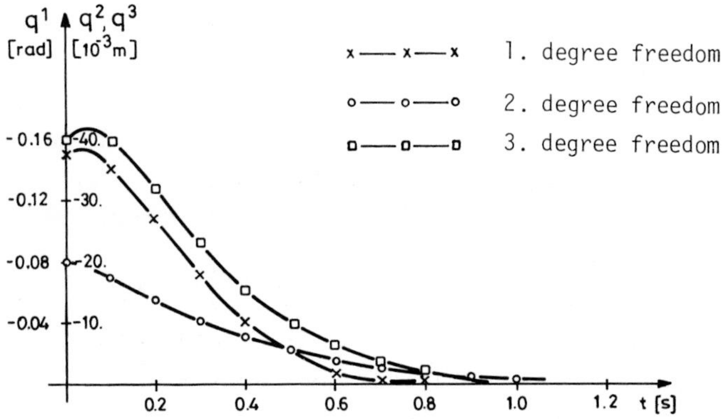

Fig. 3.30a. Simulation of manipulation system with the D^I regulator (for UMS-2)

We have thus demonstrated the synthesis of a linear regulator with two manipulation system types: the anthropomorphic (Fig. 3.1), where the coupling among d.o.f. is strong (so the linear optimal regulator requires high gains in the cross feedback loops) and the manipulator with linear d.o.f. (Fig. 3.2), where the coupling is much weaker and the regulator synthesis is simpler (so the regulator has much fewer feedback loops).

In the rest of this paragraph we shall present the synthesis of the linear optimal regulator in a discrete-time domain (para. 2.7.1) using

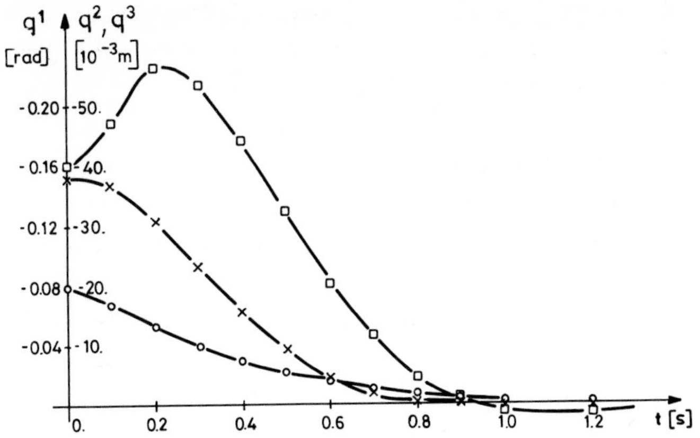

Fig. 3.30b. Simulation of manipulation system with D^{III} regulator (for UMS-2)

as an example the manipulation system UMS-2. The purpose of this is to compare the performance of the manipulation system controlled by different types of discrete regulators with the performance of the same system controlled by a continuous regulator. That is, our purpose is to investigate the influence of discrete time on the dynamic performance of the manipulation system [12]. In this way we can determine the microcomputer requirements when the control of the manipulation system should be realized by a linear optimal regulator.

In order to synthesize discrete regulators on the basis of the procedure described in para. 2.7.1, programs for forming the matrices of the discrete model and criterion have been developed. For each of above regulators for UMS-2 the corresponding discrete regulator has been calculated. The behaviour of the discrete regulators when controlling the manipulation system has been examined by the simulation of the system motion with various initial conditions.

With the simulation of the discrete regulators special attention has been paid to the choice of the sampling interval T_D, since this is one of the basic parameters upon which the behaviour of the complete system depends.

The results have been obtained in several steps.

(1) In the first step the equivalence of the continuous and discrete

regulators were proved in the case of very short sampling interval: $T_D \to 0$.

In order to show this, we consider the first region of the functional movement and weighting matrices: $Q = \mathrm{diag}(50.\ 50.\ 50.\ 1.\ 1.\ 1.)$ and $R = I_3$. In the synthesis of the discrete regulator we have chosen the sampling interval to be $T_D = 1\mathrm{ms}$. The matrices of the discrete system have been calculated and they are presented in Table 3.7. The feedback matrix of the discrete regulator is obtained in the form

$$D(T_D = 1\mathrm{ms}) = \begin{bmatrix} 7.044 & 0. & 0. & 1.504 & 0. & 0. \\ 0. & 7.060 & 0. & 0. & 1.041 & 0. \\ 0. & 0. & 7.054 & 0. & 0. & 1.078 \end{bmatrix},$$

the elements of which are very close to the corresponding elements of matrix D^I of the continual regulator.

In Table 3.8 the matrices of the discrete system are formed for a sampling period $T_D = 50$ ms.

(2) The second step compared the feedback gains of the continuous and discrete regulators. Fig. 3.31 illustrates the change of gains $d^{ij}(T_D)$ of matrix $D^I(T_D)$, synthesized according to the dynamic characteristics of the model in the first part of the functional movement. In the asymptotic case, when $T_D \to 0$, the gains $d^{ij}(T_D)$ are identical to the corresponding gains of the continuous regulator.

Figs. 3.32a and 3.32b illustrate the gains of the discrete regulator in the third region of the movement. Gains $d^{ij}(T_D)$ in Fig. 3.32b appear as the result of the coupling between the first and the third d.o.f. of the manipulator.

From Figs. 3.31, 3.32a and 3.32b it follows that the feedback gains of the discrete regulator change with the sampling period T_D (they decrease when T_D increases).

(3) In this step the simulations of the manipulator UMS-2 controlled by continuous and discrete regulators have been compared. Comparison has been performed on the complete dynamic model of the system

0.1000E+01	0.0000E+00	0.0000E+00	0.9987E-03	0.0000E+00	0.0000E+00
0.0000E+00	0.1000E+01	0.0000E+00	0.0000E+00	0.9967E-03	0.0000E+00
0.0000E+00	0.0000E+00	0.1000E+01	0.0000E+00	0.0000E+00	0.9969E-03
0.0000E+00	0.0000E+00	0.0000E+00	0.9974E+00	0.0000E+00	0.0000E+00
0.0000E+00	0.0000E+00	0.0000E+00	0.0000E+00	0.9934E+00	0.0000E+00
0.0000E+00	0.0000E+00	0.0000E+00	0.0000E+00	0.0000E+00	0.9938E+00

$\bar{\phi}^{o}(T_D)$

0.2502E-05	0.0000E+00	0.0000E+00
0.0000E+00	0.1526E-05	0.0000E+00
0.0000E+00	0.0000E+00	0.2263E-05
0.5002E-02	0.0000E+00	0.0000E+00
0.0000E+00	0.3050E-02	0.0000E+00
0.0000E+00	0.0000E+00	0.4521E-02

$\bar{D}^{o}(T_D)$

0.5000E-01	0.0000E+00	0.0000E+00	0.2498E-04	0.0000E+00	0.0000E+00
0.0000E+00	0.5000E-01	0.0000E+00	0.0000E+00	0.2494E-04	0.0000E+00
0.0000E+00	0.0000E+00	0.5000E-01	0.0000E+00	0.0000E+00	0.2495E-04
0.2498E-04	0.0000E+00	0.0000E+00	0.9975E-03	0.0000E+00	0.0000E+00
0.0000E+00	0.2494E-04	0.0000E+00	0.0000E+00	0.9934E-03	0.0000E+00
0.0000E+00	0.0000E+00	0.2495E-04	0.0000E+00	0.0000E+00	0.9939E-03

$\hat{Q}(T_D)$

0.4171E-07	0.0000E+00	0.0000E+00
0.0000E+00	0.2546E-07	0.0000E+00
0.0000E+00	0.0000E+00	0.3773E-07
0.2498E-05	0.0000E+00	0.0000E+00
0.0000E+00	0.1520E-05	0.0000E+00
0.0000E+00	0.0000E+00	0.2254E-05

$M(T_D)$

0.1000E-02	0.0000E+00	0.0000E+00
0.0000E+00	0.1000E-02	0.0000E+00
0.0000E+00	0.0000E+00	0.1000E-02

$\hat{R}(T_D)$

Table 3.7. Matrices of system for T_D = 1ms

0.1000E+01	0.0000E+00	0.0000E+00	0.4692E-01	0.0000E+00	0.0000E+00
0.0000E+00	0.1000E+01	0.0000E+00	0.0000E+00	0.4253E-01	0.0000E+00
0.0000E+00	0.0000E+00	0.1000E+01	0.0000E+00	0.0000E+00	0.4300E-01
0.0000E+00	0.0000E+00	0.0000E+00	0.8794E+00	0.0000E+00	0.0000E+00
0.0000E+00	0.0000E+00	0.0000E+00	0.0000E+00	0.7168E+00	0.0000E+00
0.0000E+00	0.0000E+00	0.0000E+00	0.0000E+00	0.0000E+00	0.7338E+00

$\bar{\phi}^{o}(T_D)$

0.6001E-02	0.0000E+00	0.0000E+00
0.0000E+00	0.3433E-02	0.0000E+00
0.0000E+00	0.0000E+00	0.5127E-02
0.2350E+00	0.0000E+00	0.0000E+00
0.0000E+00	0.1301E+00	0.0000E+00
0.0000E+00	0.0000E+00	0.1950E+00

$\bar{D}^{o}(T_D)$

0.2500E+01	0.0000E+00	0.0000E+00	0.5991E-01	0.0000E+00	0.0000E+00
0.0000E+00	0.2500E+01	0.0000E+00	0.0000E+00	0.5610E-01	0.0000E+00
0.0000E+00	0.0000E+00	0.2500E+01	0.0000E+00	0.0000E+00	0.5652E-01
0.5991E-01	0.0000E+00	0.0000E+00	0.4599E-01	0.0000E+00	0.0000E+00
0.0000E+00	0.5610E-01	0.0000E+00	0.0000E+00	0.3814E-01	0.0000E+00
0.0000E+00	0.0000E+00	0.5652E-01	0.0000E+00	0.0000E+00	0.3894E-01

$\hat{Q}(T_D)$

0.5054E-02	0.0000E+00	0.0000E+00
0.0000E+00	0.2939E-02	0.0000E+00
0.0000E+00	0.0000E+00	0.4380E-02
0.5693E-02	0.0000E+00	0.0000E+00
0.0000E+00	0.2863E-02	0.0000E+00
0.0000E+00	0.0000E+00	0.4338E-02

$M(T_D)$

0.5097E-01	0.0000E+00	0.0000E+00
0.0000E+00	0.5031E-01	0.0000E+00
0.0000E+00	0.0000E+00	0.5070E-01

$\hat{R}(T_D)$

0.76673E+00	0.12135E+00
0.76673E+00	-0.12135E+00
0.84738E+00	0.00000E+00
0.75642E+00	0.43797E-01
0.75642E+00	-0.43797E-01
0.72113E+00	0.00000E+00

Eigenvalues of system

Table 3.8. Matrices of discrete system (T_D = 50ms)

with T_D = 50 msec, 100 msec and 200 msec. The control vector was formed by superposing the nominal (programmed) control and the signals from the discrete regulator output (2.7.11).

Figs. 3.33a, b, c present the simulation results for the manipulation system with continuous and discrete regulators (T_D = 50 msec). The regulators were synthesized in the first part of the functional movement in which the system is practically noninteractive. The following weighting matrices were adopted: Q=diag(50. 50. 50. 1. 1. 1.), $\underline{R}$ = I. Fig. 3.33a gives the variation of the first state coordinate (deviation of the internal coordinate $q^1(t)$ from the nominal value $q^{1o}(t)$, $\forall t \varepsilon T$). It is evident, that the difference in tracking of the nominal trajectory with continuous and discrete regulators is negligible. A similar conclusion can be drawn for the second and third state coordinate. These figures also show the control signals for the cases of the continuous and discrete regulators.

Figs. 3.34a, b and c compare the continuous and discrete regulators for a sampling period of T_D = 100 msec. It is evident, that as $t \to \tau$ the state coordinates x(t), deviate from zero, i.e., the internal manipulator coordinates do not attain the corresponding nominal values. However, this is not required by the conditions of practical system stability but the following system state is required: $\Delta x(t) \varepsilon \bar{X}^F(t)$, $\forall t \varepsilon T_s$. If τ_s = 1,5 sec and the region $\bar{X}^F$ is defined by the deviation of the system from the nominal trajectory $|\Delta q^1| \leq 0.02$ [rad], $|\Delta q^2| \leq 0.0025$ [m] and $|\Delta q^3| \leq 0.002$ [m], the conditions of system practical stability will be satisfied. To obtain greater positioning accuracy it is necessary to adopt a smaller sampling period T_D. Figs. 3.35a, b and c demonstrate that the system looses practical stability in the case of a sampling period T_D = 200 msec.

All previous results concern the regulators synthesized according to the first part of the movement. Fig. 3.36 illustrates the results of simulating the system, with control, synthesized according to the dynamic characteristic of the system in the third regicn of the movement. The simulations are presented for T_D = 50 ms, 100 ms and 200 ms and, for the sake of comparison, the simulations with continuous regulators D^I and D^{III} are also illustrated. Obviously the conditions of practical stability are not satisfied only

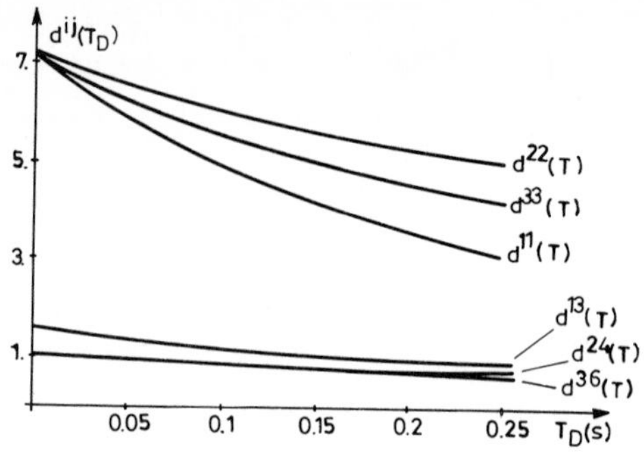

Fig. 3.31. Change in elements of feedback matrix $D^I(T_D)$

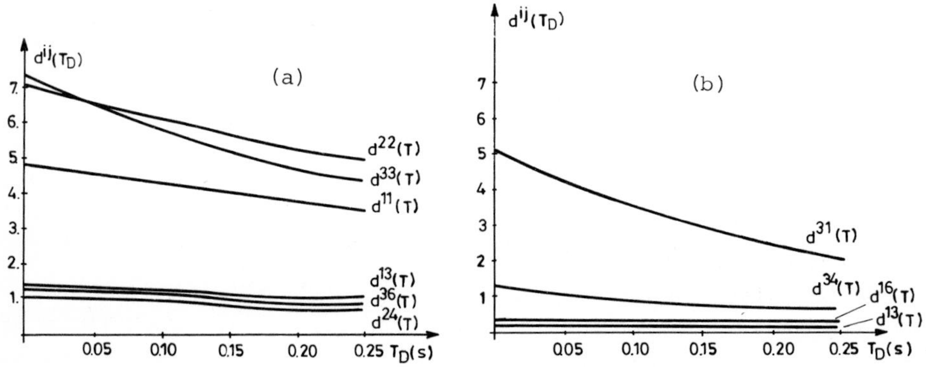

Fig. 3.32. Change of feedback matrix $D^{III}(T_D)$

in the case of the regulator D^{III} (T_D = 200 ms) on account of the long sampling period. Similar results have been obtained for the other d.o.f. of the manipulator.

It follows that by means of the linear discrete optimal regulator practical system stability can be ensured if the sampling period is between 50 and 100 msec. As the duration of one computer instruction of a contemporary microcomputer is from 1 to 2 μsec, during one sampling period between 25000 and 100000 instructions can be executed. This permits the processing of very complex sensory signals or the use of more complex control structures.

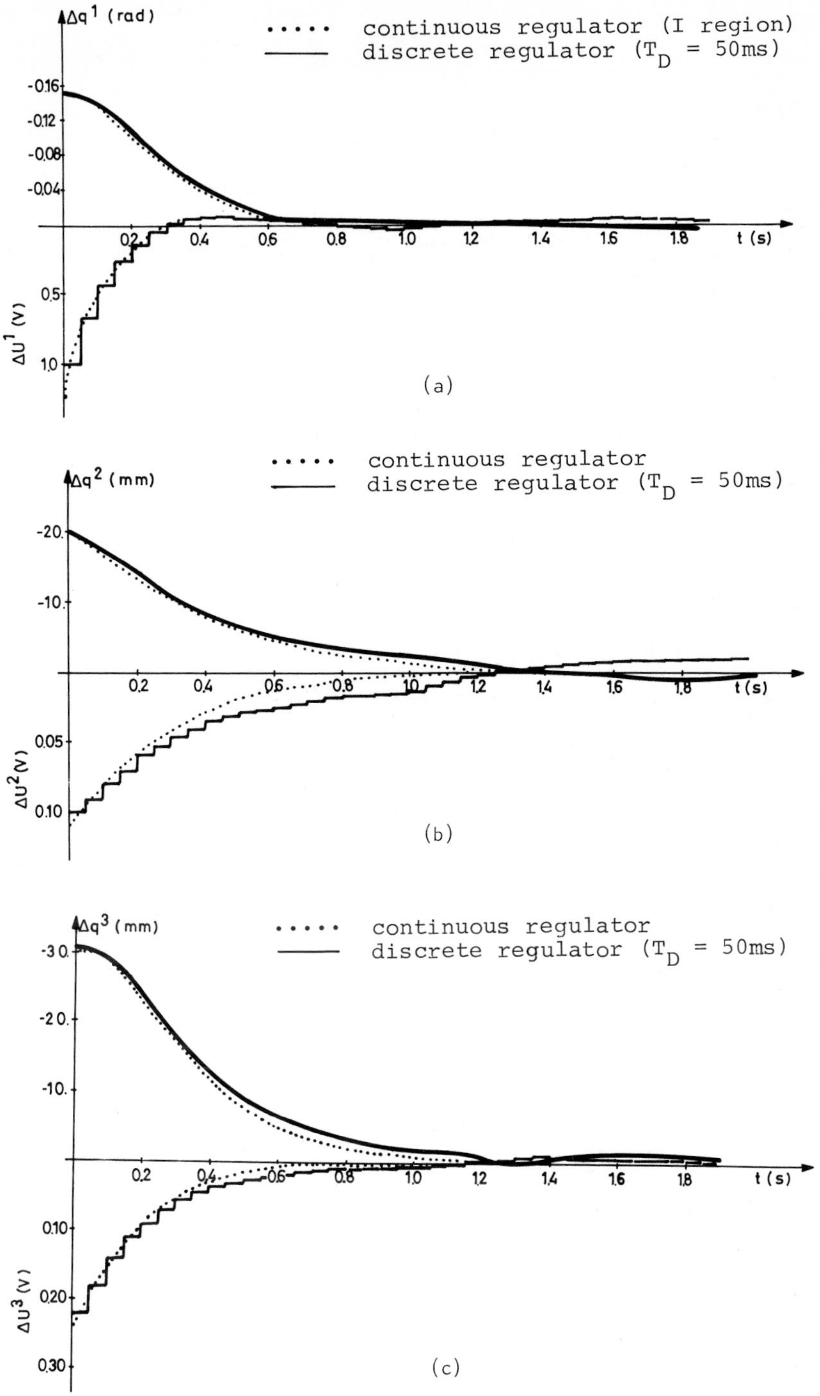

Fig. 3.33. Results of simulation of manipulator UMS-2 with continuous and discrete regulator (I region, T_D=50ms)

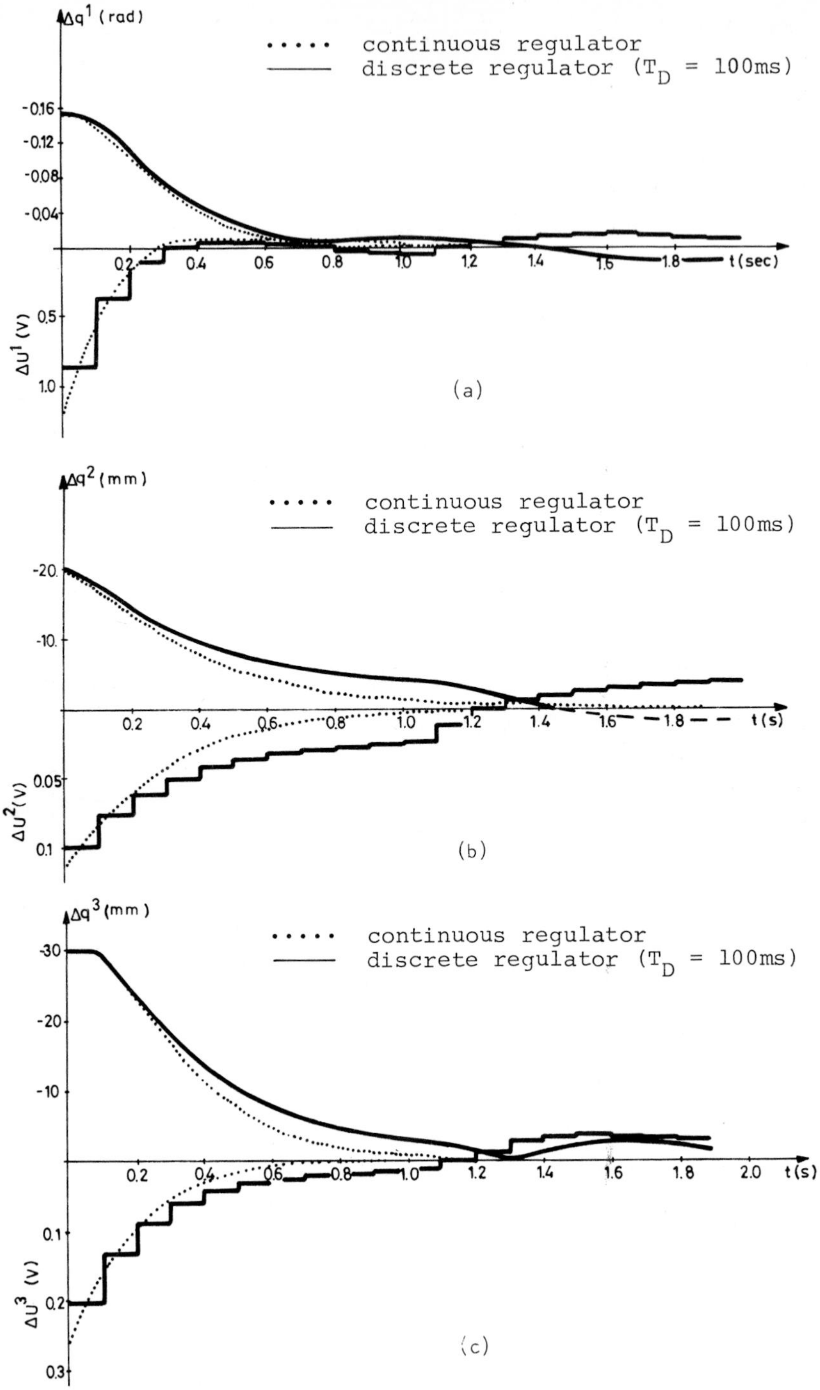

Fig. 3.34. Results of simulation of manipulator UMS-2 with continuous and discrete regulator (I region, T_D = 100ms)

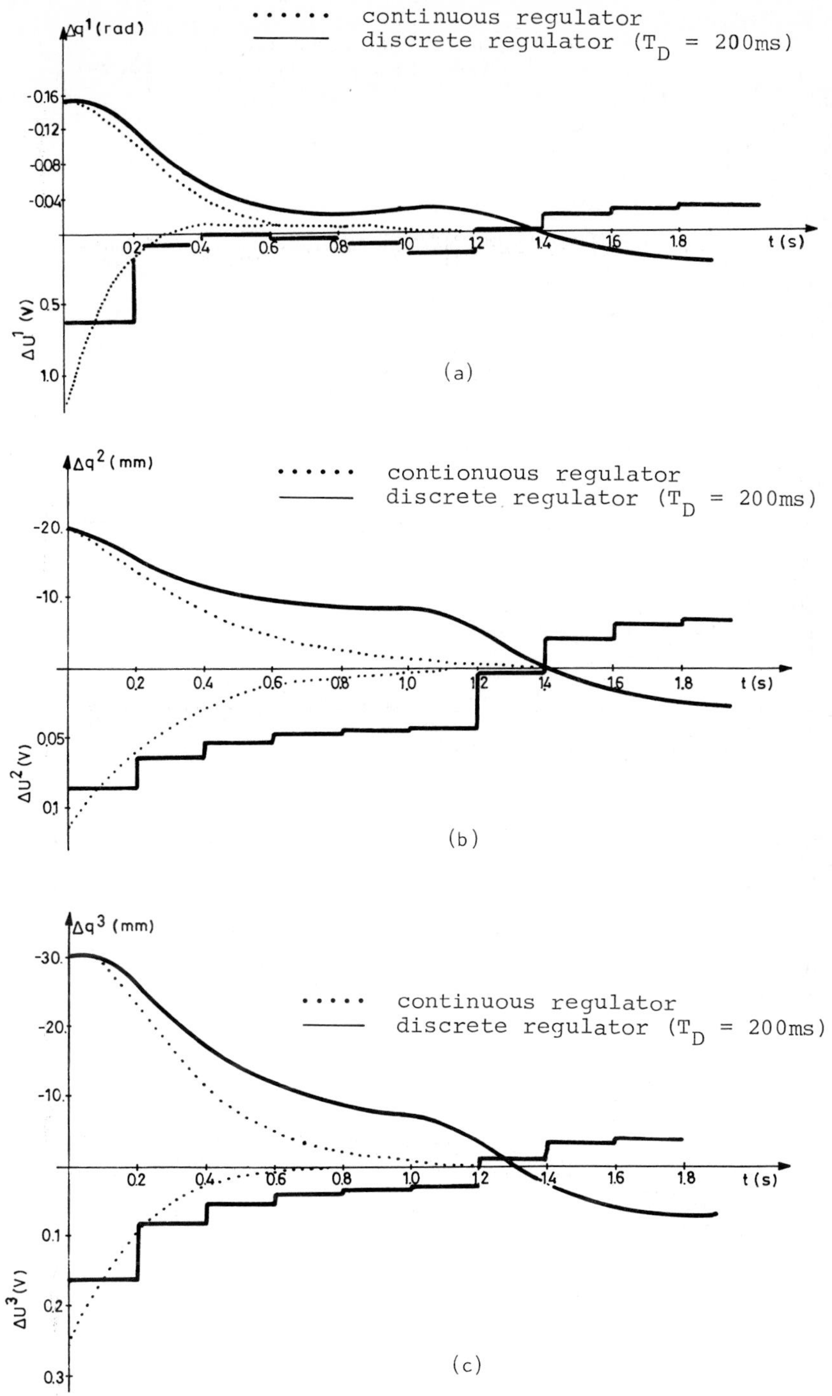

Fig. 3.35. Results of simulation of manipulator UMS-2 with continuous and discrete regulator (I d.o.f., I region, T_D = 200ms)

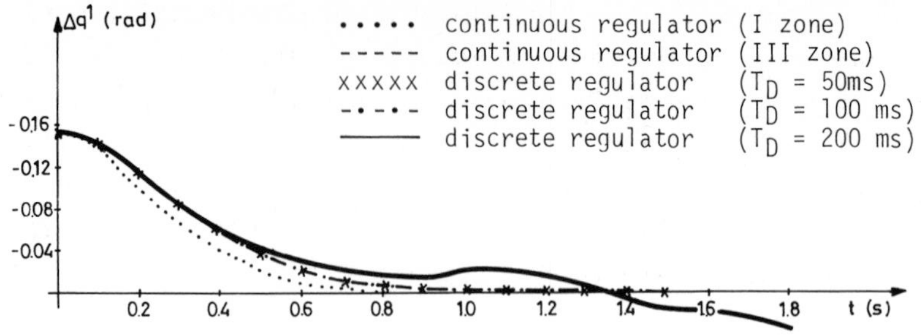

Fig. 3.36. Results of simulation of manipulator UMS-2 with continuous and discrete regulator

It should be noticed that parallel processing creates many ways of realizing different control laws. This will be the subject of para. 3.6. Linear optimal regulators can also be realized by parallel processing, but one should examine the usefulness of introducing a number of microprocessors and the relatively complex control structure required by centralized regulators.

3.2.3. Decentralized control - control synthesis at the stage of perturbed dynamics

We shall not present here the synthesis of decoupled control for tracking the nominal trajectories since this will be presented in detail in para. 3.3.2, which concerns the task of transferring the working object along a prescribed trajectory and orienting the object during the motion. Thus, we present here only the results of applying decoupled control to the minimal configuration of the manipulator in order to compare them with the results obtained using the linear optimal regulator.

Since all d.o.f. of the minimal configuration of the manipulator (n=3) are powered by actuators (3.2.1), subsystems S^i, $\forall i \varepsilon I$, in this case, are linear time-invariable systems of the order $n_i=2$. Such subsystems can be stabilized by local control in the form of linear incomplete feedback

$$\Delta u_i^L = -k_i^T \Delta y^i = -k_i^T C^i \Delta x^i, \qquad \forall i \varepsilon I \tag{3.2.19}$$

where $k_j \varepsilon R^{n_i}$ is the gain vector. By such local control free subsystems

S^i, $\forall i\varepsilon I$, given by (2.6.8) and (3.2.1) can be stabilized with the desired degree Π_i of exponential stability. The stability of the global system S can then be examined (when the coupling among subsystems S^i, given by (2.6.1), is taken into account). However, since this analysis will be presented in more detail in para. 3.3.2, we present here only the results of simulating the dynamics of the minimal configuration with control in the form

$$u^i = u^{io}(t) + \Delta u_i^L(t) = u^{io}(t) - k_i^T C^i \Delta x^i, \quad \forall i\varepsilon I \qquad (3.2.20)$$

for tracking the nominal trajectory in Figs. 3.4 - 3.5.

In order to further simplify the control structure and in order to avoid introducing tachogenerators, the feedback gains k_i are chosen as follows:

$$\frac{C_M^1 k_1}{r_R^1} = (3000,0)^T, \quad \frac{C_M^i k_i}{r_R^i} = (6000,0)^T, \quad \text{for} \quad i = 2, 3$$

This means that, in this case, the feedback by velocity is not introduced. So the chosen linear feedback gains for subsystems and the control in the form (3.2.20), yield the simulation results in Figs. 3.37 and 3.38. Fig. 3.37 shows the deviation $\Delta q^i(t)$ of the angles of the manipulator from the nominal trajectories, for the initial conditions on the bounds of the region $\bar{X}^I$ being defined in para. 2.2.2, i.e., for $\Delta q^i(0) = 0.2$, $\forall i\varepsilon I$. Fig. 3.38 shows the driving torques developed during the tracking of the nominal trajectory by control (3.2.20). According to the results of the simulation in Fig. 3.37, it can be seen that the conditions of practical stability are satisfied in this case [13].

If the results of simulating the system with a linear centralized regulator (Fig. 3.23. and Fig. 3.24) are compared with the results of simulating the system with decoupled control (Figs. 3.37 and 3.38), it follows that both approaches can satisfy the conditions of practical stability. Obviously, the control structure is far simpler in the case of the decoupled control than in the case of the linear regulator. The control in form (3.2.20) requires only six feedback loops (in the concrete case only three positional feedbak loops are introduced), while in the case of the linear regulator eighteen feedback loops would have to be introduced. Obviously the decoupled control, besides the advantages concerning the simplicity and the price of the control system, has advantages concerning reliability. On the other hand, comparing Fig. 3.24 with Fig. 3.38 it can be seen that the optimal regulator requires less energy than the decoupled control. However, since the maximum driving torques developed during the tracking of the nominal trajectory in Figs. 3.4 and 3.5. are approximately the same, both controls can be realized by actuators (D.C. motors) with the same capacity.

We have briefly described the use of decentralized control in tracking the trajectories of the manipulator, i.e., in realizing the simplest manipulation task, the transfer of the manipulator tip along the prescribed trajectory. As we said, the analysis of such control will be presented in paras. 3.3.2 and 3.3.3, while para. 3.6 deals with the realization of such control.

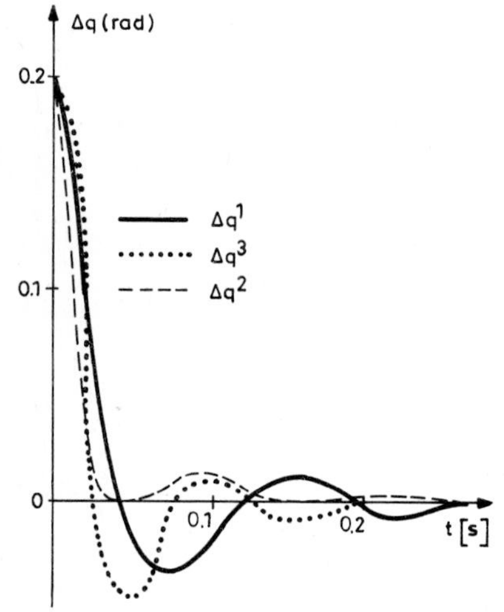

Fig. 3.37. Tracking of the nominal with decoupled control (UMS-1)

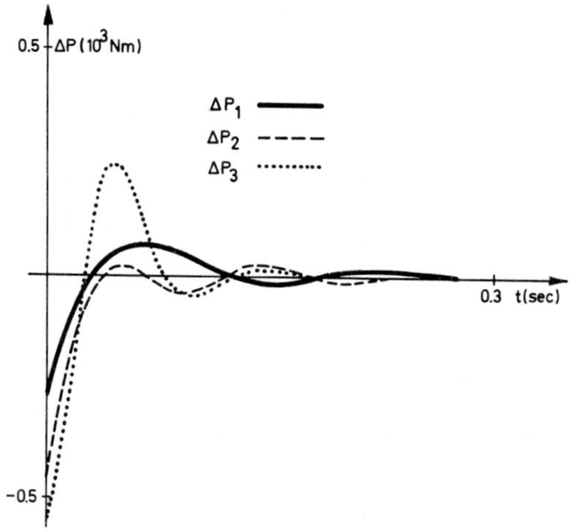

Fig. 3.38. Deviation of driving torques from the nominal for a system with decoupled control (UMS-1)

3.2.4. Asymptotic regulator - control synthesis at the stage of perturbed dynamics*)

In this paragraph we present the synthesis of the asymptotic regulator for tracking the nominal trajectories for the manipulation system UMS-2 (Fig. 3.2). As shown in para. 2.6.6 the properties of the asymptotic regulator can be used at the stage of perturbed dynamics to decouple the manipulation system. This decoupling will be illustrated with the manipulation system UMS-2, which is relatively weakly coupled. This means that the synthesis of the decoupled control is relatively simple [14].

We shall consider the functional movement presented in Figs. 3.7, 3.8. in para. 3.2.1. By analyzing the time - varying elements of the matrices $\tilde{A}^o(t)$ and $\tilde{B}^o(t)$ it has been shown that the system is most interactive (coupled) at the mid-point of the considered trajectory (Fig. 3.25, para. 3.2.2). In the concrete case, submatrices A_{21}, A_{22} and B_{21} of the matrices $\bar{\tilde{A}}^o$ and $\bar{\tilde{B}}^o$ for the region X^{III} of maximum coupling are

$$A_{21} = \begin{bmatrix} 0. & 0. & -7.246 \\ 0. & 0. & 0. \\ 0. & 0. & 1.899 \end{bmatrix} \qquad A_{22} = \begin{bmatrix} -2.83 & 0. & -3.354 \\ 0. & -6.66 & 0. \\ 0.336 & 0. & -6.190 \end{bmatrix}$$

$$B_{21} = \mathrm{diag}(1.033,\ 3.060,\ 4.535)$$

The decoupled control by means of the asymptotic regulator will be synthesized according to the following requirements.

A. The dominant eigenvalues of the manipulation system should be real and equal for all three degrees of freedom: $\sigma_i^o = -4$, i=1, 2, 3. The corresponding eigenvectors should ensure the asymptotic decoupling of the system over the separate degrees of freedom:

$$\tilde{x}_1^o = [1.\quad 0.\quad 0.\quad v\quad v\quad v]^T,$$

$$\tilde{x}_2^o = [0.\quad 1.\quad 0.\quad v\quad v\quad v]^T,$$

$$\tilde{x}_3^o = [0.\quad 0.\quad 1.\quad v\quad v\quad v]^T,$$

*) This paragraph was written by Mr N.Kirćanski, M.Sc.

where v denotes the arbitrary values not specified in advance. The positions of these values correspond to those of the generalized velocities of the manipulator angles, so the choice of these values is not free.

B. The asymptotically infinite eigenvalues of the system should be $\sigma_j^\infty/\sqrt{\rho} = 1/\sqrt{\rho}$, $j = 1,2,3$ and the corresponding eigenvectors should ensure the asymptotic decoupling of the infinite modes: $\nu_1^\infty = [1/b_1 \;\; 0 \;\; 0]^T$, $\nu_2^\infty = [0 \;\; 1/b_2 \;\; 0]^T$, $\nu_3^\infty = [0 \;\; 0 \;\; 1/b_3]^T$.

The procedure for synthesizing the regulator follows directly from the method described in para. 2.6.6. For instance, the form of the eigenvectors which corresponds to the infinite modes and the form of the weighting matrix $\tilde{W}$ (2.6.59) show that $\tilde{W} = I_n$. In this case matrix Q has the form

$$Q = \tilde{H}^T\tilde{H} = \begin{bmatrix} H_{11}^T H_{11} & H_{11}^T \\ H_{11} & I_n \end{bmatrix}$$

where $H_{11} = -\mathrm{diag}(\sigma_1^o \; \ldots \; \sigma_n^o)$. From the form of matrix $\tilde{H}$ it follows that

$$\tilde{H} = [H_{11} \mid I_n] = [\mathrm{diag}(4.0 \quad 4.0 \quad 4.0) \mid I_3].$$

From $\tilde{N} = [\nu_1^\infty \; \ldots \; \nu_n^\infty]$ and $S = \mathrm{diag}(\sigma_1^\infty \; \ldots \; \sigma_n^\infty)$, wherein σ_i^∞ are real and negative, it follows that

$$S = \mathrm{diag}(-1.0 \quad -1.0 \quad -1.0), \text{ and from } \tilde{N} = [\nu_1^\infty \; \ldots \; \nu_n^\infty] =$$

$$= \mathrm{diag}(\nu_{1d}^\infty \; \ldots \; \nu_{nd}^\infty): \tilde{N} = \mathrm{diag}(0.968 \quad 0.327 \quad 0.220).$$

It immediately follows that the weighting matrix is

$$R^o = \tilde{N}^{-T}S^{-2}\tilde{N}^{-1} = \mathrm{diag}(1.067 \quad 9.352 \quad 20.70)$$

Matrix Q has the form

$$Q = \begin{bmatrix} \mathrm{diag}(16. \quad 16. \quad 16.) & \mathrm{diag}(4. \quad 4. \quad 4.) \\ \mathrm{diag}(4. \quad 4. \quad 4.) & I_3 \end{bmatrix}.$$

Programs have been written for calculating the matrices Q and R^o by digital computer. By using these programs the asymptotic regulator was synthesized for various values of the factor ρ. Figs. 3.39 and 3.40 present variations of the eigenvector coordinates of the system, $\tilde{x}_1(\rho)$ and $\tilde{x}_3(\rho)$ (the eigenvector $\tilde{x}_2(\rho)$ is not presented because subsystem S^2 is not dynamically coupled with the other subsystems). Table 3.9 presents the system eigenvalues as functions of ρ and Fig. 3.41 shows the gain coefficients as functions of same factor ρ. It is evident, that by diminishing the value of ρ, the eigenvalues and eigenvectors approach the desired asymptotic values but, at the same time, the gain coefficients are raised. Hence, a compromised value is determined (for instance, $\rho = 0.01$) with which the system becomes decoupled and the gains are not too high.

For $\rho = 0.01$ the feedback matrix is

$$D^{III}(\rho = 0.01) = \begin{bmatrix} 37.72 & 0. & -1.28 & 10.25 & 0. & -1.344 \\ 0. & 13.08 & 0. & 0. & 2.72 & 0. \\ -1.3 & 0. & 9.27 & -0.313 & 0. & 1.97 \end{bmatrix}$$

Simulation results of the complete system behaviour under the action of such a regulator are presented in Figs. 3.42, from which it follows that the system time constants correspond to the dominant system eigenvalues and the system behaves as though it were decoupled.

As in para. 3.2.2, we shall present how discrete time influences the behaviour of the manipulation system with a linear regulator. According to the procedure in para. 2.7.2, asymptotic regulator can be synthesized in discrete-time domain. This is necessary if the asymptotic regulator has to be implemented on microcomputer. The same manipulation system UMS-2 was considered and the regulators corresponding to the regulators in a continuous-time domain were synthesized.

First, we examined the influence of the sampling frequency on the feedback gains of the asymptotic regulator. The results are presented in Fig. 3.43. Obviously the gains, when $T_D \to 0$, asymptotically approach the gains of the corresponding continuous regulator and they decrease as T_D increases.

Fig. 3.44 presents the simulation of the complete dynamic model of the manipulation system UMS-2 for a sampling period $T_D = 50$ ms and initial conditions: $\Delta x(0) = \{-0.15[\mathrm{rad}], -0.02[\mathrm{m}], -0.03[\mathrm{m}], -0.10[\mathrm{rad/s}],$

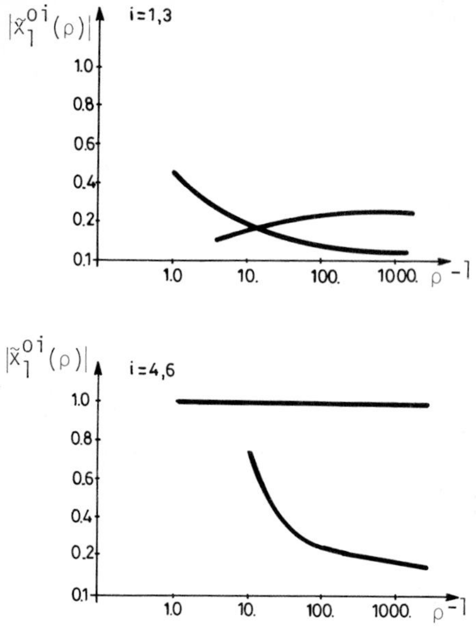

Fig. 3.39. Eigenvectors of the system as functions of ρ (for UMS-2)

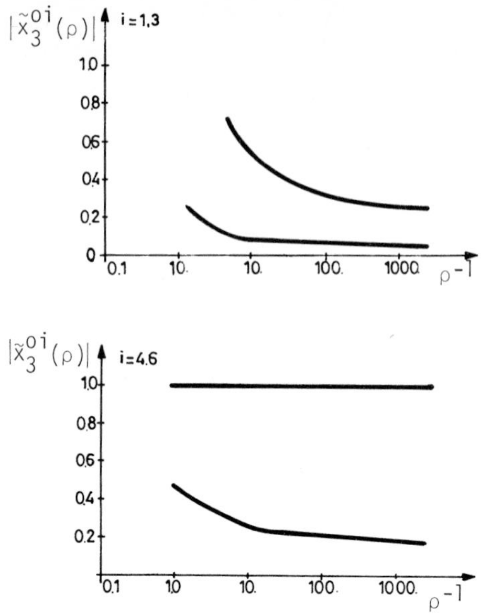

Fig. 3.40. Eigenvectors of the system as functions of ρ (for UMS-2)

ρ	σ_1	σ_2	σ_3	σ_4, σ_5	σ_6
0.1	-6.56	-1.80	-1.90	-3.56 ± j 1.45	-7.10
0.01	-4.14	-3.43	-3.48	-10.60 ± j 1.60	-11.5
0.002	-4.02	-3.87	-3.90	-22.60 ± j 1.80	-23.0
0.0001	-4.00	-3.99	-3.99	-99.90 ± j 1.84	-100.00

TABLE 3.9 Eigenvalues of the system as functions of ρ for manipulator UMS-2 (linearized model of a system with asymptotic regulator)

0.025[m/s], -0.04[m/s]}. These results are compared in the same figure with the corresponding results of tracking by the continual regulator. Obviously, the difference in tracking the nominal trajectory by the continuous and the discrete regulators is negligible. By a similar analysis it can be shown that for longer sampling periods the tracking of the nominal trajectories becomes worse.

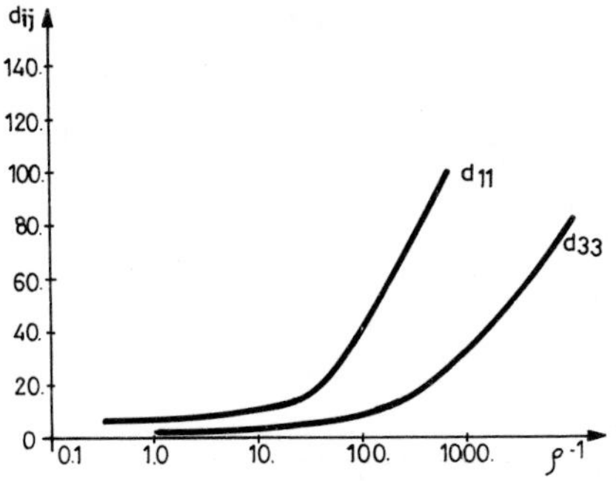

Fig. 3.41. Dependance of gains on factor ρ

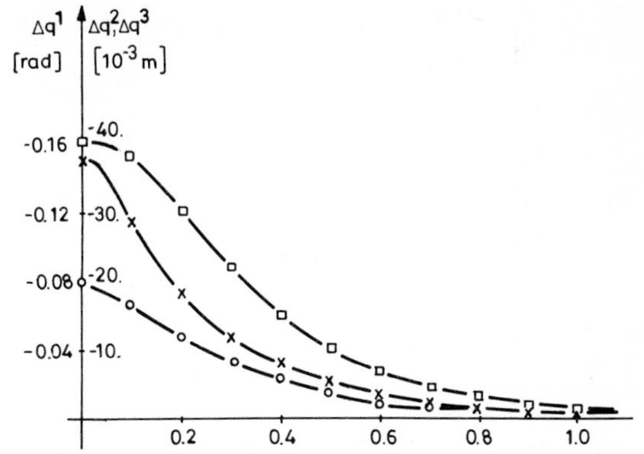

Fig. 3.42. Simulation of manipulator UMS-2 with asymptotic regulator

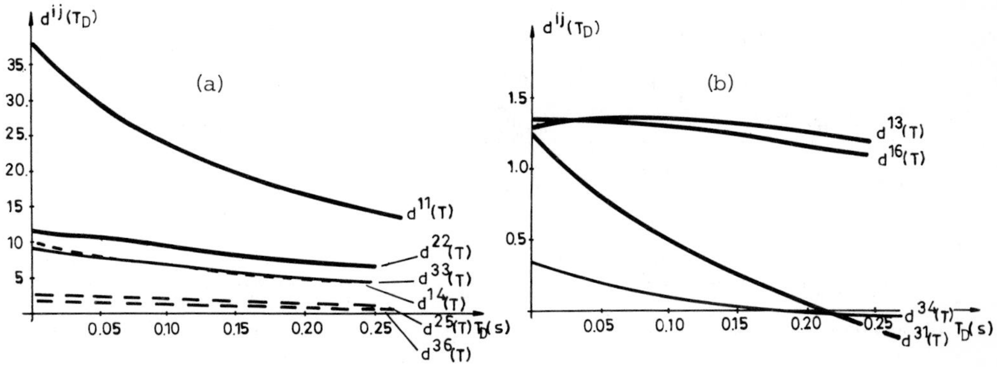

Fig. 3.43. Change of asymptotic regulator matrix elements $D^{III}(T_D)$ (ρ = 0.01, UMS-2)

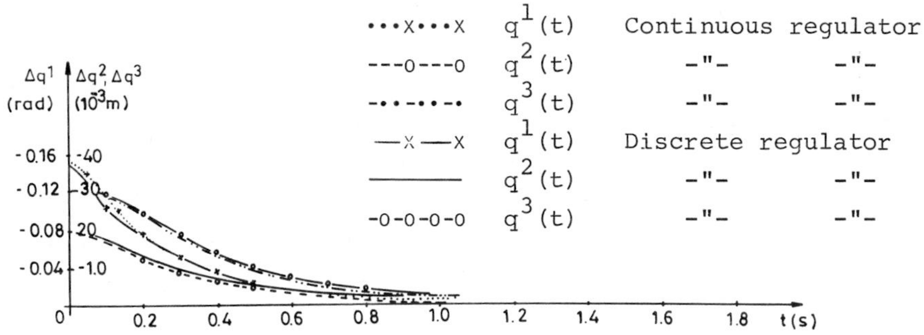

Fig. 3.44. Simulation of manipulator UMS-2 with asymptotic regulator (T_D = 50 ms)

3.3. Transfer of Working Object with Desired Orientation Along Prescribed Trajectory

The task considered in this paragraph is an extension of the task discussed in para. 3.2. Here, in addition to transferring the manipulator tip along a prescribed trajectory in space, it is necessary to achieve a particular orientation of the working object during the transfer, i.e., to ensure that some axis of the working object has a defined orientation with respect to the axes of the absolute coordinate system. In order to realize this task three degrees of freedom of the minimal configuration are not sufficient. This is demonstrated in Fig. 3.4 where the d.o.f. of the gripper are fixed. Manipulators with n=6 d.o.f. are therefore considered (Figs. 3.1 - 3.3). Here the real D.C. motors, for which $L_R \neq 0$, are considered, so that the actuator models S^i ($\forall i \in I$) are given by a linear stationary system of order $n_i = 3$, given by (3.1.2). The system model $S(S^M + S^i, \forall i \in I)$ is obtained in the form (2.2.1), where the state vector is of the order $N = 6x3 = 18$, $x = (x^{1T}, x^{2T}, \ldots, x^{6T})^T$, $x^i = (q^i, \dot{q}^i, i_R^i)^T$, $\forall i \in I$. The output of the system S is given by $y = (q^1, \dot{q}^1, \ldots, q^6, \dot{q}^6)^T$.

The problem of transferring the object with a desired orientation is defined as follows. The manipulator initial state $x^o(0)$ and terminal state $x^o(\tau_s)$ are given. The initial state defines the initial position of the manipulator, the position of the object and its orientation, while $x^o(\tau_s)$ defines the position and orientation of the object, which should be accomplished in time τ_s. We shall suppose that the trajectory, along which the object (gripper) should move, and the change of orientation of the object from $x^o(0)$ to $x^o(\tau_s)$ are defined.

3.3.1. Nominal dynamics synthesis

We now consider the problem of the nominal trajectory $x^o(t)$, $\forall t \in T$ and the programmed input $u^o(t)$ synthesis. The trajectories of the six angles and six angular velocities are unique when the trajectory and the orientation of the working object are defined in space. In paras. 2.3 and 3.2.1 we discussed the difficulties in synthesizing the "optimal" trajectories. However, if at the level of nominal dynamics, the system S is decoupled to two functional subsystems $\tilde{S}_o$ and $\tilde{S}_x$ so that the subsystem $\tilde{S}_o$ consists of the first three d.o.f. with their actuators (minimal configuration of manipulator) and the subsystem $\tilde{S}_x$ consists of the three d.o.f. of the gripper with their actuators, then the problem of nominal trajectory synthesis $q^o(t)$ and $\dot{q}^o(t)$, $\forall t\varepsilon T$ becomes simple [1, 15]. In this case, the state vector is divided into the state vectors of the two subsystems $x_o = (x^{1T}, x^{2T}, x^{3T})^T$ and $x_x = (x^{4T}, x^{5T}, x^{6T})^T$, $x_o \varepsilon R^9$, $x_x \varepsilon R^9$. The input vector $u(t)$ is also split into two input vectors for subsystems, $u_o = (u^1, u^2, u^3)^T$, $u_x = (u^4, u^5, u^6)^T$, $u_o \varepsilon R^3$, $u_x \varepsilon R^3$.

The control task also can be split into two subtasks. Subsystem $\tilde{S}_o$ should ensure the motion of the object along the desired trajectory and subsystem $\tilde{S}_x$ should ensure the desired orientation of the object. This means that the trajectories of the subsystem $\tilde{S}_o$ are synthesized so that $x_o^o(0) \varepsilon X_o^I$ implies $x_o^o(t, x^o(0)) \varepsilon X_o^F$, $\forall t \varepsilon T_s$ and $x_o^o(t, x^o(0)) \varepsilon X_o^t$, $\forall t\varepsilon T$. In addition, the nominal trajectories are calculated for the functional subsystem $\tilde{S}_x$, so that $x_x^o(0) \varepsilon X_x^I$ implies $x_x^o(t, x^o(0)) \varepsilon X_x^F$, $\forall t \varepsilon T_s$ and $s_x^o(t, x^o(0)) \varepsilon X_x^t$, $\forall t \varepsilon T$, assuming that $X^I = X_o^I \times X_x^I$, $X^t = X_o^t \times X_x^t$ and $X^F = X_o^F \times X_x^F$. By decoupling the system S in this way the synthesis of nominal trajectories becomes simple.

Since the manipulator tip trajectory is defined, three angles and three angular velocities of the subsystem $\tilde{S}_o$ are uniquely defined. The angles and angular velocities can be calculated as in para. 3.2.1 since the task for the subsystem S_o is reduced to that of realizing the manipulator tip motion along a prescribed trajectory in space. The trajectories of angles, velocities and accelerations can be calculated using eqs. (3.2.3) - (3.2.6) or by optimization using dynamic programming.

When the nominal trajectories $x_o^o(t)$, $\forall t\varepsilon T$ for the subsystem $\tilde{S}_o$ are calculated, we have to synthesize the nominal trajectories for the gripper subsystem $\tilde{S}_x$. This subsystem should provide the desired orientation of the working object during its transfer along the prescribed trajectory in space. The desired orientation of the object can be involved by prescribing the trajectory of the object tip with respect to the trajectory of the manipulator tip (gripper joint). Here again optimization by dynamic programming, as described in para. 3.2.1 may, in principle, be used to optimize some parameters of the trajectory.

The object tip coordinates in the Cartesian coordinate system are functions of all six manipulator angles:

$$(S_{P_i}) = f_{P_i}(q^1, q^2, \ldots, q^6), \qquad i = 1,2,3, \tag{3.3.1}$$

where $(S_{Pi})^T = (x_p, y_p, z_p)^T$ are the coordinates of the object tip. Since with (3.2.3) - (3.2.6) the trajectories of the first three angles are defined by q^{io}, i = 1, 2, 3, we get

$$(S_{P_i}) = f_{P_i}(q^{10}, q^{20}, q^{30}, q^4, q^5, q^6), \quad i = 1,2,3, \tag{3.3.2}$$

i.e., the coordinates of the object tip are functions of the gripper angles. Thus, there exists a unique relation between the coordinates of the object tip (or gripper) and the three angles of the gripper.

Based on (3.3.1) one can write

$$[A_{P1} \,\vdots\, A_{P2}] \begin{bmatrix} \Delta q_o^o \\ --- \\ \Delta q_x^o \end{bmatrix} = \Delta \vec{S}_P, \tag{3.3.3}$$

where the elements of the matrices A_{p1} and A_{p2} (3×3) are given by

$$a_{ij}^{P1} = \frac{\partial f_{P_i}}{\partial q^j}; \qquad a_{ij}^{P2} = \frac{\partial f_{P_i}}{\partial q^{j+3}}, \qquad i, j = 1,2,3$$

Since the orientation of the object is defined, the object tip trajectory $\vec{S}_p(t)$ is defined as well as the trajectory of the object tip velocity $\dot{\vec{S}}_p(t)$. As with the calculation of the minimal configuration trajectories we shall observe small increments of the object tip movement along the trajectory $\vec{S}_p^o(t)$, during which the matrices A_{p1} and A_{p2} do not change significantly. Let us assume the same time intervals $\Delta t_\ell = t_\ell - t_{\ell-1}$ that we assumed in the synthesis of the minimal configuration trajectories. When the positions of the manipulator and the gripper at the instant t_ℓ are known, the matrices $A_{p1}(t_\ell)$, $A_{p2}(t_\ell)$ can be calculated and since $\dot{q}_o^o(t_\ell)$ are already known, we get

$$\dot{q}_x^o(t_\ell) = A_{P2}(t_\ell)^{-1}(\dot{\vec{S}}_P^o(t_\ell) - A_{P1}(t_\ell)\dot{q}_o^o(t_\ell)) \tag{3.3.4}$$

assuming that the matrix A_{p2} is nonsingular. One thus calculates the gripper angular velocities and the gripper angles trajectories. If the matrix A_{p2} is singular, the singular points should be carefully studied. The matrix A_{p2} could be singular if the desired object orientation requires fewer than three angles of the gripper to be specified. In this case, one d.o.f. of the gripper can be fixed, so the desired orientation of the object can be defined by the other d.o.f. One thus calculates the nominal angles $q^o(t)$ and nominal velocities $\dot{q}^o(t)$ for both functional subsystems $\tilde{S}_o$ and $\tilde{S}_x$. If a distribution (profile) of the manipulator tip velocity is defined, then the accelerations $\ddot{q}^o(t)$ for the six d.o.f. are defined. Now, the driving torques $P^o(t)$, $\forall t \varepsilon T$ can be calculated according to the model of the mechanical part of the system S^M (1.1.30). The nominal rotor current $i_R^{io}(t)$, $\forall t \varepsilon T$ and $\forall i \varepsilon I$ can then be calculated. The nominal trajectories for all state coordinates $x^o(t)$ of the system S can thus be calculated. Using the model S (3.2.1), the programmed control $u^o(t)$, $\forall t \varepsilon T$ can then be calculated.

Figs. 3.45 - 3.46 present the results of the nominal trajectories and programmed inputs synthesis for a particular task for the manipulator UMS-1 (Fig. 3.1). The working object should move from the position A to the position B (Fig. 3.45e). The object position at the point A is defined by the manipulator coordinates. The gripper joint $\vec{S}_A(0) = (0.425, 0.167, 0.571)^T$, while the gripper center of gravity coordinates are $\vec{S}_{PA}(0) = (0.455, 0.203, 0.630)^T$. The position of the object at the point B is given by the manipulator tip coordinates $\vec{S}_B^o(\tau) = (0.368, 0.365, 0.355)^T$, while the coordinates of the gripper c.o.g. are $\vec{S}_{PB}^o(\tau) = (0.391, 0.410, 0.410)^T$ [m]. The manipulator angles corresponding to position A are $q^o(0) = (0, 1.1, 0.5, 0, 0, 0)^T$ [rad], while the manipulator angles for the point B are $q^o(\tau) = (0.08, 0.94, 1.25, 0., 0.78, -1.50)^T$ [rad]. The movement should be performed in $\tau = \tau_s = 0.9$s. The working object should be moved in such way that the tip of manipulator should move along a straight line between points $A(\vec{S}_A^o(0))$ and $B(\vec{S}_B^o(\tau))$ and the gripper c.o.g. should also move along a straight line between its two terminal positions $A(\vec{S}_{PA}^o(0))$ and $B(\vec{S}_{PB}^o(\tau))$.

Manipulator tip acceleration is determined according to (3.2.7), with $a_{max} = 1.48$ m/sec^2 (distance = 0.30 m, $\tau = 0.9$ sec). The conditions for the object orientation are defined so that only two gripper d.o.f. are sufficient to define it. The matrix A_{p2} in (3.3.4) is therefore singular. In this case one chooses $q^4(t) = 0$, $\forall t \varepsilon T$ and the orientation of the object is achieved by the other two d.o.f. of the subsystem $\tilde{S}_x$. The angular velocities $\dot{q}^5$ and $\dot{q}^6$ of the fifth and sixth d.o.f. are calculated according to (3.3.4) but, in this case, the matrix $A_{p2}(t)$ has dimension 2×2, and $A_{p1}(t)$ has dimension 2×3 [7].

Fig. 3.45e presents the movement in three quarter projection in three points on the nominal trajectory. Fig. 3.45a shows the distance of the manipulator tip from the initial position A in the time domain. Fig. 3.45b shows the desired manipulator tip velocity (3.2.7) during the motion. The tip coordinates in the Cartesian absolute coordinate system are presented in Fig. 3.45c. The angular velocities in the initial and terminal positions have to be zero i.e., $\dot{q}^o(0) = 0$ and $\dot{q}^o(\tau) = 0$. Differences between the gripper c.o.g. coordinates and the manipulator tip coordinates are presented in Fig. 3.45d, representing the gripper orientation change during the motion. Fig. 3.46a shows the nominal trajectories of the subsystem $\tilde{S}_o$ angles. Fig. 3.46b shows the nominal trajectories of the gripper (subsystem $\tilde{S}_x$) angles. The initial and terminal conditions motor rotor currents are defined by $i_R^i(0) = i_{STA}^i$ and $i_R^i(\tau) = i_{STB}^i$, $\forall i \varepsilon I$, where i_{STA}^i and i_{STB}^i are currents which hold the manipulator at the points A and B.

With the nominal trajectories for all state coordinates $x^o(t)$, calculated on the basis of the mathematical model (1.1.30), (1.2.1), the nominal driving torques are calculated and presented in Figs. 3.46c. and 3.46d. Finally, using the model S the programmed control $u^o(t)$, $\forall t \varepsilon T$ is calculated. This is shown in Figs. 3.46e - 3.46f.

It should be mentioned that the values of the parameters of the manipulator UMS-1 are modified with respect to their values given in Table 3.1. The new values of the parameters, for which the control has been synthesized for the manipulation task, are given in Table 3.10. One

MEMBER	1	2	3	4
MASS (kg) m_i	0	4.7	6.	6.
LENGTH (m) ℓ_i	0.135	0.33	0.35	0.15
$J_{ix}(10^{-2}kgm^2)$		0.11	0.18	0.18
$J_{iy}(10^{-2}kgm^2)$		0.22	0.36	0.24
$J_{iz}(10^{-2}kgm^2)$	0.92	0.22	0.36	0.24

Table 3.10. Manipulator UMS-1 parameters

should remember that the mass of the gripper includes the mass of the working object. The actuators are the same D.C. motors, data for which are given in Table 3.2.

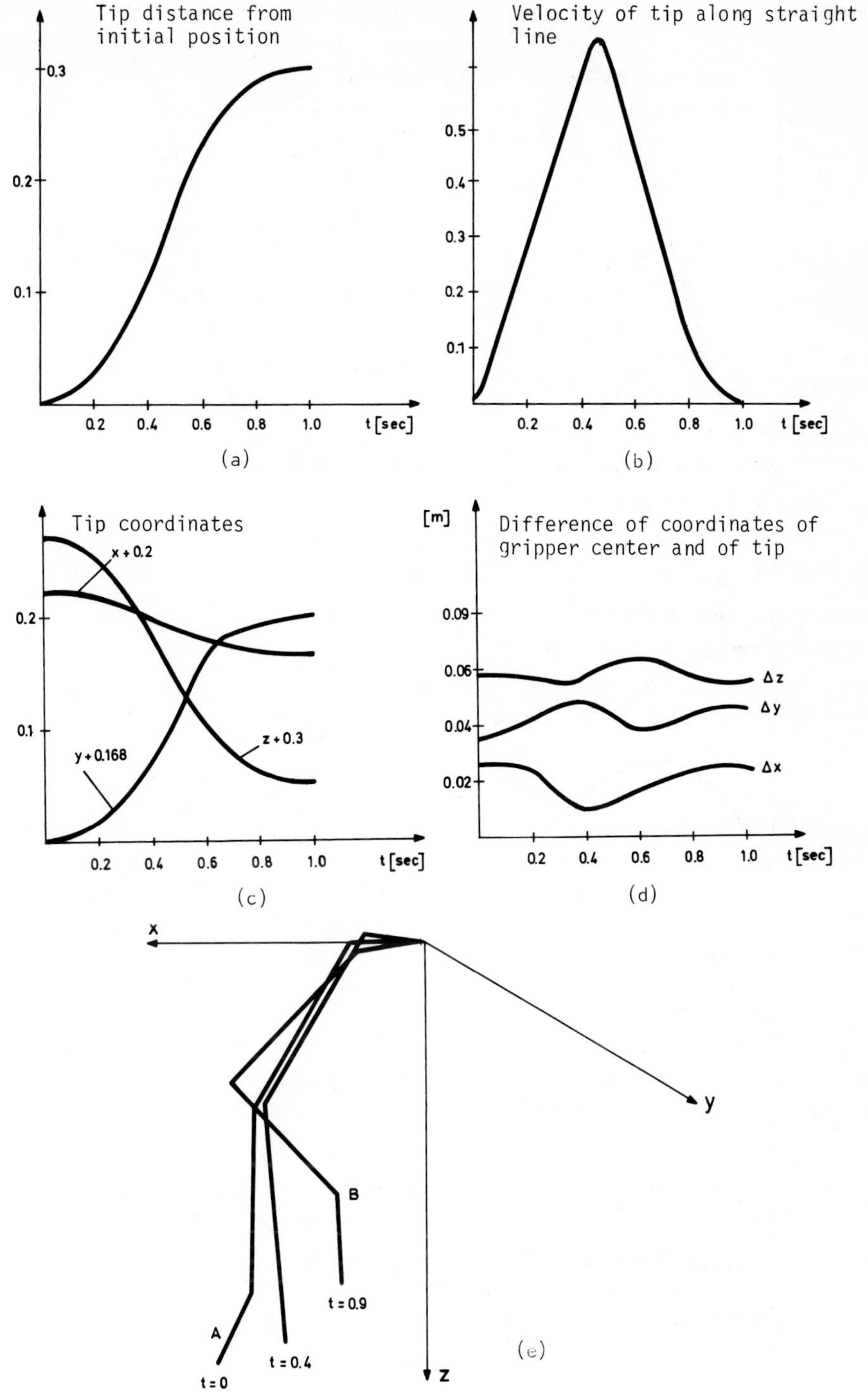

Fig. 3.45. Nominal dynamics: coordinates of tip and gripper center of gravity (UMS-1)

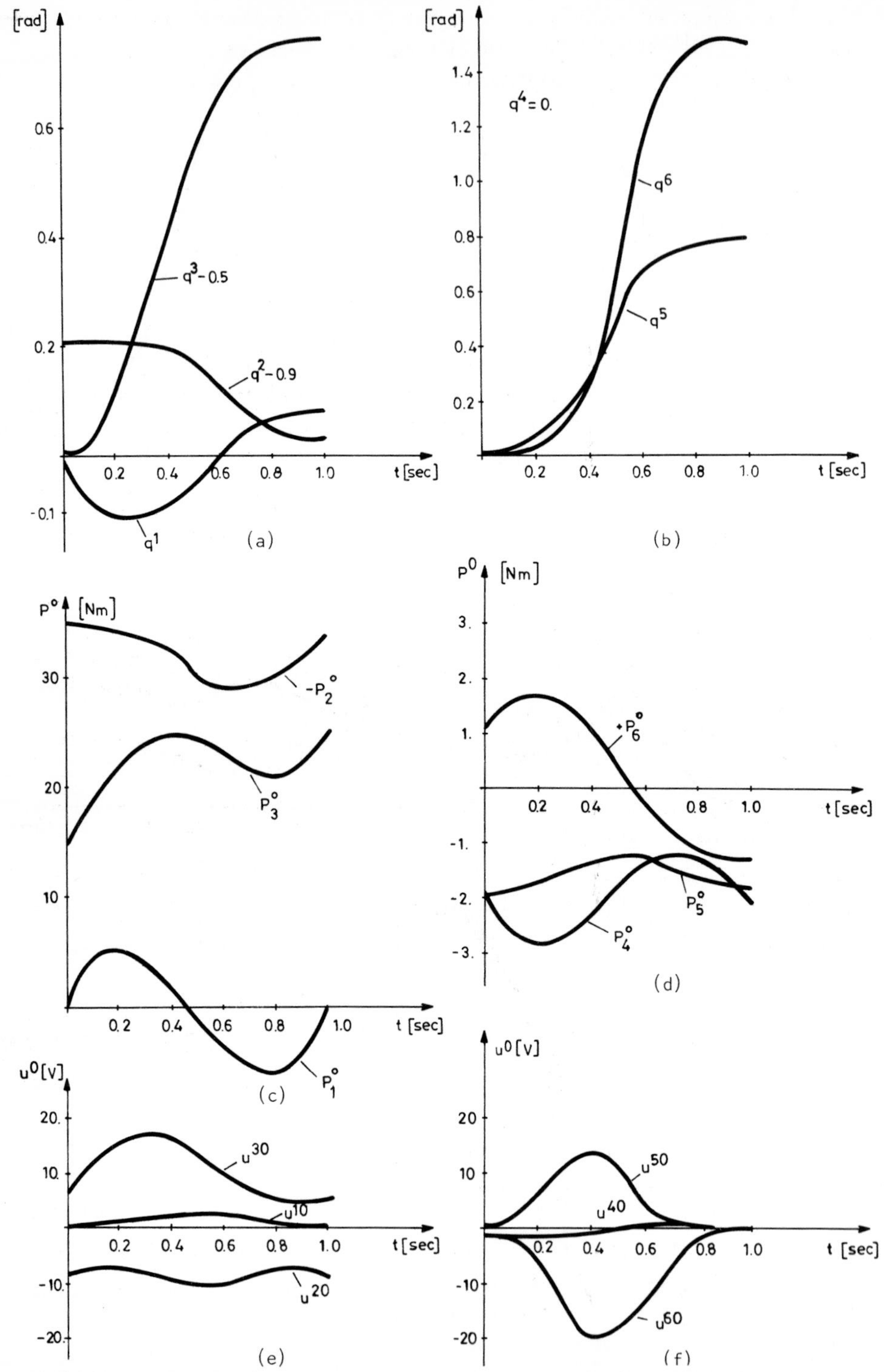

Fig. 3.46. Nominal dynamics: trajectories of angles, driving torques and inputs (UMS-1)

For the sake of comparing various manipulator configurations, the nominal trajectories for the manipulation system UMS-2 (Fig. 3.2, data in Table 3.3) have been synthesized. We have considered the same task, namely transfer of the gripper along the prescribed trajectory with a particular orientation of the working object. We have considered the movement with dynamic requirements as in the case of the manipulator UMS-1 (Fig. 3.45) in order to compare the performances of various manipulators with various control laws, synthesized at the stage of perturbed dynamics [8] (see para. 3.3.2).

The manipulator tip (point D in Fig. 3.2) should move along the straight line between the point $\vec{S}_A(0) = (0., 0.45, 0.45)^T[m]$ and the point $\vec{S}_B(\tau) = (0.288, 0.45, 0.534)^T[m]$. During the motion, the gripper should be oriented in such a way that its main axis is vertical. The acceleration of the tip is given by (3.2.7), where $a_{max} = 1.5\ m/s^2$ (see Fig. 3.47a).

According to this procedure, the change of internal coordinates of the manipulator is determined (Fig. 3.47b). From the Figure it can be seen that the vectors with generalized coordinates at the beginning and at the end of the trajectory are given as: $q^o(0) = \{0[rad], 0.05[m], 0.[m], 0.[rad], -1.57[rad], 0.[rad]\}$ and $q^o(\tau) = \{0.57[rad], 1.134[m], 0.084[m], 0.[rad], -1.57[rad], 0.[rad]\}$, $\tau = 0.9$ sec. The nominal driving torques $P^o(t)$ have been calculated on the basis of model (1.1.30) and are presented in Figs. 3.47c. and 3.47d*). According to (1.2.1) and the mathematical models of the actuators, the parameters of which are given in Table 3.4., the nominal (programmed) control $u^o(t)$ has been determined (Figs. 3.47e. and 3.47f).

For the third example of nominal trajectory synthesis for the manipulation task in question, we have synthesized the nominal trajectories for the system UMS-3B (presented in Fig. 3.3). Data on the mehanism are given in Table 3.5. and data on actuators are given in Table 3.6. Since the manipulator is intended for painting we have considered the task of painting a plane surface, so the manipulator tip trajectory has been chosen as in Fig. 3.48a. The tip of the paint gun should move along the trajectory ABCDEFG. During the motion the gun should keep its initial orientation in space. The part AB of the trajectory repre-

---

*) It should be mentioned that in the synthesis it has been assumed that the manipulator carries an object of mass 4.5 kg.

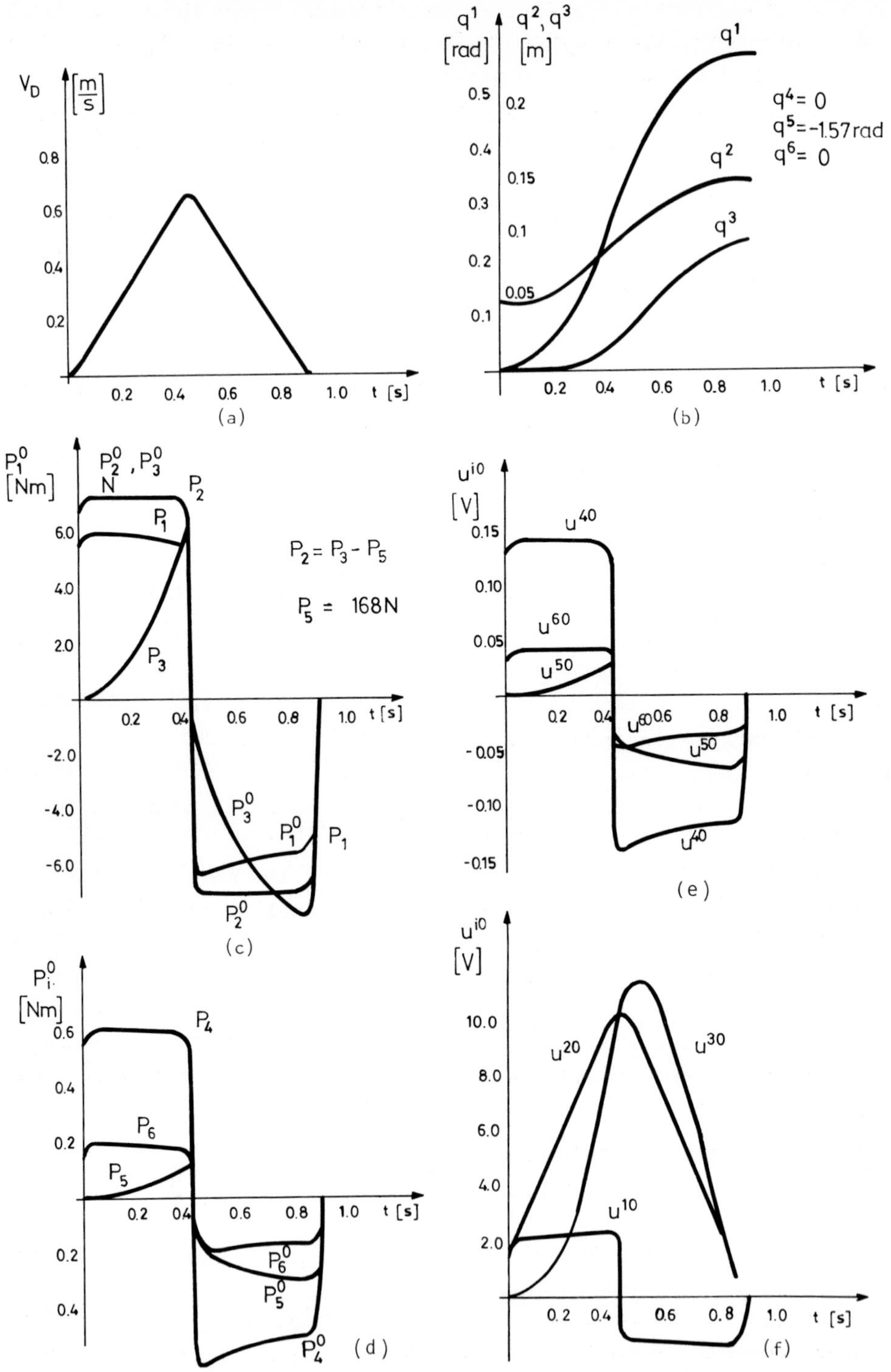

Fig. 3.47. Nominal dynamics (UMS-2)

sents the gun being brought to the initial point for painting, and BCDEFG represents the trajectory during the painting. The initial position of the manipulator is in Fig. 3.3, and the gun is aimed at point A. The painting trajectory is approximately a repetition of the trajectory BC. Thus, we present the results of the nominal synthesis of the part of the trajectory ABC. The part AB of the trajectory should be covered by the tip of the gun in 1 sec with the velocity profile as in Fig. 3.48b. (i.e. with the acceleration as in (3.2.7)). The painting for BC is performed with a constant velocity of 0.3[m/s]. The velocity profile for BC is thus given as in Fig. 3.48c.

The nominal trajectories of the generalized coordinates q^i have been calculated according to the algorithm previously described and are shown in Fig. 3.48d. The nominal driving torques have been calculated on the basis of the model of the mechanical part of the system (1.1.30) and are presented in Figs. 3.49a and 3.49b. The nominal control (currents of the hydraulic servo-valves) has been calculated, on the basis of complete model (2.2.1) and is presented in Figs. 3.49c and 3.49d.

3.3.2. Decentralized control - control synthesis at the stage of perturbed dynamics

When the nominal trajectory $x^o(t)$, $\forall t \varepsilon T$ and programmed control $u^o(t)$, $\forall t \varepsilon T$ have been calculated, the control synthesis at the stage of perturbed dynamics must ensure that $\forall x(0) \varepsilon X^I$ implies $x(t, x(0)) \varepsilon X^t$, $\forall t \varepsilon T$ and $x(t, x(0)) \varepsilon X^F$, $t \varepsilon T_s = \{t: t = \tau = \tau_s\}$. The model of the deviation from the nominal is considered at the stage of perturbed dynamics.

Since all d.o.f. of the manipulators considered are powered by the corresponding actuators (3.1.2), (1.2.1) the system S can be viewed as a set of subsystems S^i, $\forall i \varepsilon I$, given by (2.6.8), which are interconnected by driving torques ΔP_i given by (2.6.1). As stated in para. 2.6, the control is chosen in the form

$$\Delta u^i(t) = \Delta u_i^L(t) + \Delta u_i^G(t), \qquad \forall i \varepsilon I, \tag{3.3.5}$$

where $\Delta u_i^L(t)$ is chosen as a local control possessing the information on state coordinates $\Delta x^i(t)$ of subsystem S^i only, while Δu_i^G is chosen as a global control, as a function of the coupling ΔP_i. First we shall

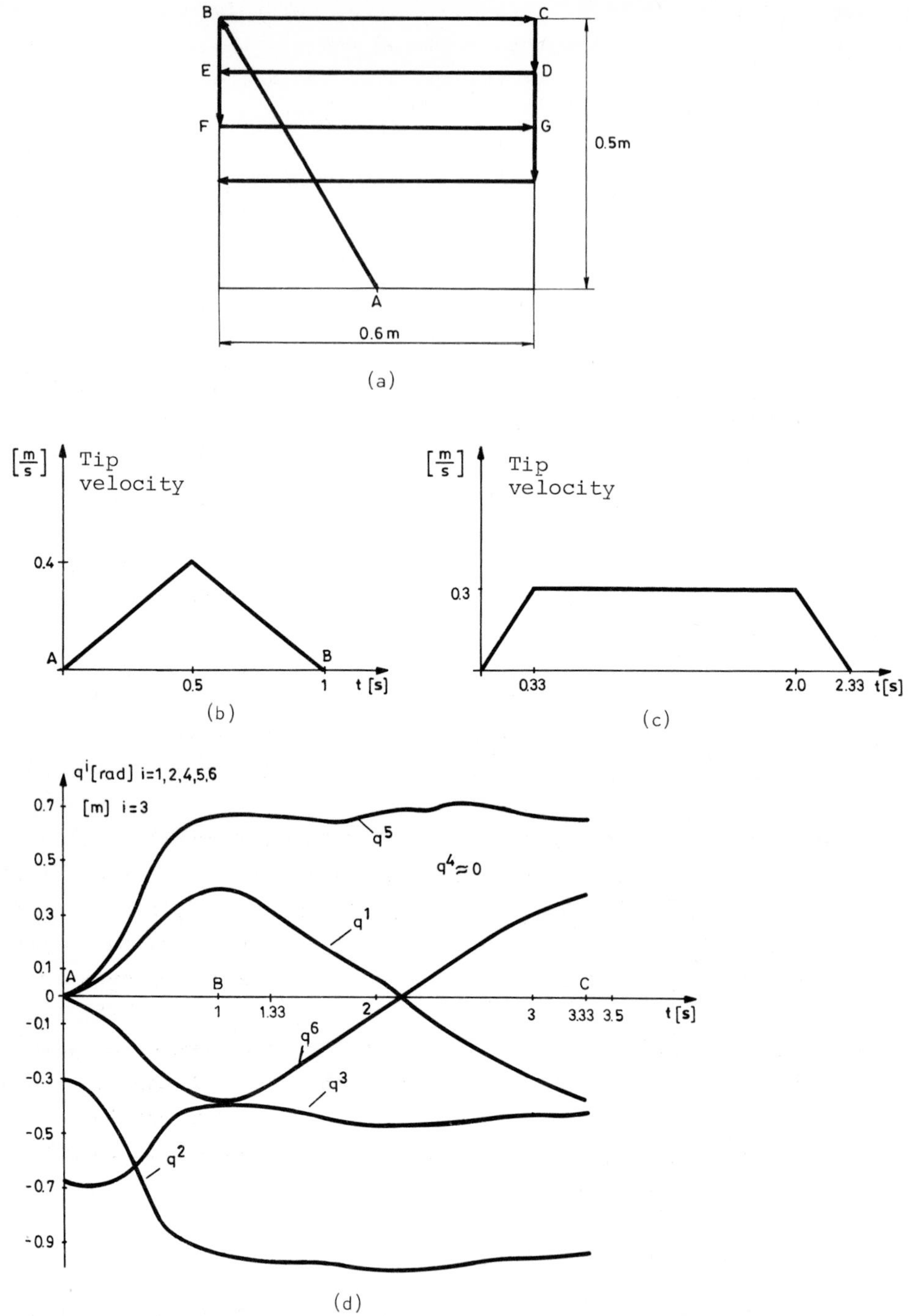

Fig. 3.48. Nominal trajectories for painting by manipulator UMS-3B

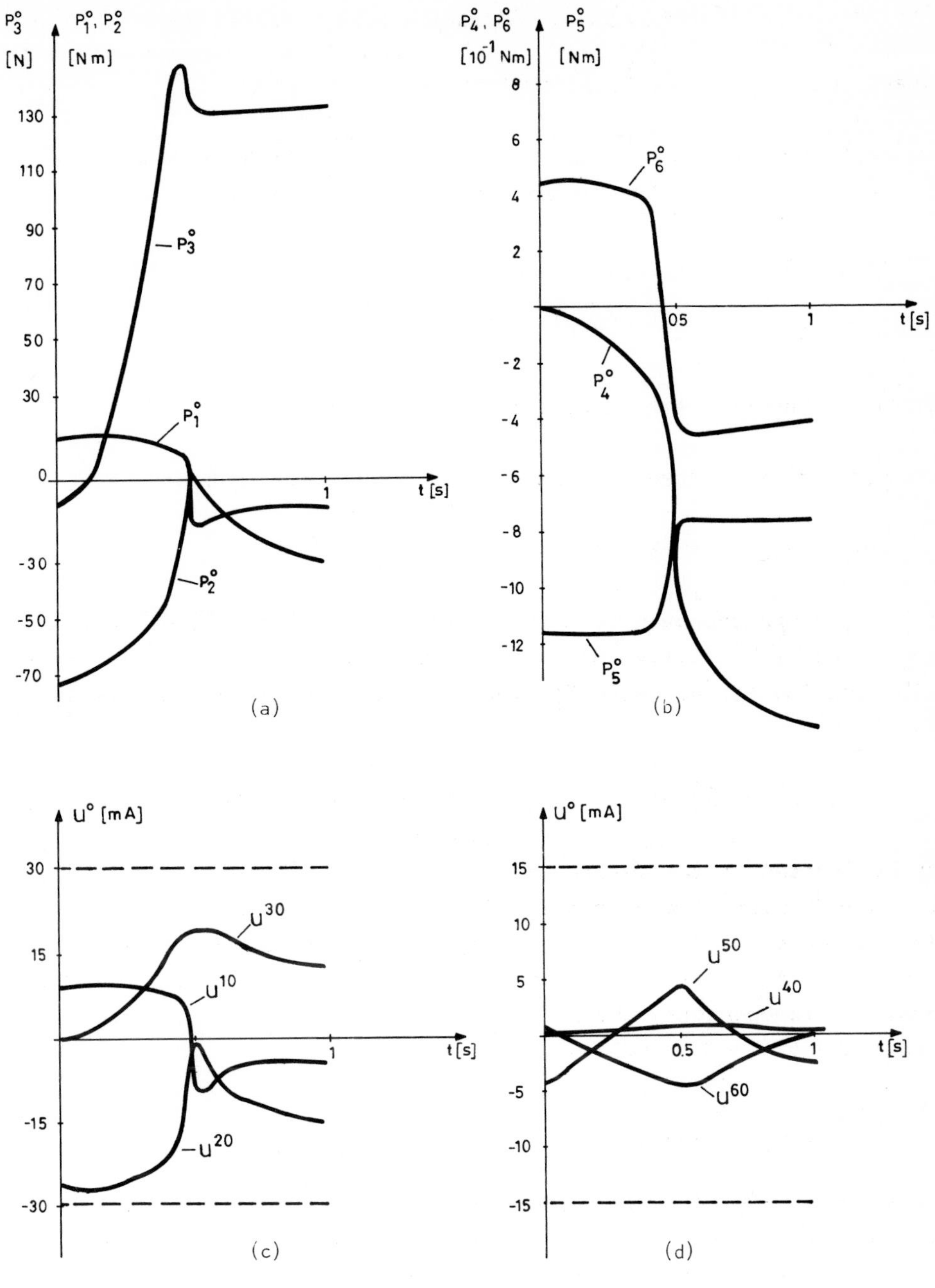

Fig. 3.49. Nominal generalized forces and programmed control for manipulator UMS-3B

consider the stabilization of the manipulation system powered by D.C. motors and we shall apply the local control in the form of output feedback.

The local control law should be synthesized in order to stabilize the subsystems S^i independently of ΔP_i^* (with chosen $\bar{H}_{ii}$, $\forall i\varepsilon I$), where

$$S^i : \Delta\dot{x}^i = \bar{A}^i\Delta x^i + \bar{b}^i N(t, \Delta u_i^L) \qquad \forall i\varepsilon I$$

$$\Delta y^i = C^i\Delta x^i, \ \Delta x^i(0) \text{ are given} \tag{3.3.6}$$

around the nominal trajectory, i.e. around the point $\Delta x^i(t) = 0$.

Local control is chosen in the form of linear output feedback:

$$\Delta u_i^L = -k_i^T\Delta y^i = -k_i^T C^i\Delta x^i, \qquad \forall i\varepsilon I, \tag{3.3.7}$$

where $k_i \varepsilon R^2$ is the gain vector. By considering the form of matrix $\bar{A}^i$ and of vector $\bar{b}^i$ (3.1.2) with this incomplete feedback, by an appropriate choice of the gains k_i, the poles of the closed loop subsystems S^i,

$$S^i : \Delta\dot{x}^i = (\bar{A}^i-\bar{b}^i k_i^T C^i)\Delta x^i, \quad \forall i\varepsilon I, \ \Delta x^i(0) \text{ are given} \tag{3.3.8}$$

may be located in the desired positions $-\sigma_1^i \pm j\omega^i$, $-\sigma_2^i$ or $-\sigma_1^i$, $-\sigma_2^i$, $-\sigma_3^i$ in the left part of the complex plane ($\sigma_p^i > 0$, for $p = 1, 2, 3$). Two eigenvalues of the closed loop subsystem matrix may be chosen arbitrarily, while the third one is defined by the choice of the first two. Thus, all independent subsystems S^i are set to be exponentially stable with the degree of stability

$$\Pi_i^1 = \min_{p=1,2,3} \sigma_p^i \tag{3.3.9}$$

However, the subsystem inputs are constrianed (2.4.2). Thus, the subsystem S^i is exponentially stable with degree Π_i^1 only in the some finite region in the subsystem S^i state space.

That is, for the chosen k_i the exponential stability of the subsystem S^i with degree Π_i^1 is guaranteed only in the finite region in the state space R^{n_i} in which the following holds:

$$|k_i^T C^i \Delta x^i| \leq \bar{u}_m^i = u_m^i - \max_{t\in T} u^{oi}(t), \quad \forall i\in I, \tag{3.3.10}$$

However, even for the states outside the region defined by (3.3.10), the subsystem S^i is exponentially stable with degree of stability $\Pi_i < \Pi_i^1$ [16]. λ is introduced with

$$\lambda = \begin{cases} 1 \quad \text{for} \quad |k_i^T C^i \Delta x^i| \leq \bar{u}_m^i, \\ \bar{u}_m^i / |k_i^T C^i \Delta x^i| \quad \text{otherwise.} \end{cases}$$

Then, the independent subsystem (3.3.6) may be written in the form

$$S^i : \Delta\dot{x}^i = (\bar{A}^i - \bar{b}^i k_i^T C^i)\Delta x^i + \bar{b}^i (1-\lambda) k_i^T C^i \Delta x^i, \quad \forall i\in I \tag{3.3.11}$$

Subsystem S^i, according to equation (3.3.11), is exponentially stable with degree of stability Π_i^1 in the finite region in which

$$\mathrm{Re}(\sigma(\bar{A}^i - \lambda \bar{b}^i k_i^T C^i)) \leq -\Pi_i^1, \quad \forall i\in I, \tag{3.3.12}$$

where $\sigma(\ldots)$ denotes the eigenvalue of the matrix in the brackets. For the defined Π_i, λ_g may be calculated from (3.3.12). This defines the region in which the subsystem is exponentially stable with degree Π_i:

$$|k_i^T C^i \Delta x^i| \leq \frac{|\bar{u}_m^i|}{\lambda_g}, \quad \forall i\in I \tag{3.3.13}$$

By condition (3.3.13), the region in the subsystem S^i state space R^{n_i} is bounded with respect to the state coordinates Δq^i and $\Delta\dot{q}^i$. The third coordinate, motor rotor current Δi_R^i, is restricted by the chosen D.C. motor characteristics. Thus, one may write

$$|\bar{k}_i^2 \Delta\dot{q}^i + \bar{k}_i^3 \Delta i_R^i| \leq \bar{k}_m^i \rightarrow |\bar{k}_i^T \Delta x^i| \leq \bar{k}_m^i, \tag{3.3.14}$$

where $\bar{k}_i = (0, \bar{k}_i^2, \bar{k}_i^3)^T$ and $\bar{k}_m^i$ are the constants defined by the motor specifications.

The region X_i in the state space R^{n_i} is thus restricted in such a way that the subsystem S^i is exponentially stable with degree of stability Π_i. Thus,

$$\Delta x^i(0)\in X_i \ \text{.AND.}\ \Delta u_i = -k_i^T C^i \Delta x^i \rightarrow ||\Delta x^i(t)|| \leq M_i e^{-\Pi_i t} ||\Delta x^i(0)|| \tag{3.3.15}$$

where $M_i > 0$. The region $X_i \subset R^{n_i}$ is bounded by

$$X_i = \{\Delta x^i: |k_i^T C^i \Delta x^i| < \frac{|\bar{u}_m^i|}{\lambda_g} \text{ .AND. } |\bar{k}_i^T \Delta x^i| \leq \bar{k}_m^i\}, \quad \forall i \varepsilon I \tag{3.3.16}$$

In order to investigate the stability of the composite system S, we introduce Liapunov´s functions for free subsystems. We therefore introduce the following non-singular transformation [17]

$$\Delta x^i = \hat{T}_i \Delta\hat{x}^i. \tag{3.3.17}$$

By applying the transformation $\hat{T}_i$, the subsystem S^i (3.3.11) is transformed into

$$S^i : \Delta\dot{\hat{x}}^i = \Lambda^i \Delta\hat{x}^i + \hat{T}_i^{-1}\bar{b}^i(1-\lambda)k_i^T C^i \hat{T}_i \Delta\hat{x}^i, \tag{3.3.18}$$

where $\Lambda^i = \hat{T}_i^{-1}(\bar{A}^i - \bar{b}^i k_i^T C^i)\hat{T}_i$. The transformation matrix $\hat{T}_i$ (3×3) should transform the subsystem matrix into a quasidiagonal or diagonal form:

$$\Lambda^i = \begin{bmatrix} -\sigma_1^i & \omega_1^i & 0 \\ -\omega_1^i & -\sigma_1^i & 0 \\ 0 & 0 & -\sigma_2^i \end{bmatrix} \quad \text{or} \quad \Lambda^i = -\text{diag}(\sigma_1^i, \sigma_2^i, \sigma_3^i), \ \forall i \varepsilon I$$

Liapunov´s functions for the subsystem S^i is chosen in the form

$$v_i(\Delta\hat{x}^i) = (\Delta\hat{x}^{iT}\hat{H}^i\Delta\hat{x}^i)^{1/2} \quad \forall i \varepsilon I, \tag{3.3.19}$$

where $\hat{H}^i$ is a 3×3 matrix of the form $\hat{H}^i = \text{diag}\{\theta_1^i, \theta_2^i, \theta_3^i\}$, $\theta_j^i > 0$, $j = 1,2,3$. If the subsystem state is observed in the region X_i^1, defined by (3.3.16) when $\lambda_g = 1$, the following relationship holds

$$\dot{v}_i(\Delta\hat{x}^i) = \Delta\hat{x}^{iT}(\Lambda^i\hat{H}^i + \hat{H}^i\Lambda^i)\Delta\hat{x}^i/2v_i \leq -\Pi_i^1 v_i(\Delta\hat{x}^i) \leq$$
$$-\Pi_i^1 \min_{j=1,2,3}(\theta_j^i)||\Delta\hat{x}^i||, \quad \forall i \varepsilon I \tag{3.3.20}$$

This choice of Liapunov functions (by introducing the transformation $\hat{T}_i$) permits an exact estimation of the exponential stability of degree Π_i^1 for the free subsystem [18]. Condition (3.3.20) is satisfied in the region X_i^1. The regions X_i^1 can be approximated by the regions

$$\tilde{X}_i^1 = \{\Delta x^i: v_i(\Delta x^i) \leq v_{io}\}, \qquad \forall i\varepsilon I, \tag{3.3.21}$$

where $v_{io} > 0$ are constants. The regions $\tilde{X}_i^1 \subseteq X_i^1$ should approximate the stability regions X_i^1 as well as possible. The elements θ_j^i of the matrix $\hat{H}^i$ should be chosen so as to maximize the volume of the region $\tilde{X}_i^1$ which approximates the exponential stability region X_i^1. For the case in which the subsystem is observed in a region wider than X_i^1 (for $\lambda_g = 1$), i.e. if the region is considered for $\lambda < \lambda_g$, then

$$\dot{v}_i(\Delta\hat{x}^i) \leq -\Pi_i v_i(\Delta\hat{x}^i), \; \Pi_i < \Pi_i^1, \qquad \forall i\varepsilon I \tag{3.3.22}$$

for $\Delta\hat{x}^i \varepsilon X_i \supseteq X_i^1$ given by (3.3.16). It is obvious that, for $\lambda < 1$, one obtains a wider region in the state-space R^{n_i}, for which the subsystem is exponentially stable with a smaller degree of stability [19].

We now examine the asymptotic stability of the global system S. Since the coupling among subsystems is given by ΔP_i^*, the numbers ξ_{ik}, $\forall i,k\varepsilon I$ should be determined which satisfy condition (2.6.23) on $\tilde{X}\times T$, where $\tilde{X}$ is defined by (2.6.22). The numbers ξ_{ij} should first be determined for the case when the global control is not introduced. With these ξ_{ik} one checks whether condition (2.6.24) is satisfied, where the elements of matrix G are given by $G_{ij} = -\Pi_i \delta_{ij} + \xi_{ij}$, while the v_{io} are defined by (3.3.21). If condition (2.6.24) is satisfied, the region $\tilde{X}$, defined by (2.6.22), is the approximation of the region in which the global system S is asymptotically stable, i.e., the system S is asymptotically stable around the nominal trajectory and control in the region $\tilde{X}$, which is the product of the regions $\tilde{X}_i$ approximating the regions X_i of exponential stability of the free subsystems S^i. Finally, by (2.6.26), one can estimate the degree of exponential shrinkage of the region $\tilde{X}(t)$, which contains the state of the system for $t\varepsilon T$. The region $\tilde{X}(t)$, which should contain the state of the system, is given by

$$\tilde{X}(t) = \{\Delta x(t): \max_{i\varepsilon I} \frac{v_i(\Delta x^i(t))}{v_{io}} \leq \max_{i\varepsilon I} \frac{v_i(\Delta x^i(0))}{v_{io}} e^{-\eta t}, \; \forall t\varepsilon T\} \tag{3.3.23}$$

Thus, the region $\tilde{X}(t)$ shrinks exponentially with degree η. One can now determine according to (2.6.28) whether the conditions of manipulator practical stability around the nominal trajectory are satisfied.

Let us consider the manipulation system UMS-1 and the stabilization around the nominal trajectories given in para. 3.3.1 (Figs. 3.45 and

3.46).

First, in order to achieve exponential stability of degree Π_i^1 for the independent subsystems, the feedback gains k_i for the subsystems S^i (D.C. motors) - are chosen according to (3.3.7). Because of the form of the matrices A^i and vectors b^i and because the output vector has been chosen as $\Delta y^i = (\Delta q^i, \Delta \dot{q}^i)^T$, the independent subsystems can be stabilized with degree of stability Π_i^1. This can at most be $\Pi_i^* = 31.3$ with the feedback gains $k_i = (300,45)^T$. However, these gains are too high, so to realize them in practice may be very difficult. On the other hand, with gains so high the region X_i of exponential stability for the subsystem is very narrow. Thus, two cases of subsystem exponential stability are considered:

$$\Pi_{i1} = 7.7, \quad k_{i1} = (50,2)^T, \quad \forall i \varepsilon I,$$

$$\Pi_{i2} = 21, \quad k_{i2} = (130,3)^T, \quad \forall i \varepsilon I$$

In both cases the Liapunov functions are chosen according to (3.3.19), where the $\hat{H}^i$ are taken in the form $\hat{H}^i = \text{diag}\ (1, 1, 1)$. With these Liapunov functions, the constants v_{io} in (3.3.21) are calculated. One thus defines the regions $\tilde{X}_i$, which estimate the regions X_i of subsystem exponential stability with stability degree Π_{i1}, or Π_{i2} for the case $\lambda_g = 1$. The coupling ΔP_i around the nominal trajectory $x^o(t)$ and the nominal driving forces $P^o(t)$ are then investigated in the region $\tilde{X}\times T$. ΔP_i is investigated by calculating the driving forces in the manipulator joints in the region $\tilde{X}\times T$. These calculations are based on the mathematical manipulator model, implemented in a digital computer. Thus, the coefficients ξ_{ik} are calculated in both cases.

The matrices G and the vectors Gv_o are given in Tables 3.11 and 3.12. The values of the vector v_o defining the regions $\tilde{X}_i$ (3.3.21), which estimate the region of exponential stability for $u_m^i = 20V$, $\forall i \varepsilon I$, and $u_m^i = 27V$, $\forall i \varepsilon I$, are also given in the tables.

Finally it is shown that condition (2.6.24) is satisfied in the second case ($\Pi_{i2} = 21$), so the asymptotic stability of the system S can be guaranteed in the region $\tilde{X}$. However, condition (2.6.24) is not satisfied when $\Pi_{i1} = 7.7$, so the system asymptotic stability in the region $\tilde{X}$ cannot be guaranteed. This demonstrates that by increasing the free subsystem degree of stability Π_i, the composite stability of the total system S is achieved more easily but the region $\tilde{X}$ in which the system S is asymptotically stable decreases in size. That is, by increasing Π_i, the gains have to increase, so the regions X_i are smaller. In order to achieve asymptotic stability of the system S in the wider region $\tilde{X}$, with lower degree of stability Π_i, one introduces the global control Δu_i^G, with respect to the coupling ΔP_i. The force transducers are introduced in order to measure the driving torques P_i on the motor shafts, so force feedbacks are introduced. In order to simplify the control structure, global control is chosen in the form of "linear" feedback with respect to ΔP_i [19, 20]:

$$\Delta u_i^G(t) = K_i^G \cdot |\Delta P_i| \cdot \operatorname{sgn}(\Delta u_i^L(t)), \qquad i \in I \tag{3.3.24}$$

The choice of the feedback sign (3.3.24) (equal to the sign of $\Delta u_i^L(t) = -k_i^T \Delta y^i$) follows from the form of the vectors $\bar{b}^i$ and $\bar{f}^i$ and from the condition $(\mathrm{grad}\, v_i)^T \bar{b}^i \Delta u_i^G(t) < 0$ which should be satisfied. However, since the local control is chosen in the form of incomplete feedback, it is not always true that $(\mathrm{grad}\, v_i)^T \bar{b}^i \Delta u_i^L < 0$. In the local control feedback by rotor current is not introduced, so, when the current mode determines the subsystem behaviour, the condition $(\mathrm{grad}\, v_i)^T \bar{b}^i \Delta u_i^L < 0$ is not satisfied. However, the current mode of the subsystem is very stable (the open-loop subsystem poles are $\sigma = 0, -27.30, -72.71$), so the subsystem is stable when the current mode is dominant. The global control (3.3.24) cannot influence the behaviour of the subsystem when the current mode is dominant, so the chosen form of the global control is satisfactory in the case under consideration. In order to take this form of global control into account when analyzing the stability of the global system the elements of matrix G are not determined according to (2.6.23) but from the condition

$$(\mathrm{grad}\, v_i)^T [\bar{A}^i \Delta x^i + \bar{b}^i N(t, \Delta u_i^L + \Delta u_i^G) + \bar{f}^i \Delta P_i] \leq \sum_{j=1}^{n} G_{ij} v_j,$$

$$\forall i \in I, \forall (t, \Delta x) \in T \times \tilde{X}$$

It is clear that the chosen form of the global control (3.3.24) is "suboptimal", since it does not consider the nature of the effect of the coupling on the subsystem stability. The effect of coupling may be stabilizing or destabilizing. The chosen form (3.3.24) amplifies the stabilizing effect of the coupling in both cases. The global gains K_i^G are therefore chosen to be consistent with obtaining the practical stability of the system but are kept as low as possible. The choice of

TABLE 3.11: $\Pi_i = 7.7$

$$G = \begin{bmatrix} -0.24665E+02 & 0.90383E+02 & 0.42645E+02 & 0.53593E+01 & 0.10211E+02 & 0.49262E+01 \\ 0.16357E+03 & 0.44846E+01 & 0.29652E+02 & 0.12139E+02 & 0.20930E+01 & 0.50289E+01 \\ 0.54502E+02 & 0.78712E+02 & -0.72756E+01 & 0.73548E+01 & 0.52599E+01 & 0.57823E+01 \\ 0.14043E+02 & 0.21005E+02 & 0.35074E+01 & -0.32471E+02 & 0.36406E+01 & 0.35279E+01 \\ 0.18882E+02 & 0.14759E+00 & 0.43552E+01 & 0.78227E+00 & -0.28967E+02 & 0.21784E+01 \\ 0.26619E+01 & 0.90625E+01 & 0.66404E+01 & 0.47446E+01 & 0.10504E+01 & -0.31941E+02 \end{bmatrix}$$

$$v_o = [0.24887E+02 \quad 0.18462E+02 \quad 0.16893E+02 \quad 0.24869E+02 \quad 0.23638E+02 \quad 0.26105E+02]^T$$

$$Gv_o = [0.41872E+04 \quad 0.44152E+04 \quad 0.15934E+04 \quad -0.18825E+03 \quad -0.18329E+03 \quad -0.34975E+03]^T$$

TABLE 3.12: $\Pi_i = 21$

$$G = \begin{bmatrix} -0.16548E+02 & 0.24562E+01 & 0.49177E+01 & 0.14003E+01 & 0.13206E+01 & 0.82907E+00 \\ 0.72032E+01 & -0.16652E+02 & 0.74186E+01 & 0.97855E+00 & 0.60734E+00 & 0.54229E+00 \\ 0.45561E+01 & 0.63169E+01 & -0.21491E+02 & 0.48627E+00 & 0.11649E+01 & 0.35127E+00 \\ 0.48802E+00 & 0.88452E+00 & 0.69431E+00 & -0.20636E+02 & 0.33326E+00 & 0.57757E+00 \\ 0.49493E+00 & 0.52139E+00 & 0.12379E+01 & 0.26029E+00 & -0.20908E+02 & 0.16077E+00 \\ 0.10976E+01 & 0.88585E+00 & 0.66316E+00 & 0.18637E+00 & 0.16861E+00 & -0.21449E+02 \end{bmatrix}$$

$$v_o = [0.91588E+02 \quad 0.63829E+02 \quad 0.75412E+02 \quad 0.90333E+02 \quad 0.91619E+02 \quad 0.92482E+02]^T$$

$$Gv_o = [-0.52126E+03 \quad -0.15196E+03 \quad -0.45931E+03 \quad -0.15956E+04 \quad -0.16223E+04 \quad -0.17797E+04]^T$$

TABLE 3.13: $\Pi_i = 7.7, \ \Delta u_i^G = 0.232|\Delta P_i|\mathrm{sgn}(\Delta u_i^L) \ i = 1, 2, 3$

$$G = \begin{bmatrix} -0.31830E+02 & 0.56296E+00 & 0.11003E+00 & 0.84391E+01 & 0.99622E+01 & 0.32916E+01 \\ 0.54236E+01 & -0.39088E+02 & 0.10400E+02 & 0.12183E+02 & 0.24954E+01 & 0.45845E+01 \\ 0.32300E-01 & 0.16732E+00 & -0.30337E+02 & 0.74320E+01 & 0.87344E+01 & 0.56848E+01 \\ 0.89812E-02 & 0.13366E-01 & 0.20279E-01 & -0.33311E+02 & 0.31862E+01 & 0.69634E+01 \\ 0.70058E-02 & 0.10002E-01 & 0.25636E+00 & 0.81864E+00 & -0.29003E+02 & 0.18141E+01 \\ 0.11014E-00 & 0.77288E-02 & 0.23659E-01 & 0.46964E+01 & 0.18265E+01 & -0.31889E+02 \end{bmatrix}$$

$$v_o = [0.24887E+02 \quad 0.18462E+02 \quad 0.16893E+02 \quad 0.24869E+02 \quad 0.23538E+02 \quad 0.25105E+02]^T$$

$$Gv_o = [-0.69572E+03 \quad -0.71121E+03 \quad -0.32799E+03 \quad -0.13075E+03 \quad -0.11603E+03 \quad -0.32211E+03]^T$$

the gains is iterative. They are iteratively increased until the practical stability of the global system is achieved. The lowest K_i^G satisfying the practical stability of the system is $K_i^G = 0.232$, $i \in I$, in this particular case. However, since the effect of the coupling on the gripper d.o.f. is weaker, it is sufficient to introduce global control for the first three d.o.f. only. Global control is introduced when the subsystems are exponentially stabilized with a degree of stability Π_{i1}. Since the control is constrained, it is necessary to consider the system in the narrower region $\tilde{X}_1 \subseteq \tilde{X}$, so local control is always beyond the maximum (3.3.10). That is, the maximum amplitude of the control $\bar{u}_m^i$ at the perturbed level is separated into two parts:

$$\bar{u}_m^i = \tilde{u}_m^{iL} + \tilde{u}_m^{iG} \tag{3.3.25}$$

Local exponential stability of the subsystem is then considered in the regions X_{1i} given by (3.3.16), where $\tilde{u}_m^{iL} \leq \bar{u}_m^i$ is substituted for $\bar{u}_m^i$. These regions are then approximated by the regions $\tilde{X}_{1i} \subseteq X_{1i}$, defined by (3.3.21). Global control in the form given by Eq. (3.3.24) is introduced for S^i, $i = 1,2,3$, but its amplitude is constrained by $\tilde{u}_m^{iG}$. The coupling among subsystems, together with global control, is then investigated and the elements of the matrix G are determined. Thus, we calculate the matrix G and vector Gv_o, shown in Table 3.13. Coupling is investigated in the region $\tilde{X}_1 \times T$, where $\tilde{X}_1 = \tilde{X}_{11} \times \tilde{X}_{12} \times \tilde{X}_{13} \times \tilde{X}_4 \times \tilde{X}_5 \times \tilde{X}_6$. In Table 3.13. the values of v_{io} are also given for $|u^{io} + \Delta u_i^L| \leq 20$ and $i = 1,2,3$. One thus shows that, in this case, condition (2.6.24) is satisfied. This means that, when subsystems are exponentially stable with degree Π_{i1} and global control (3.3.24) is introduced, the asymptotic stability of the composite system S in the region $\tilde{X}_1$ can be guaranteed.

Suppose that the conditions for practical stability around the nominal trajectory $x^o(t)$, $\forall t \in T$, are given by the following ($\tau = 0.9$ sec).

$$\bar{X}^T = \{x(0): ||\Delta y^i(0)|| \leq 0.26, \text{ for } i=1,3,4,5,6, ||\Delta y^2(0))|| \leq 0.18\}$$

$$\bar{X}^t(t) = \{\Delta x(t): ||\Delta y^i(t)|| \leq 0.26, \text{ for } i=1,3,4,5,6, ||\Delta y^2|| \leq 0.18, \forall t \in T\}$$

$$\bar{X}^F = \{\Delta x(\tau): ||\Delta y^i|| < 0.05, \forall i \in I\}; \ |\Delta i_R^i| \text{ constrained by } (3.3.14).$$

Here, $||\Delta y^i|| = (\Delta q^{i2} + c_i \Delta \dot{q}^{i2})^{1/2}$, where $c_i = 0.1$, $\forall i \in I$.

It is shown that the estimated region of asymptotic stability of the composite system S, for the case in Table 3.13, satisfies all conditions for practical stability (2.6.18). This result is verified by simulation of the manipulator dynamics. Tracking of the nominal trajectory $x^{o}(t)$, $\forall t \in T$ from Figs. 3.45 - 3.46. is simulated for initial conditions taken on the boundary of the region X^{I}, $\Delta q(0) = (-0.26, 0.18, -0.22, 0.26, 0.26, 0.26)^{T}$, $\Delta\dot{q}(0) = 0$, $\Delta i_{R}^{i}(0) = 0$, $\forall i \in I$. Tracking of the nominal trajectories are simulated for 4 different forms of control and the results are presented in Fig. 3.50. In control law I, the programmed control $u^{o}(t)$ is not introduced and local gains are chosen for the case $\Pi_{i1} = 7.7$. This demonstrates that when programmed control is not introduced, coupling among subsystems is too strong, so the system cannot track the nominal trajectory with the desired accuracy. This substantiates the statement in ch. 2 that the tracking of the nominal trajectory by local control only (direct decentralization of the system) in the case of active mechanisms with significant coupling among actuators may be insufficient. Control law II, with both the programmed and local control, also does not satisfy the imposed conditions for $\Pi_i = \Pi_{i1}$. For the third form of control, local control Δu_i^L for $\Pi_i = \Pi_{i2}$ (without the programmed control) is considered. Again in this case not all conditions are satisfied because the initial conditions are outside the region in which the subsystems are exponentially stable with degree Π_{i2} and the conditions in Table 3.12 are not satisfied. Finally, control law IV should be in the form

$$u^{i}(t) = u^{oi}(t) - 50(q^{i}-q^{oi}(t)) - 2(\dot{q}^{i}-\dot{q}^{oi}(t)) +$$

$$0.232|P_i-P_i^{o}| \text{ sgn } (\Delta u_i^{L}) \text{ for } i = 1,2,3, \tag{3.3.26}$$

$$u^{i}(t)=u^{oi}(t)-50(q^{i}-q^{oi}(t))-2(\dot{q}^{i}-\dot{q}^{oi}(t)) \text{ for } i=4,5,6$$

and the conditions for practical stability are satisfied.

However, the control law can be simplified with minimal loss of accuracy of tracking by the introduction of global feedback with respect to the torque P_i and by omitting the programmed control $u^{oi}(t)$; i.e., if we take

$$u^{i}(t)=-50\Delta q^{i} - 2\Delta\dot{q}^{i} - 0.232|P_i| \text{ sgn } (\Delta u_i^{L}) \text{ for } i=1,2,3 \tag{3.3.27}$$

and u^{i} for i = 4,5,6 the same as before. This case nearly coincides

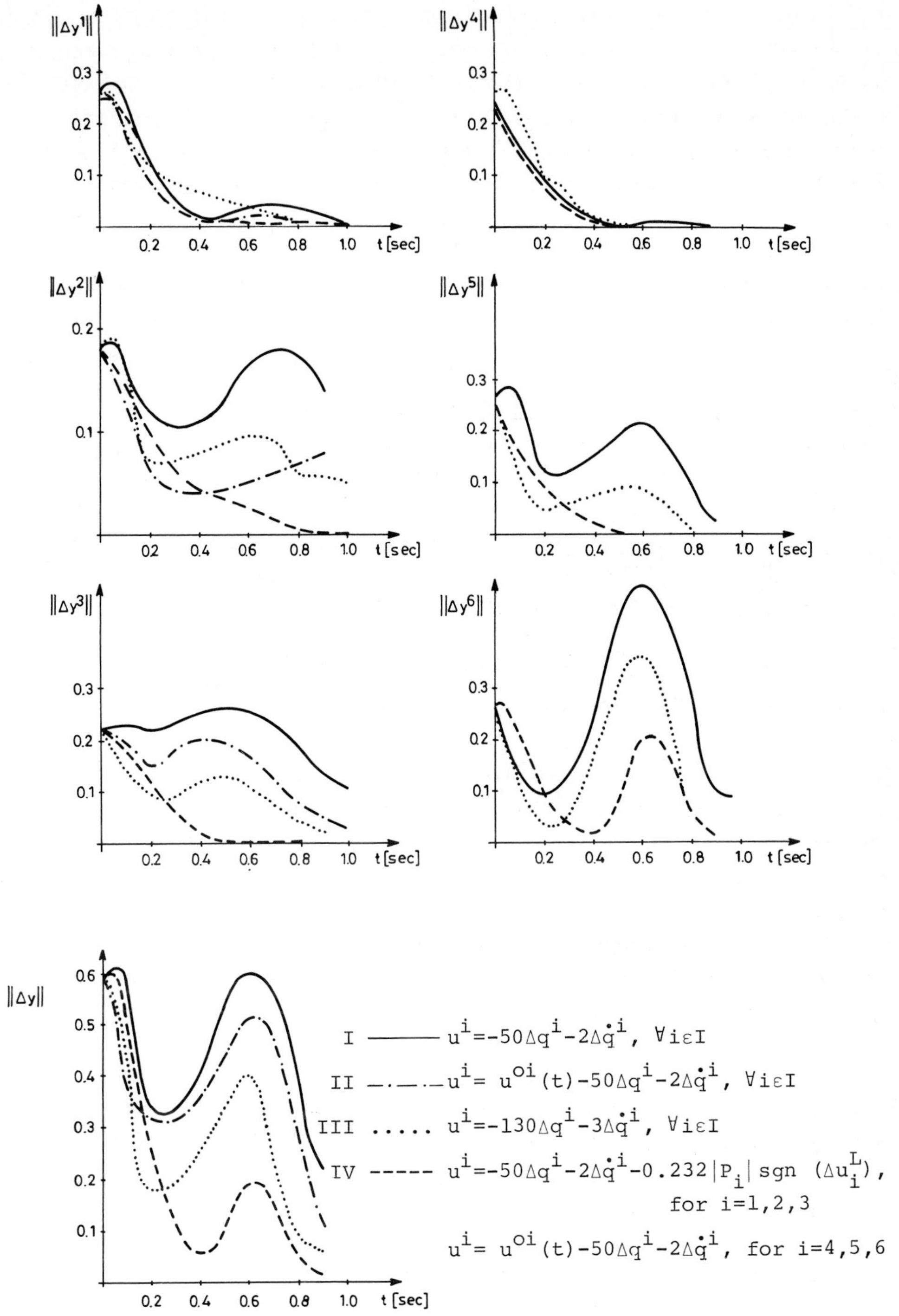

Fig. 3.50. Norms of deviation from nominal trajectories (UMS-1)

with case IV. The control (3.3.27) significantly simplifies the synthesis because it tracks the nominal trajectories quite satisfactorily but it is not necessary to calculate programmed inputs through calculation of the nominal dynamics. This allows on-line implementation of such a control law. On the other hand, this control is much more reliable and insensitive to parameter variation than the programmed control (as will be shown in the following).

Fig. 3.50 shows the norm of the output deviation from the nominal trajectories of each subsystem and of the total dynamical system $||\Delta y|| = (\sum_{i=1}^{n} ||\Delta y^i||^2)^{1/2}$. In Fig. 3.51, the input is shown (voltage on the D.C. motor rotor) developing during the tracking for all four cases. In Fig. 3.52, the driving torques during the tracking of the nominal trajectories are shown. Fig. 3.53 shows, for all four cases, the energy which must be supplied to the actuators in order to realize the tracking as shown in Fig. 3.50. From Fig. 3.53, it may be seen that when the subsystems are locally strongly stabilized, energy consumption is greater than when subsystems are locally weakly stabilized but global control is introduced.

In the example considered all imposed conditions for practical stability are satisfied - all manipulator d.o.f. can deviate within given values from the nominal trajectories. In practice the condition that the deviation from the nominal trajectory in one d.o.f. does not influence the other d.o.f. is often imposed. Obviously, this condition can be satisfied only if the coupling among subsystems is relatively weak. Thus, this condition for practical stability can be satisfied with decentralized control in which the subsystems (d.o.f.) are stabilized independently, and even more, when feedback with respect to coupling is also introduced.

Let us consider the same task of transferring an object along a prescribed trajectory with a desired orientation during the motion (as in Figs. 3.45 - 3.46). However, let us suppose that, instead of the previous conditions for practical stability, the regions of desired practical stability are given in the form

$$\bar{X}^I = \{\Delta x(0): ||\Delta y^i(0)|| \le 0.05 \text{ for } i=1,2,4,5,6, ||\Delta y^3(0)|| < 0.22\}$$

$$\bar{X}^t(t) = \{\Delta x(t): ||\Delta y^i(t)|| < 0.05, \text{ for } i=1,2,4,5,6, ||\Delta y^3(t)|| < 0.22, \forall t \in T\}$$

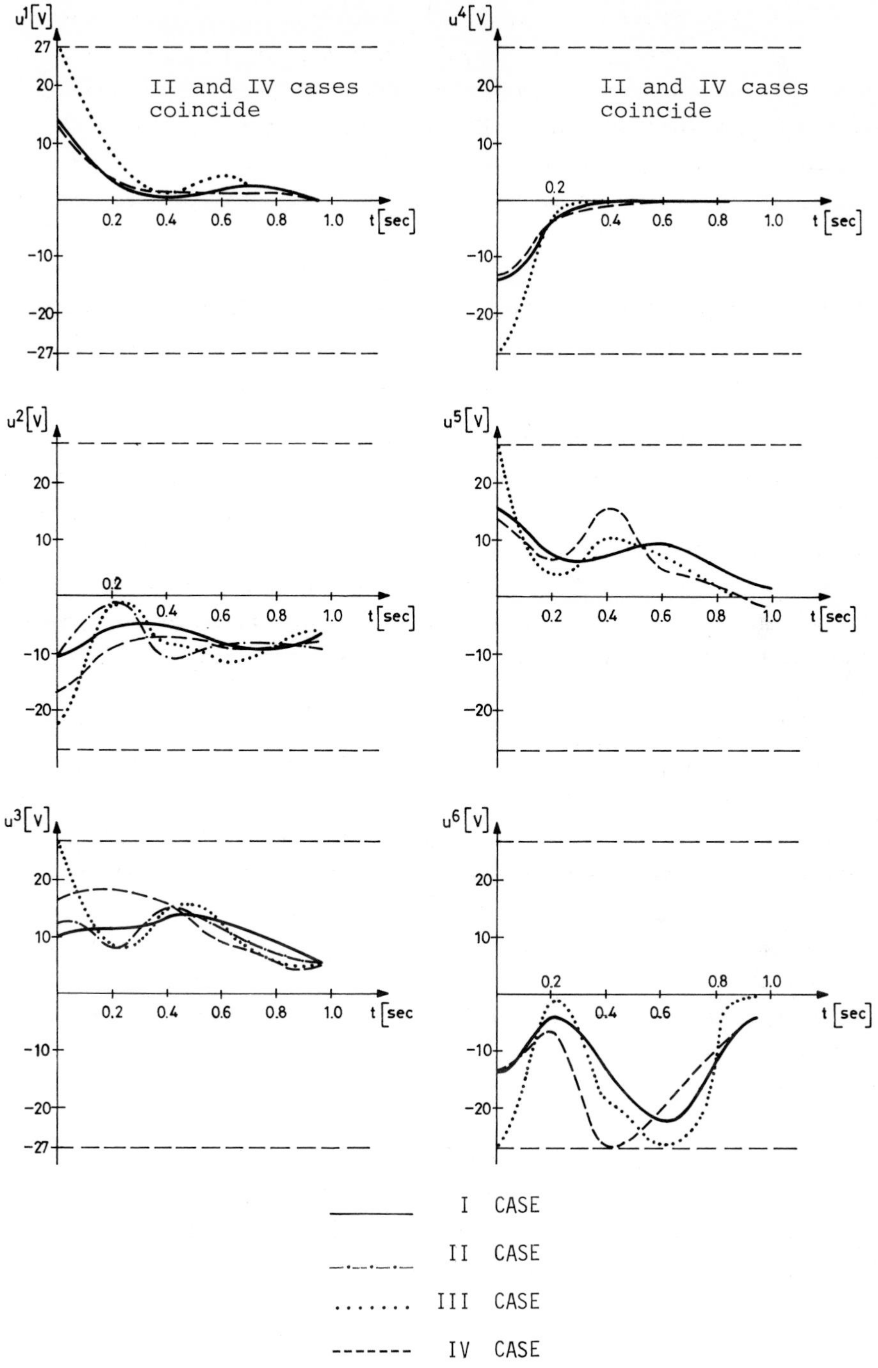

Fig. 3.51. Actuator inputs for tracking presented in Fig. 3.50 (UMS-1)

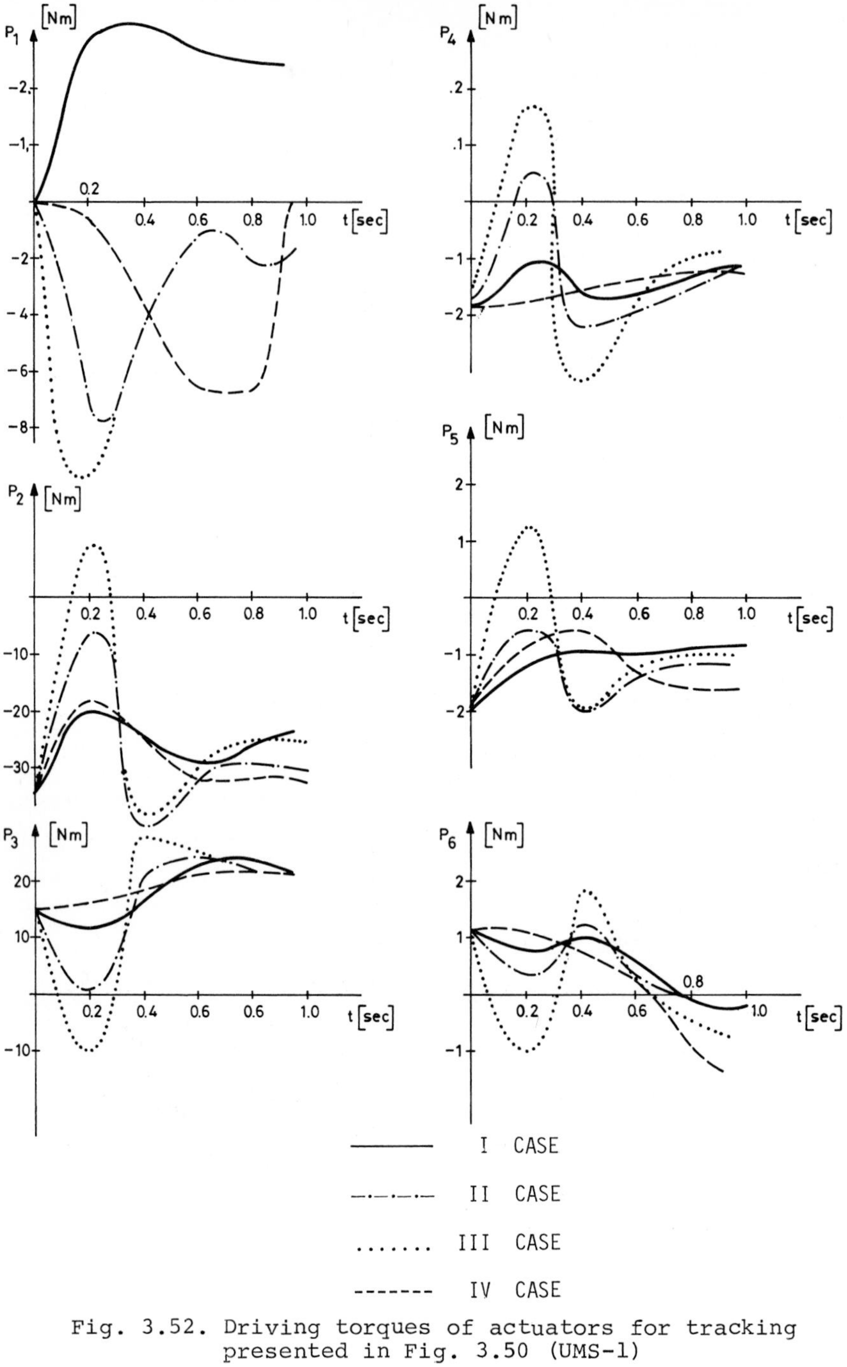

Fig. 3.52. Driving torques of actuators for tracking presented in Fig. 3.50 (UMS-1)

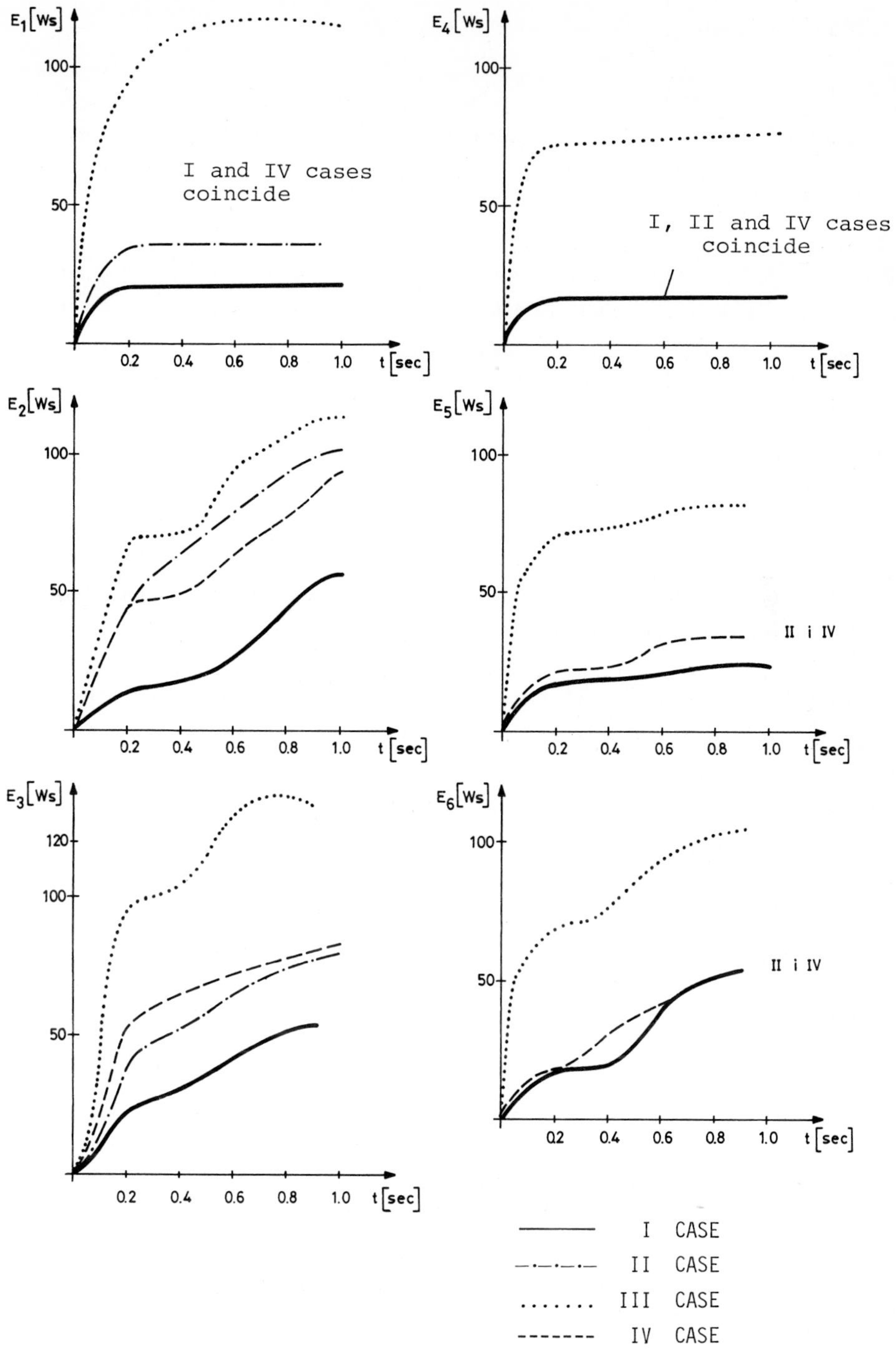

Fig. 3.53. Energy consumptions for tracking presented in Fig. 3.50 (UMS-1)

$$\bar{X}^F(\tau) = \{\Delta x(\tau): ||\Delta y^i(\tau)|| < 0.01, \quad \forall i \varepsilon I\}$$

In this case, deviation from the nominal trajectory is permitted for the subsystem S^3 only, while other d.o.f. must not deviate from their nominal trajectories. Two characteristic cases are simulated; in both cases the subsystems are locally stabilized with exponential stability of degree $\Pi_i = \Pi_{i1} = 7.7$. In case I neither programmed control nor global control is introduced. In case II, the control is of the form (3.3.27) for i = 1,2,3 and of the form (3.3.16) for i = 4,5,6. The results of the simulation for these two cases are shown in Fig. 3.54. In Fig. 3.54 the norm of the output deviation from the nominal trajectories for the subsystems and for the global system S are presented for both control laws. Initial conditions are $\Delta q(0) = (0, 0, -0.22, 0, 0, 0)^T$, $\Delta \dot{q}^i(0) = 0$, and $\Delta i^i_R(0) = 0$, $\forall i \varepsilon I$. It can be seen that, in case I, the conditions for practical stability are not satisfied: the coupling among subsystems is too strong and thus, local control cannot reduce the effect of coupling. However, in case II, the conditions for practical stability are satisfied.

Efficiency of the suggested control structure is thus demonstrated for manipulator control. The validity of the aggregate-decomposition method for the composite system stability is also verified. Results presented in Tables 3.11 - 3.13 demonstrate the possibility of estimating the regions of initial conditions for which the system S is asymptotically stable (around the nominal trajectory) for different forms of control laws, for different exponential degrees of stability for the subsystems, with or without global control. In this way, using the analysis of the aggregate stability of the system, the parameters of the decentralized control structure can be synthesized. We shall now briefly present some aspects of this approach to control synthesis [19].

The dynamic analysis of the manipulator and the introduction of programmed control (nominal manipulator dynamics) or global control becomes necessary when we want to use the potential of actuators maximally and when we want to satisfy the demands of very fast and precisely positioned manipulators. That is, if we consider a relatively slow motion of manipulators with sufficiently powerful actuators and if precise positioning is not necessary, kinematic synthesis of nominal trajectories (without calculating the nominal driving torques) and local control only would be sufficient. In this case, local control is suf-

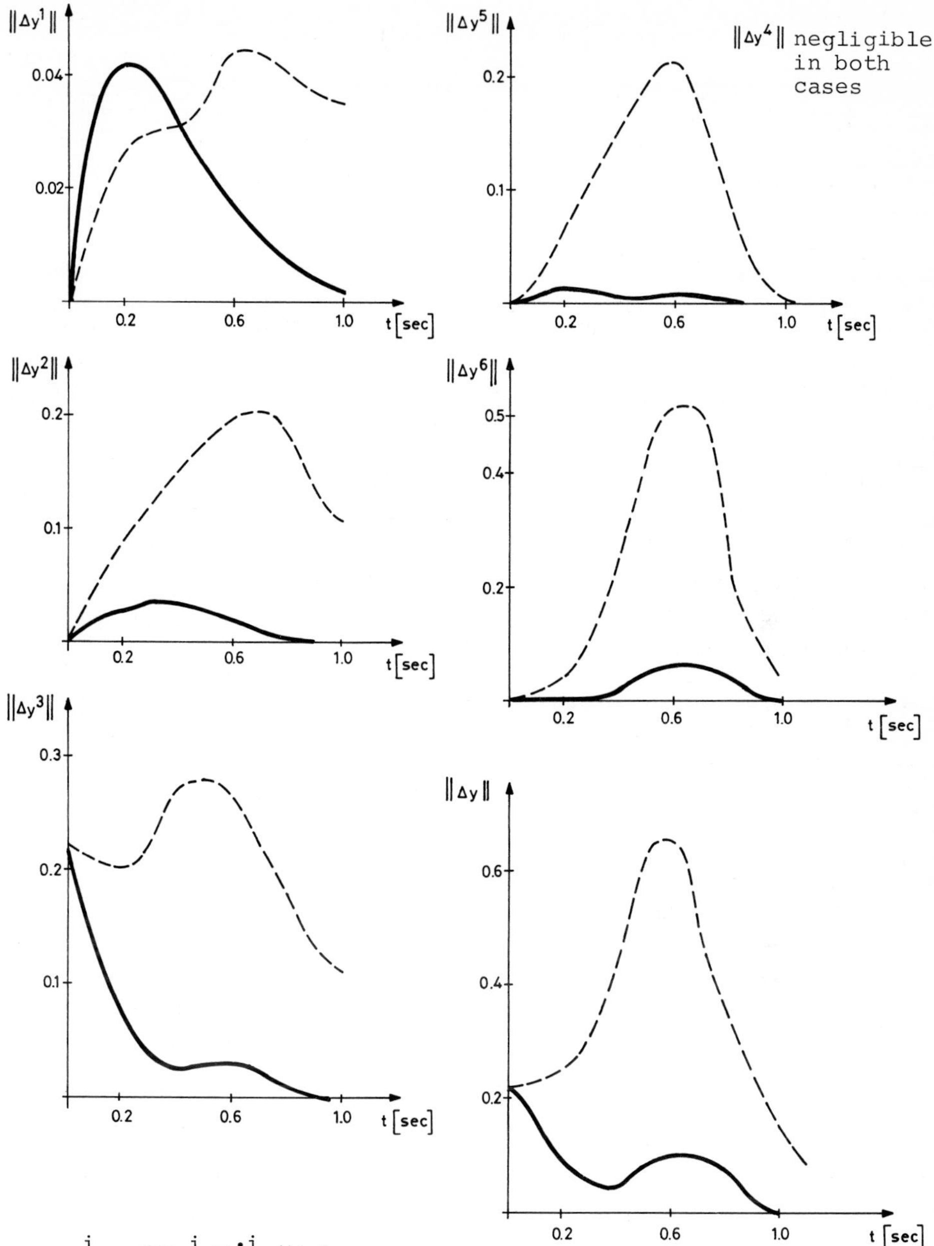

------ $u^i = -50\Delta q^i - 2\Delta\dot{q}^i \quad \forall i \varepsilon I$

——— $u^i = -50\Delta q^i - 2\Delta\dot{q}^i + 0.232|P_i|\ \mathrm{sgn}\ (\Delta u_i^L)$ i=1,2,3

$= -50\Delta q^i - 2\Delta\dot{q}^i + u^{oi}(t)$ i = 4,5,6

Fig. 3.54. Effects of coupling on tracking of nominal trajectories from Figs. 3.45, 3.46 (UMS-1)

ficient to reduce the coupling among subsystems. However, if we consider fast motion, manipulator dynamics (coupling) influences the system performance and if we do not want to use more powerful (heavier and more expensive) actuators, it is necessary to take dynamic coupling into account.

In order to demonstrate this explicitly, we consider the same manipulator motion as in Fig. 3.45, but with the time interval for the motion shorter than in Fig. 3.45. In order to do this, we increase the permissable manipulator tip acceleration. Three cases were considered. Case I: $a_{max} = 1.48$ m/s^2, case II: $a_{max} = 2.25$ m/s^2 and case III: $a_{max} = 3$ m/s^2. All other conditions are the same as before (same initial and terminal states and the same straight-line trajectory of the manipulator tip and the gripper). Thus, the nominal trajectories for all three cases are calculated and presented in Fig. 3.55. In Fig. 3.55a the change of tip velocity is presented. Fig. 3.55b shows the nominal trajectories of the angles $q^o(t)$. Figs. 3.55c and 3.55d show the nominal driving torques and nominal control. The time intervals for the motion are $\tau_I = 0.9$ sec for case I (Fig. 3.45), $\tau_{II} = 0.76$ sec and $\tau_{III} = 0.66$ sec. Obviously, if the manipulator velocity is increased, the nominal torques also increase. Now, the tracking of the nominal trajectories for all three cases is considered. The tracking is realized in two ways.

(a) By decentralized control, which does not take into account the coupling among subsystems ($K_i^G = 0$, $\forall i \in I$ and without programmed control).

(b) By local and global control in the form of force feedback.

The tracking of the three nominal trajectories with both control laws is simulated, with the initial conditions as in Fig. 3.54. The subsystems are locally exponentially stable with stability of degree $\Pi_i = \Pi_{i1}$. Thus, control (b) is of the form (3.3.27) for i = 1,2,3. However, since the coupling is stronger in faster motions, due to the non-linear characteristic of the moment equation (1.1.30), it is necessary to introduce global control even for the gripper d.o.f. in order to obtain better tracking of the nominal movements in cases II and III. Thus, $K_i^G = 1.78$ for i = 4,5,6 is introduced for cases II and III. The results of the simulation for these cases are presented in Fig. 3.56. It can be verified that the difference in tracking between controls

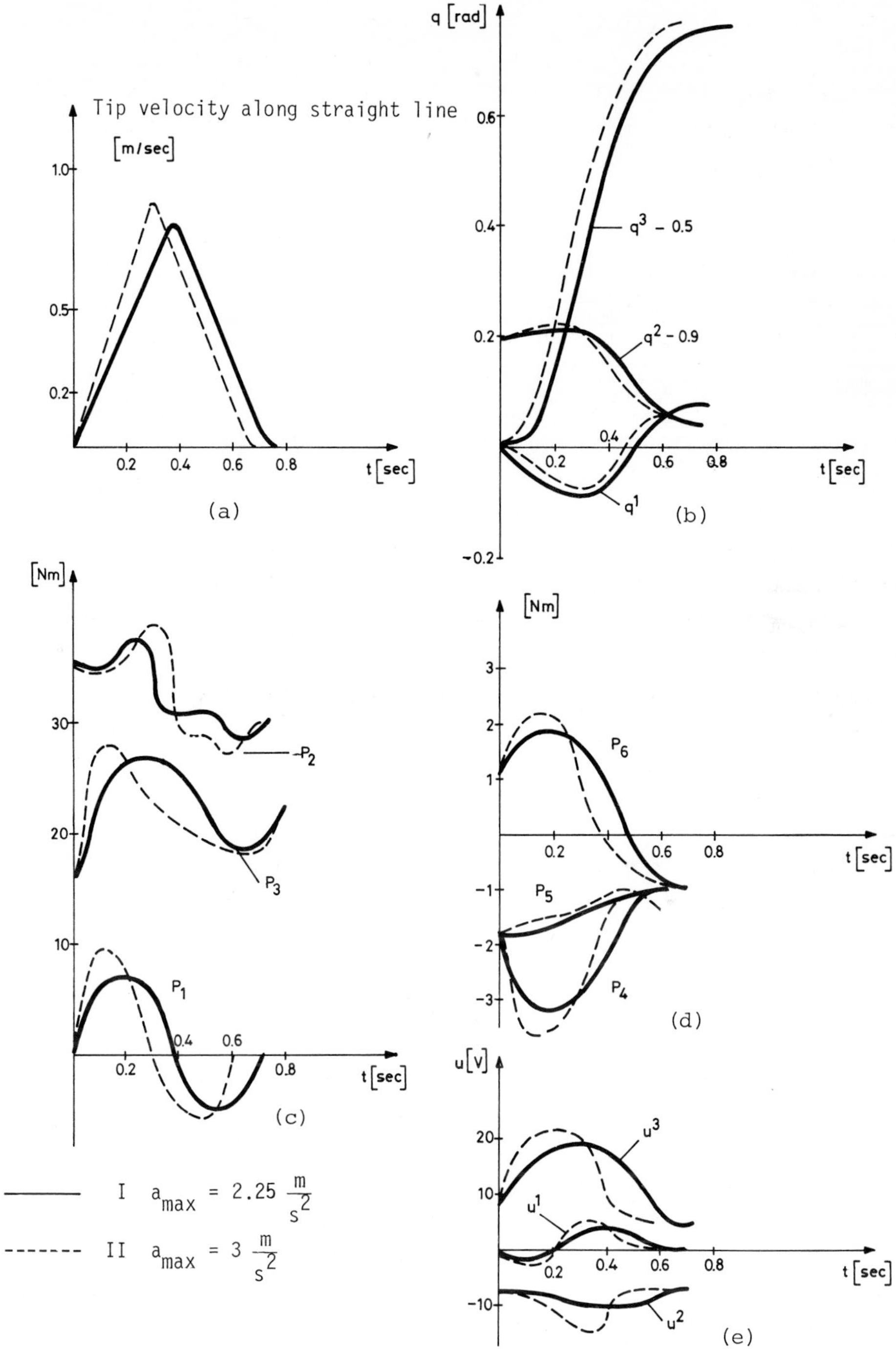

Fig. 3.55. Nominal trajectories: Accelerated motions from Fig. 3.45 (UMS-1)

$$u^i = -50\Delta q^i - 2\Delta\dot{q}^i \quad \forall i \in I$$

$$u^i = -50\Delta q^i - 2\Delta\dot{q}^i +$$
$$+0.232|P_i|\,\text{sgn}(\Delta u_i^L) \quad i=1,2,3$$

$$u^i = -50\Delta q^i - 2\Delta\dot{q}^i +$$
$$+1.78|P_i|\,\text{sgn}(\Delta u_i^L) \quad i=4,5,6$$

$$a_{max} = 2.25 \left[\frac{m}{s^2}\right]$$

$$a_{max} = 3 \left[\frac{m}{s^2}\right]$$

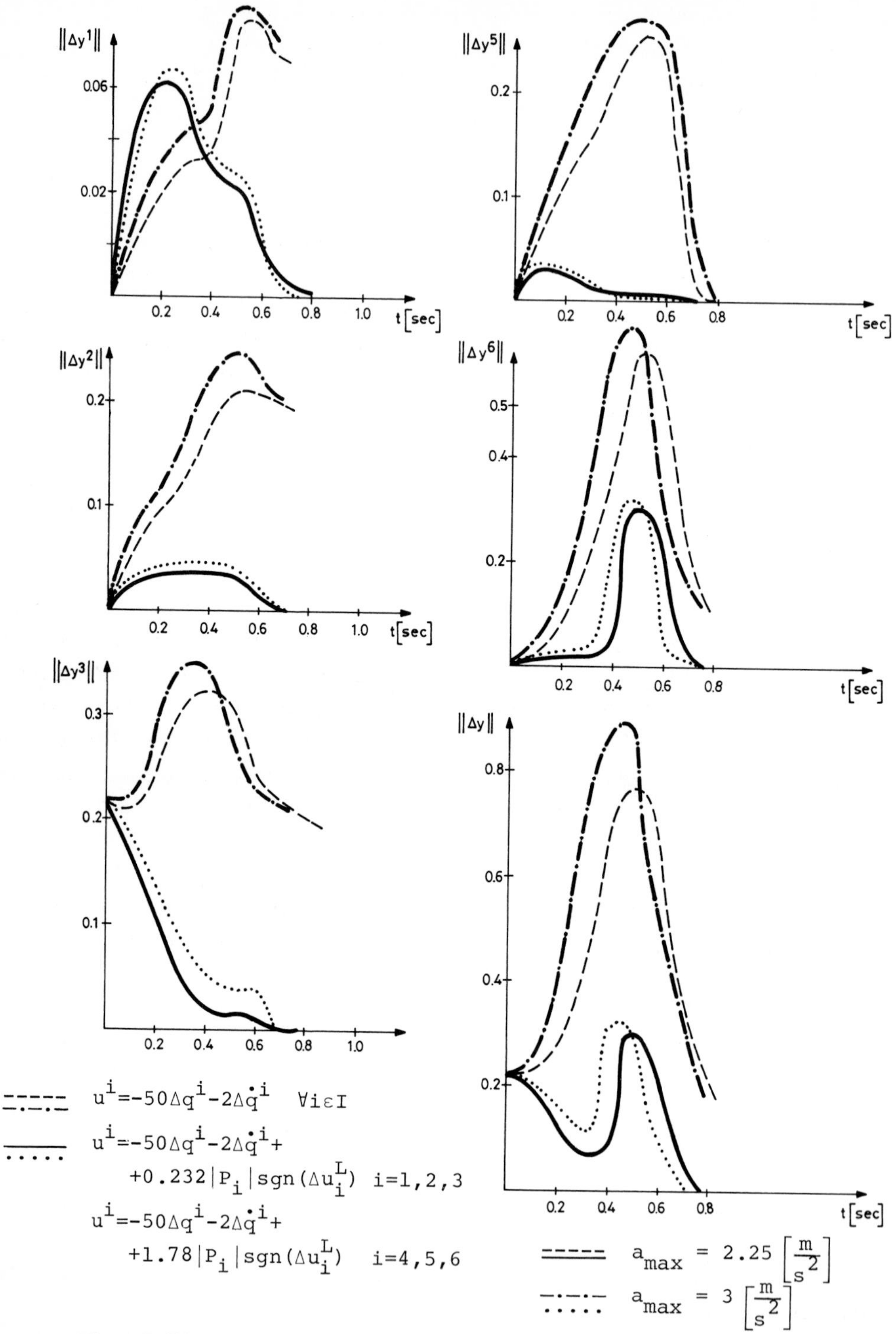

Fig. 3.56. Tracking of trajectories from Fig. 3.55 (UMS-1)

(a) and (b) is greater if the nominal motion is faster. This means that we have to take care of the dynamics of the system if the level of dynamic influence is so high that the actuators cannot reduce it by linear decentralized output feedback of the subsystems.

In order to further demonstrate this fact, the tracking of these two nominal trajectories are simulated when very weak local control is applied. That is, we consider the case when the subsystems are locally exponentially stable of degree $\Pi_i = \Pi_{i3} = 1.5$, by the choice of the gain vector $k_i = (10,0)^T$, i.e., the feedback with respect to angular velocity is not introduced. The tracking of the nominal trajectory for cases II and III is simulated with the control in the form

(a) decentralized control $u^i(t) = -10\ \Delta q^i(t),\ \forall i \in I$,

(b) decentralized control with global control with respect to the torque $u^i(t) = -10\ \Delta q^i(t) + K_i^G |P_i|\ \mathrm{sgn}\ (\Delta u_i^L),\ \forall i \in I$.

The initial conditions are as in Fig. 3.50. The results of the simulation show that the difference in the tracking of the nominal trajectories for these two forms of control increases with an increase in motion velocity. For case III, decentralized control cannot stabilize the system around the nominal trajectory, while the control which considers the manipulator dynamics (through global or nominal programmed control) produces sufficiently accurate tracking even with this very fast motion (Fig. 3.57).

Finally, some advantages of global control (with force feedback) are demonstrated and compared with the application of programmed nominal control.

It is obvious that programmed control is much more sensitive to parameter variation than is global control, where the actual load (torque) on the motor shaft is directly measured. This is the reason why programmed control is much less robust (reliable) than control in the form of global feedback with respect to load. The industrial manipulator parameters can often vary due to a change in the task conditions and environment. Thus, the torques can vary from the nominal values even if there is no deviation from the nominal initial conditions. The simulation in Fig. 3.58. shows the sensitivity of the two control laws to parameter variation. We consider the nominal trajectory from Figs.

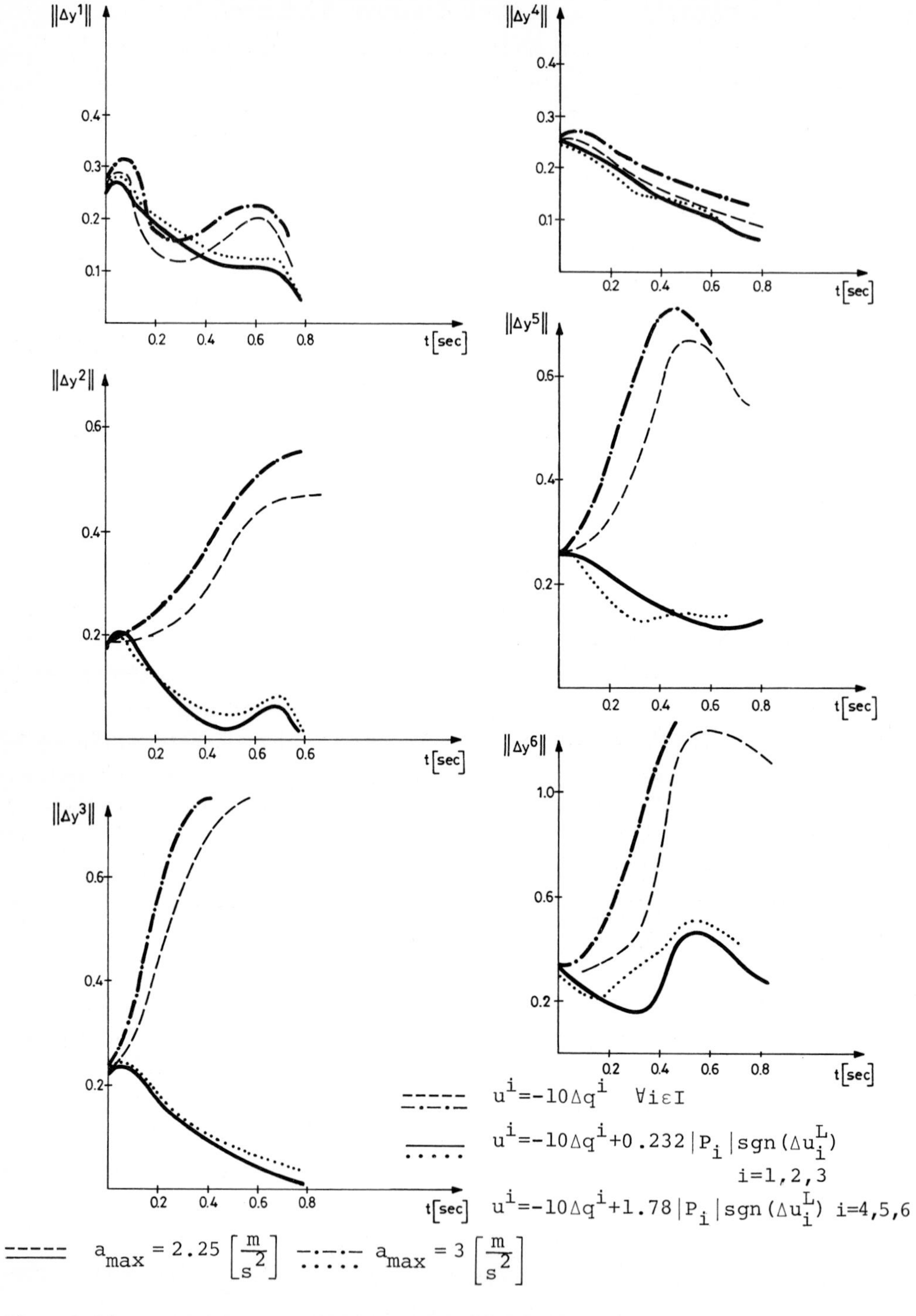

Fig. 3.57. Decoupled control with and without global feedback; tracking of nominal from Fig. 3.55.(UMS-1)

3.45 - 3.46 and observe the tracking of the nominal trajectory for the initial deviation of the state $\Delta x(0)$ from the nominal values as in Fig. 3.50. Two control laws are considered.

(a) Control in the form (3.3.27).

(b) Control in the form (3.3.26) with $K_i^G = 0$ (programmed and local control only).

We observe the tracking in two cases: I, when there is no parameter variation from nominal values and, II, when the system parameters vary significantly from their nominal values. In case II it is assumed that the working object varies significantly from the nominal case from which we calculated the nominal dynamics. The object is three times heavier than in the nominal case and the distance between the c.o.g. and the gripper joint is twice the nominal distance. The simulation results in Fig. 3.58 show significant differences between cases I and II for the control in the form of (b), while for control (a) the difference is negligible. This shows that the system performance is much more sensitive to parameter variation if programmed control is applied, while the sensitivity of the system with global control is less. That is, from the difference in tracking of the nominal trajectories with and without parameter variation, we can identify the sensitivity function of the system to parameter variation. In Fig. 3.58a the norms of the output deviation from the nominal trajectory for the subsystems are presented for the two cases and two control laws. In Fig. 3.58 the corresponding driving torques for the subsystem S^2 are presented. The nominal driving torque $P_2^o(t)$ and its change during tracking for both types of control are presented.

More precise analysis of sensitivity requires the construction of the sensitivity model of the system and analysis of the sensitivity functions [21]. It is clear that if the parameters of the system change drastically and the actuators implemented are not too powerful, high sensitivity of programmed control might lead to unsatisfactory tracking, i.e., it might happen that the conditions for practical stability are not satisfied. In this case, it is possible to pass to the system stabilization in the stage of large perturbation (as mentioned in ch. 2).

In order to compare various manipulation configurations when decentralized control is applied, the synthesis of local and global control for

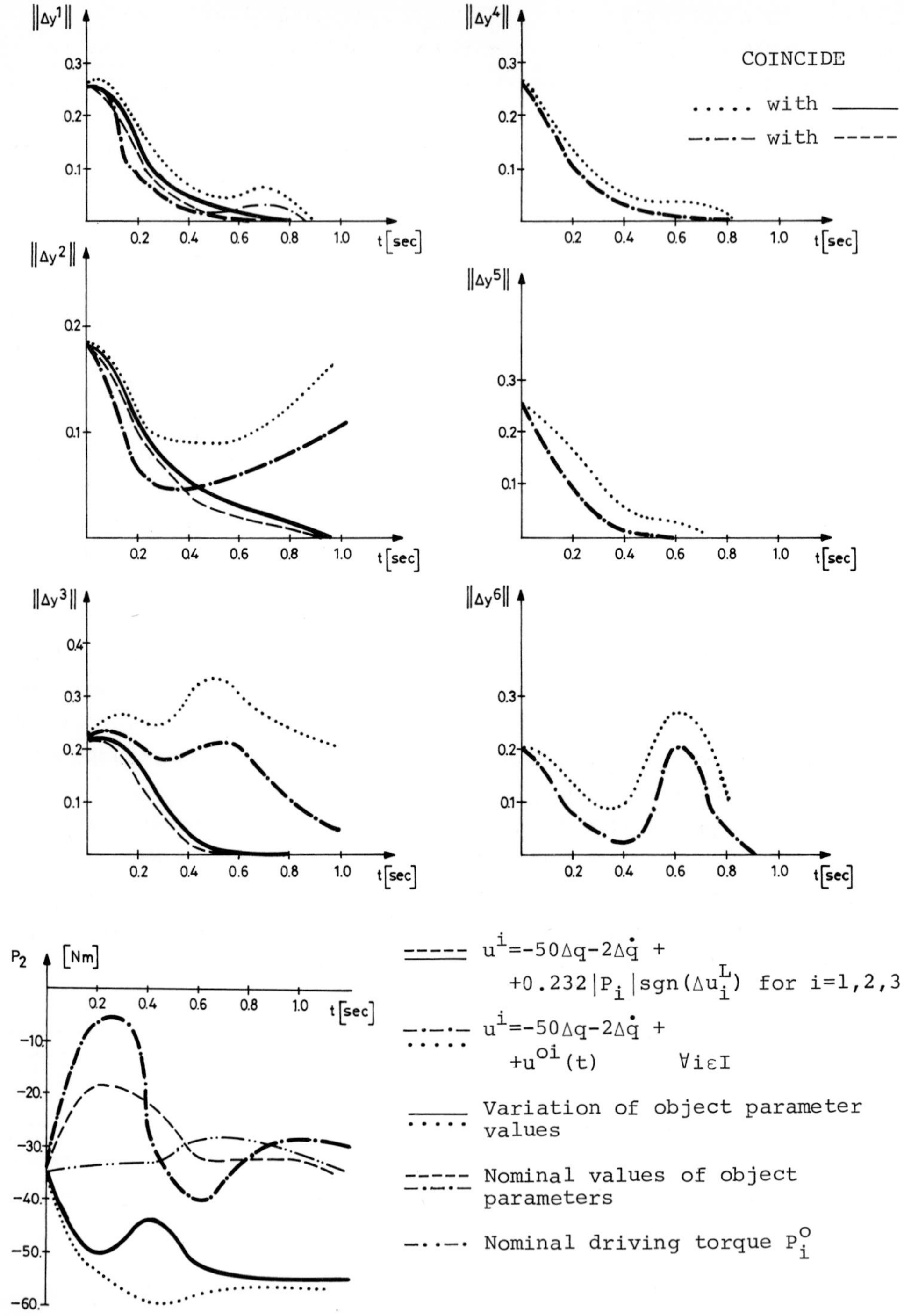

Fig. 3.58. Sensitivity to parameter variations (UMS-1)

the manipulator UMS-2 (Fig. 3.2, Tables 3.3 and 3.4) has been carried out. In the above analysis of the manipulator UMS-1, where the coupling is strong, the convenience of introducing the global control has been shown. The manipulator UMS-2 is an example of those systems where the influence of coupling is rather weak. Thus, the application of decentralized control is much simpler and it is ussually not necessary to introduce programmed, centralized control and/or global control.

Let us consider the stabilization of the system UMS-2 around the nominal trajectory $x^o(t)$ synthesized in para. 3.3.1 and presented in Fig. 3.47 [8].

At the stage of perturbed dynamics the decentralized control is to be synthesized. First, for the subsystems S^i (i.e. for D.C. motors) the local gains k_i are determined in order to obtain the degree Π_i of exponential stability. The following values for the parameters are chosen.

$$\Pi_i = 5.5 \text{ and } k_1 = (53.4,\ 21.4)^T\ [\text{V/rad, V/rad/sec}]$$

$$k_2 = (226,\ 7.59)^T\ [\text{V/m, V/m/sec}]$$

$$k_3 = (291,\ 9.77)^T\ [\text{V/m, V/m/sec}]$$

$$k_{4,5,6} = (34.3,\ 1.)^T\ [\text{V/rad, V/rad/sec}]$$

In order to investigate the significance of global control for this type of manipulator, the global control Δu_i^G (with respect to the coupling ΔP_i) is introduced. In order to simplify the control law, the global control is chosen to be linear with respect to the coupling ΔP_i, i.e., in the form (3.3.24).

In the concrete case the following global gains are chosen: $K_1^G = 0.906$ [V/Nm], $K_2^G = 0.012$[V/N], $K_3^G = 0.054$[V/N], $K_{4,5,6}^G = 0.232$[V/Nm].

The system UMS-2 with various control laws has been simulated for the following initial conditions: $\Delta q(0) = (0.26, -0.01, 0.01, 0.26, 0.25, 0)^T$, $\Delta \dot{q}^i(0) = 0.$, $\Delta i_R^i(0) = 0$, $\forall i \varepsilon I$.

Fig. 3.59 shows the deviations of the subsystem outputs from the nominal. The six control laws have been simulated and the results are presented. Here, $\bar{u}^{io}$ denotes nominal control for local subsystems where nominal coupling (load) is not taken into account (see para. 3.3.3). For manipulators where the coupling is weak, centralized nominal control might be substituted by local nominal control which does not compensate for the nominal load, thus significantly simplifying the control implementation. From Fig. 3.59 it is clear that the global control does not significantly change the performance of the system, since the coupling with this system is weak and thus, the global control can usually be omitted.

Finally we present, the synthesis of local and global control for the system UMS-3B (Fig. 3.3, data in Tables 3.5, 3.6). We shall consider the stabilization of the manipulator UMS-3B around the nominal trajectory synthesized in para. 3.3.1 and presented in Figs. 3.48 and 3.49 [22].

As already stated, the manipulator UMS-3B is somewhere between the two manipulation systems previously considered with respect to the effect of coupling on the global system stability. We shall consider stability, as in two previous cases, using the incomplete local feedback (and in the next paragraph we shall consider the stabilization using local linear optimal regulators). However, here we only introduce positional local feedback loops, i.e., it is assumed that the subsystem outputs are given by $y^i = \ell^i$, (or $y^i = q^i$). This is the most common case in practice with hydraulic actuators. Thus, only positional feedbacks are introduced.

First, for the hydraulic actuators (3.1.3) of the subsystems S^i, the local gains have been chosen in order to stabilize local subsystems, i.e., to ensure that the free subsystem poles are $\sigma < -5.0$. The gains are: $K_1 = K_2 = K_3 = 90$ [mA/m], $K_4 = K_5 = K_6 = 37$ [mA/rad], ($\Pi_i = 5.$). The degree of the exponential stability has been chosen according to the requirements of the practical stability of the system: $\bar{X}^I = \{\Delta x(0): ||\Delta y^i(0)|| \leq 0.1, |\Delta p^i| < 9.8\text{[bar]}, \forall i \in I\}$, $\bar{X}^t(t) = \{\Delta x(0): ||\Delta y^i(t)|| \leq 0.1 \exp(-2.5t), |\Delta p^i(t)| \leq 9.8\text{[bar]}, \forall i \in I, \forall t \in T\}$ and $\bar{X}^F = \{\Delta x(\tau) : ||\Delta y^i(\tau)|| \leq 0.1 \exp(-2.5\tau), |\Delta p^i(\tau)| < 9.8\text{[bar]}, \forall i \in I\}$ for $\tau = \tau_s = 0.65$s, where Δy^i has been defined previously (for the first two subsystems, the output norm is taken as $||\Delta y^i|| = (\Delta \ell^{i2} + 0.1 \Delta \dot{\ell}^{i2})^{1/2}$, $i = 1, 2$). It should be noticed, that although the nominal motion of the manipulator UMS-3B, as synthesized in para. 3.3.1, is much longer, simulation is observed on the first 0.65 sec since we are required to

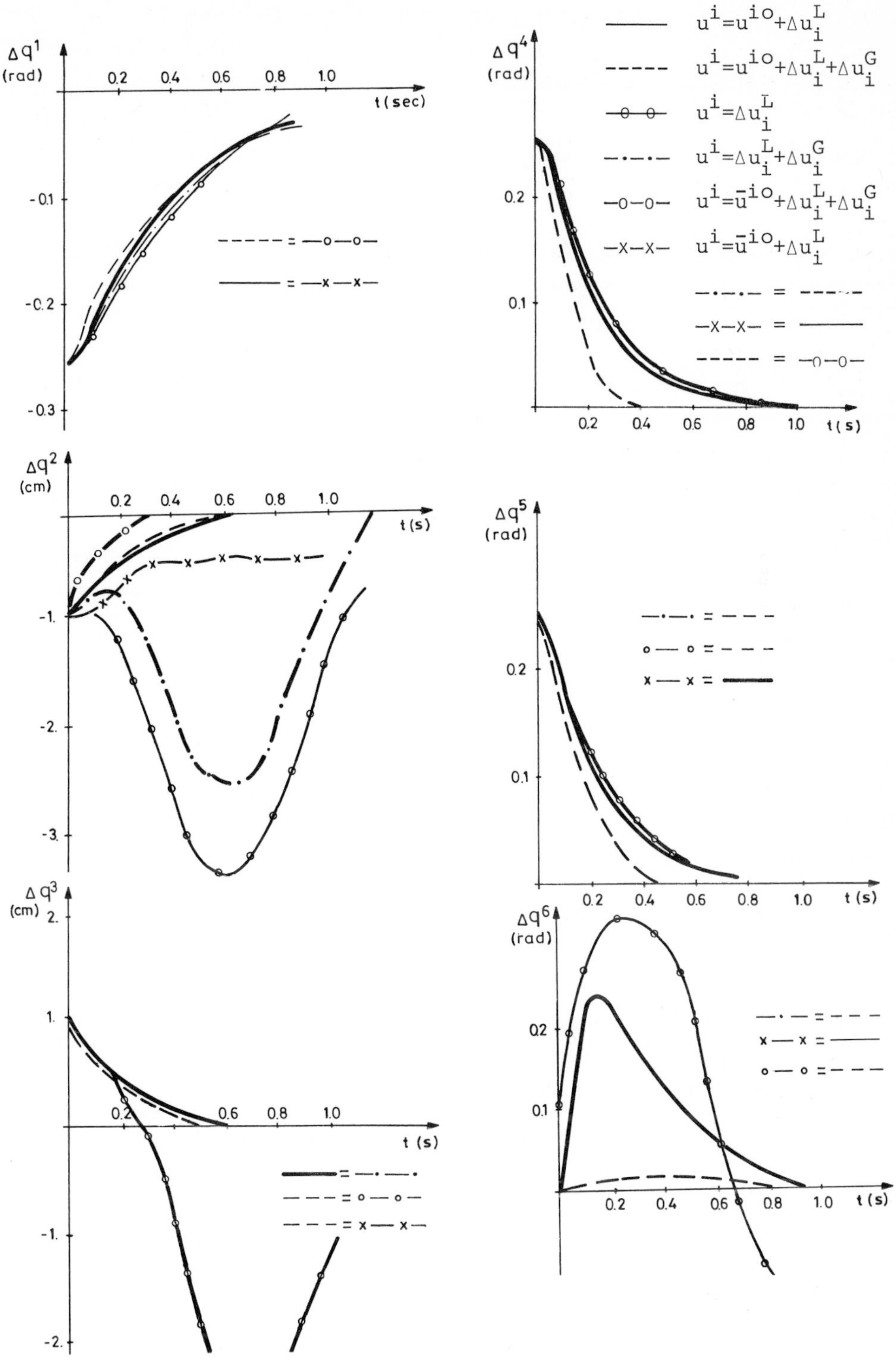

Fig. 3.59. Deviation of UMS-2 from nominal dynamics

stabilize the system during the motion on the segment AB (Fig. 3.48a) in order to ensure exact tracking of the nominal on the segment BC (when the painting is performed).

However, the stability analysis, and the simulation (Fig. 3.60) show that the practical stability of the system cannot be guaranteed if only programmed and local controls are applied (although the local gains have been chosen so as to obtain the degrees $\Pi_i = 5$ of exponential stability of the subsystems whilst the required degree of exponential shrinkage of the regions of practical stability is 2.5). Thus, global control in the form of force feedback (3.3.24) should be introduced. With the stability analysis of the global system, as previously described, the global gains have been chosen as $K_i^G = 0.12$ [mA/Nm], for $i = 1,2,4,5,6$ and $K_3^G = 0.22$ [mA/N]. It has been shown that this global control law stabilizes the system. However, the other method of stabilizing the system is to choose the local gains so as to increase the degree of exponential stability of the local subsystems. If we take $\Pi_i = 10$, $\forall i \in I$, the local gains are $K_1 = K_2 = K_3 = 179.$, $K_4 = K_5 = K_6 = 73.5$ and the system is practically stable under only nominal and local control.

The tracking of the nominal has been simulated with the three control laws mentioned and the results are presented in Figs. 3.60 - 3.63. The simulation has been caried out for the initial conditions: $\Delta\ell^1(0) = 0.$, $\Delta\ell^2(0) = 0.05$ [m], $\Delta\ell^3(0) = -0.05$ [m], $\Delta q^4(0) = \Delta q^5(0) = \Delta q^6(0) = 0.1$ [rad], $\Delta\dot{\ell}^i(0) = 0.$, $i=1,2,3$, $\Delta\dot{q}^i(0) = 0.$, $i=4,5,6$, $\Delta p^i(0) = 0.$, $\forall i \in I$. In Fig. 3.60. the norms of subsystems outputs are presented. Fig. 3.61 shows the corresponding generalized forces, developed during the tracking. Fig. 3.62 shows the inputs (currents), which should be generated during the tracking, and Fig. 3.63 shows the energy consumption of the actuators. Fig. 3.60 presents the simulations of tracking with another three control laws. In order to demonstrate the case most commonly encountered in practice (when the desired trajectories are realized by local control only), the tracking of the nominal by local control only ($\Pi_i = 10$) has been simulated (programmed control has not been introduced at all). The tracking by local control and by nominal programmed control, synthesized at the level of local subsystems (neglecting the coupling), has also been simulated. These two simulations show the effect of coupling among subsystems on global system stability. Finally, as the sixth case, the tracking by the local control (from the previous case) and by the global control in the form (3.3.24), has

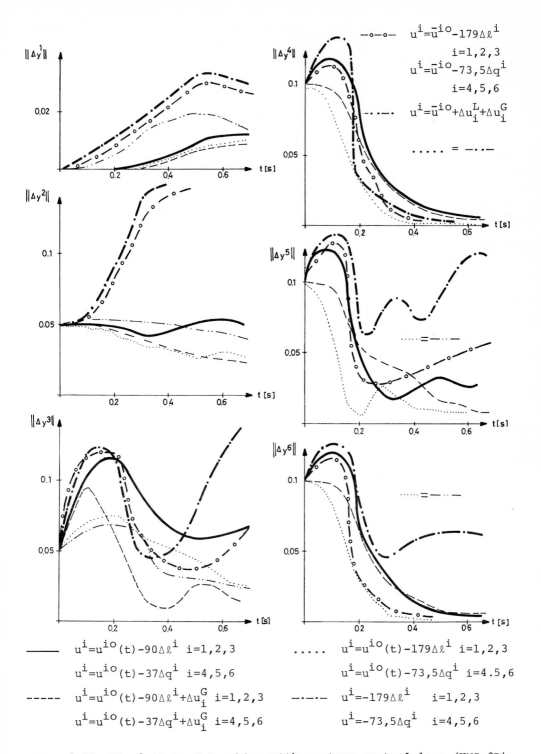

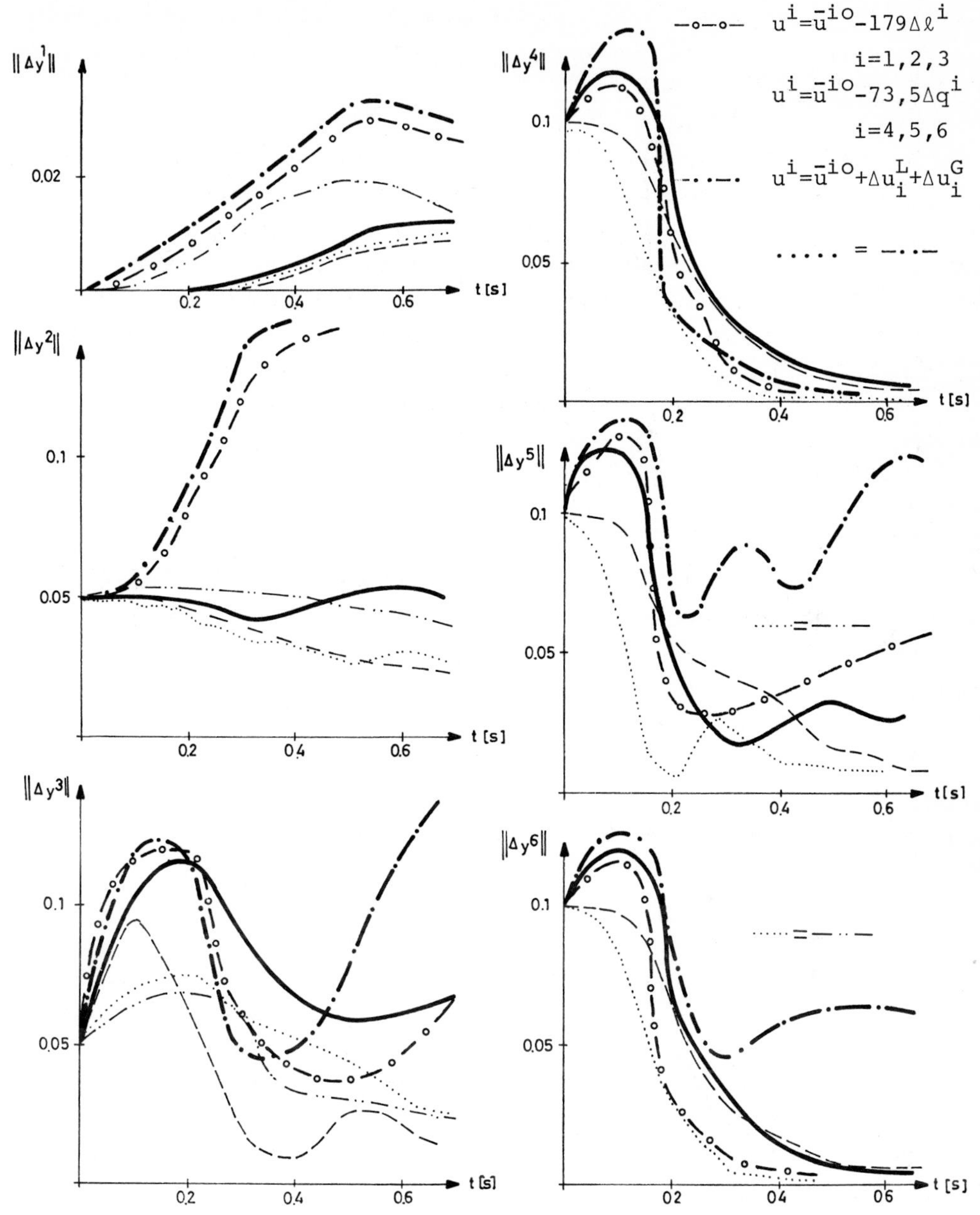

—— $u^i=u^{io}(t)-90\Delta\ell^i$ $i=1,2,3$

$u^i=u^{io}(t)-37\Delta q^i$ $i=4,5,6$

- - - - $u^i=u^{io}(t)-90\Delta\ell^i+\Delta u_i^G$ $i=1,2,3$

$u^i=u^{io}(t)-37\Delta q^i+\Delta u_i^G$ $i=4,5,6$

. . . . . $u^i=u^{io}(t)-179\Delta\ell^i$ $i=1,2,3$

$u^i=u^{io}(t)-73{,}5\Delta q^i$ $i=4.5,6$

—·—·— $u^i=-179\Delta\ell^i$ $i=1,2,3$

$u^i=-73{,}5\Delta q^i$ $i=4,5,6$

Fig. 3.60. Simulation of tracking with various control laws (UMS-3B)

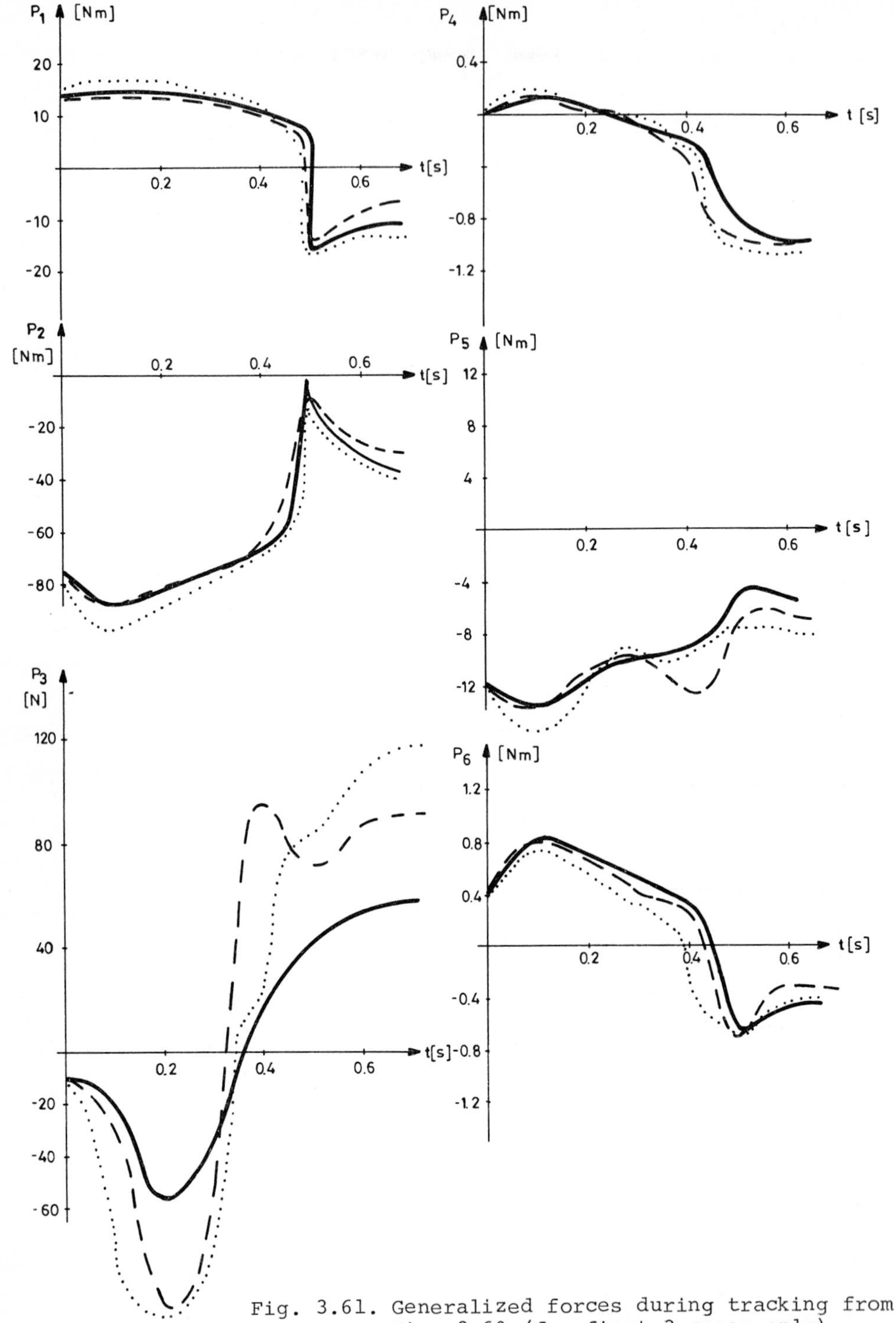

Fig. 3.61. Generalized forces during tracking from Fig. 3.60 (for first 3 cases only)

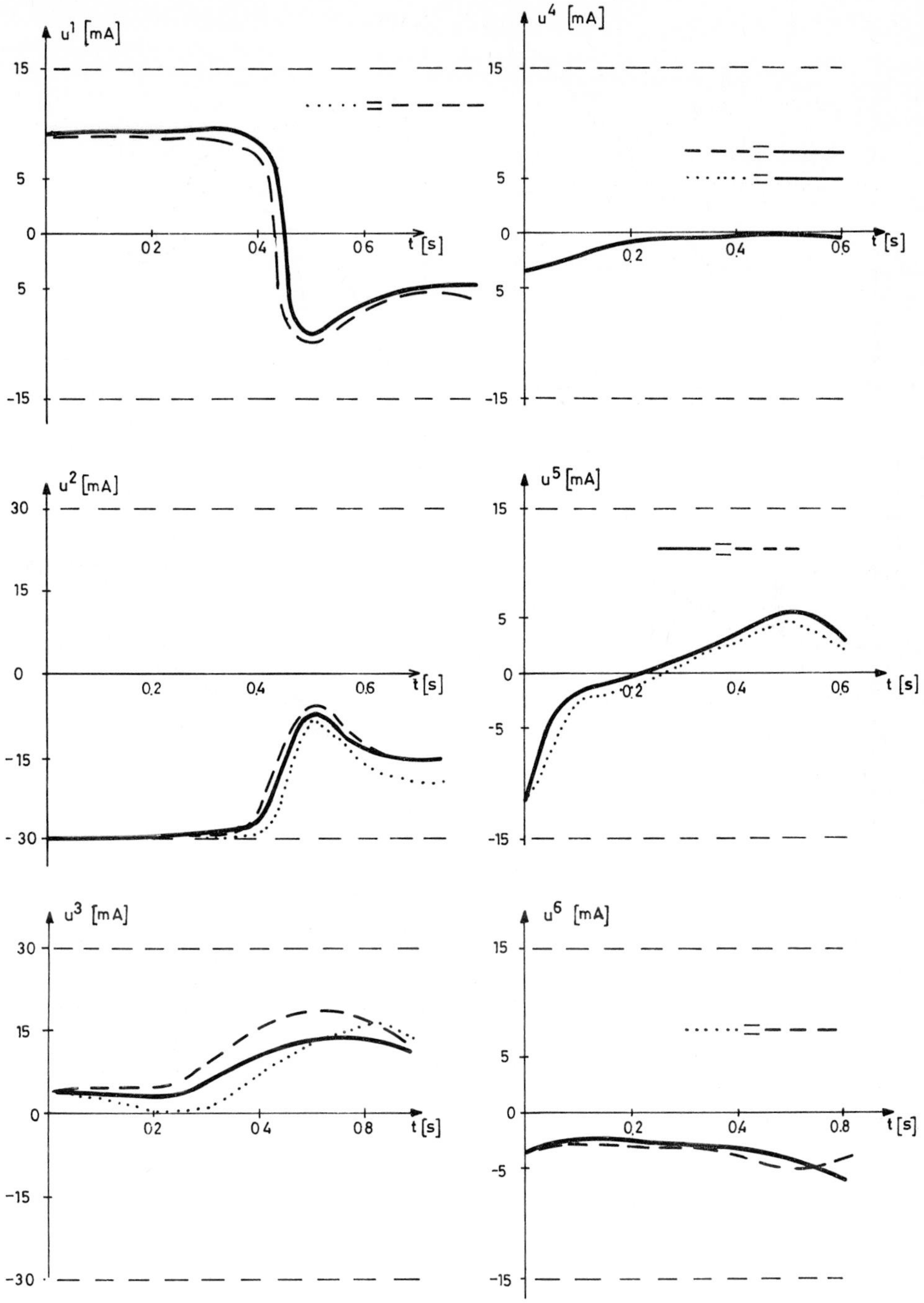

Fig. 3.62. Control during tracking from Fig. 3.60 (UMS-3B)

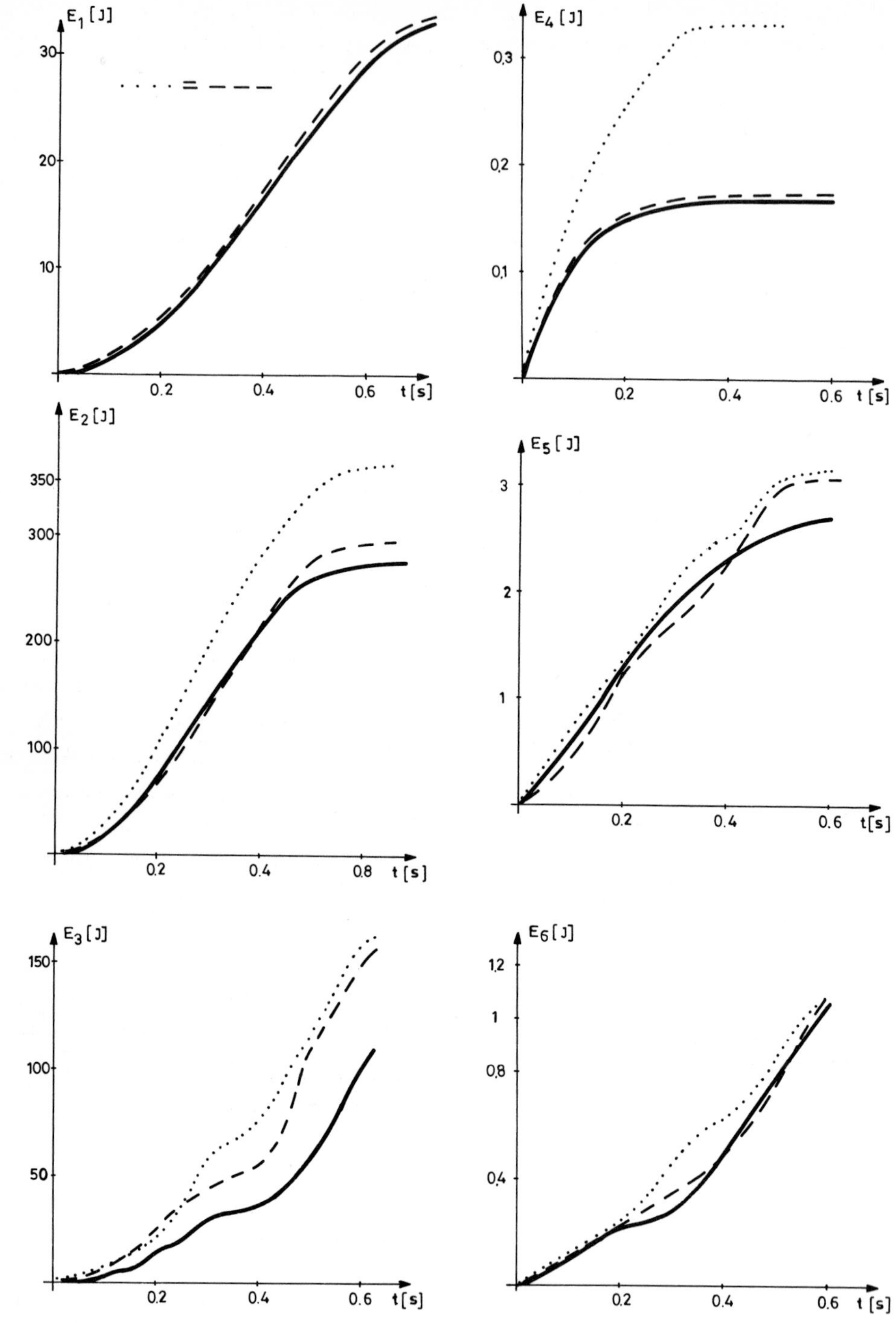

Fig. 3.63. Energy consumption for realizing tracking from Fig. 3.60

been simulated (with the global gains as above). The simulation shows how global control can improve the tracking of the nominal trajectory. Although the effect of coupling with this manipulator is not negligible, the system can be stabilized by local control (and programmed centralized control) without applying of global control. However, the energy analysis (Figs. 3.62 and 3.63) shows that higher local gains are less convenient, from the point of view of energy consumption, than lower local gains with global control. However, global control requires more expensive equipment, as we mentioned earlier.

3.3.3. Suboptimality analysis of decentralized control and various global control laws

In the previous paragraph we presented the results of decentralized control synthesis when local control is in the form of output feedbacks and the global control in the form of "linear" feedback by load. The purpose was to investigate the possibilities of a common control synthesis (of local feedbacks) and to investigate the effect of the dynamics of the mechanism on the manipulator performance. In practice the local subsystems are commonly stabilized by incomplete feedbacks (local servo-feedbacks). In the previous paragraph it was shown how the aggregation-decomposition method can be used for global system stability analysis and how the regions of the system asymptotic stability can be estimated. It was shown how the parameters of the global control can be synthesized. The local and global control can be synthesized in a few iterations until a satisfactory system performance is achieved.

For the sake of a more precise investigation of the relation between local and global control, we shall present two examples of decentralized control synthesis using minimization of the standard quadratic criteria (by introducing complete feedback with respect to the subsystem state). Global control synthesis will also be presented in terms of investigating the suboptimality of control. The influence of the model distribution on the syboptimality of the decentralized control will be also presented [20].

We consider the six d.o.f. manipulator UMS-1 (Fig. 3.1). Data on the manipulator and actuators are given in Tables 3.10 and 3.2. The control task is as follows. The nominal trajectory $x_o^o(t)$, $\forall t \varepsilon T$, is given for

$\tau = 1$ sec. The nominal programmed control $u^o(t)$, $\forall t \in T$, realizing $x^o(t)$, $\forall t \in T$ when $x(0) = x^o(0)$ and no perturbation acts on the system is synthesized on the basis of the centralized model (2.2.1) of the system S. The tracking of $x^o(t)$ should be realized at the second stage of the control synthesis. The regions of practical stability in this particular case are given by $\bar{X}^I = \{\Delta x(0): ||\Delta y^i(0)|| \leq 0.25, |\Delta i_R^i(0)| < 5A, \forall i \in I\}$ and $\bar{X}^t(t) = \{\Delta x(t): ||\Delta y^i(t)|| \leq 0.25 \exp(-2.5t), |\Delta i_R^i(t)| < 5A, \forall i \in I, \forall t \in T\}$; and $\bar{X}^F = \{\Delta x(\tau): ||\Delta y^i(\tau)|| \leq 0.25 \exp(-2.5\tau), |\Delta i_R^i(\tau)| \leq 5A, \forall i \in I\}$, for $\tau = \tau_s$.

First, local control is synthesized for each subsystem S^i (2.6.8) by minimizing the criterion (2.6.12) with $\Pi = 2.5$. The weighting matrix Q_i is presented in Table 3.14. In Fig. 3.64 local gains versus weighting elements r_i are presented. Since the actuator input is constrained ($u_m^i = 27$ V, $\forall i \in I$) the subsystem exponential stability regions X_i are determined, together with their approximations $\tilde{X}_i$ (3.3.21). Then, according to (2.6.23), the numbers ξ_{ij} are determined using the set model of the system S on a digital computer. The dependance of ξ on r_i is also presented in Fig. 3.64. The condition (2.6.24) is satisfied for all r_i in Fig. 3.64. However, the conditions for practical stability (2.6.28) are satisfied only for $r_i \leq 1.0$. The dependance of η (calculated according to (2.6.27)) on r_i is presented in Fig. 3.64 together with the estimation of the max $J(x(0))$, according to (2.6.31). It can be seen that the estimation of the suboptimality of local control (2.6.13) when applied to the global system S is minimal for $r_i \simeq 0.5$. In Table 3.14 the numbers v_{io} determining the $\tilde{X}_i$ are presented, together with the vector Gv_o for $r_i = 0.5$ and the synthesized local gains.

In order to further reduce the suboptimality of the local control the global control is introduced, which should take into account the dynamics of the mechanical part S^M of the system. This global control is taken in the form (2.6.34). Three different realizations of $\Delta\bar{P}_i^*$ are considered. In the case (Ga), global control is introduced in the form of force feedback and driving torques in the manipulator joints should be directly measured. The dependance of the estimation (2.6.37) of the suboptimality of the control (2.6.13), (2.6.34) on the global gain K^G is shown in Fig. 3.65. The "optimum" choice of the global gain, $K^G = 0.112$, for which the estimation of the suboptimality is minimal, is calculated according to (2.6.40).

In the second case (Gb), the coupling $\Delta\bar{P}^*$ among subsystems is calculated according to approximate relations. That is, the coupling among the subsystems are driving torques consisting of inertial forces, centrifugal and gravity moments and so on. In (Gb.1) we shall approximate coupling, taking into account only gravitational moments, i.e. the global control is realized by on-line calculation:

$$\Delta\bar{P}_i^* = \vec{e}_i \sum_{j=i}^{n} [\vec{r}_{ji} \times \vec{G}_j] - \vec{e}_i^{\,o} \sum_{j=i}^{n} [\vec{r}_{ji}^{\,o} \times \vec{G}_j], \tag{3.3.28}$$

where $\vec{r}_{ji}^{\,o} \in R^3$ denotes the nominal value (as a function of time) of the position vector of the center of gravity of the j-th manipulator member with respect to the i-th joint center, $\vec{e}_i^{\,o} \in R^3$ denotes the nominal value of the i-th joint axis vector and $\vec{G}_j \in R^3$ denotes the gravity force of the j-th member. The calculation of (3.2.28) is simple. Since $\vec{G}_j$ = const and $|\vec{r}_{ji}| \sim ||x||$ the form (3.3.28) corresponds to the general form (2.6.42), i.e. the ΔP_i^* in (3.3.28) is linear in state. Thus, suboptimality of the control (2.6.13), (2.6.34), (3.3.28) can be estimated by (2.6.45). The dependance of the estimation of the suboptimality of the chosen control on K^G is also presented in Fig. 3.65*).

In the case (Gb.2) $\Delta\bar{P}_i^*$ is realized by on-line calculation according to the approximate equations

$$\Delta\bar{P}_i^* = \vec{e}_i \sum_{j=i}^{n} [\vec{r}_{ji} \times (\vec{\bar{F}}_j(t, q, \ddot{q}^i) + \vec{G}_j)] - \vec{e}_i^{\,o} \sum_{j=i}^{n} [\vec{r}_{ji}^{\,o} \times (\vec{F}_i^{\,o}(t, q^o, \ddot{q}^{oi}) + \vec{G}_j)], \tag{3.3.29}$$

where $\vec{\bar{F}}_j \in R^3$ denotes the j-th member inertial force due to $\ddot{q}^i$. According to (3.3.29) we calculate only the contribution of the i-th joint acceleration to the i-th joint load. This means that by (3.3.29) we calculate exactly the terms $H_{ii}^o(t, \Delta q)\Delta\ddot{q}^i$. It can easily be seen that the form (3.3.29) corresponds to the form given by (2.6.29). However in this case the coupling is not completely taken into account, i.e., the numbers $\xi_{ijk}^{(1)}$ and $\xi_{ij}^{(2)}$ are less than the corresponding numbers in (2.6.29). The estimation of the maximum criterion cost with the control (2.6.13), (2.6.34), (3.3.29) changes with K^G as shown in Fig. 3.65.

From Figs. 3.64 and 3.65 we can easily find how much the particular system suboptimality is decreased by introducing of various forms of

---

*) In this case $K^G>1$ may be used, since (3.3.28) cannot compensate for the whole coupling ΔP_i for $K^G=1$. Thus ξ_{ij}^* might be positive even for $K_G>1$.

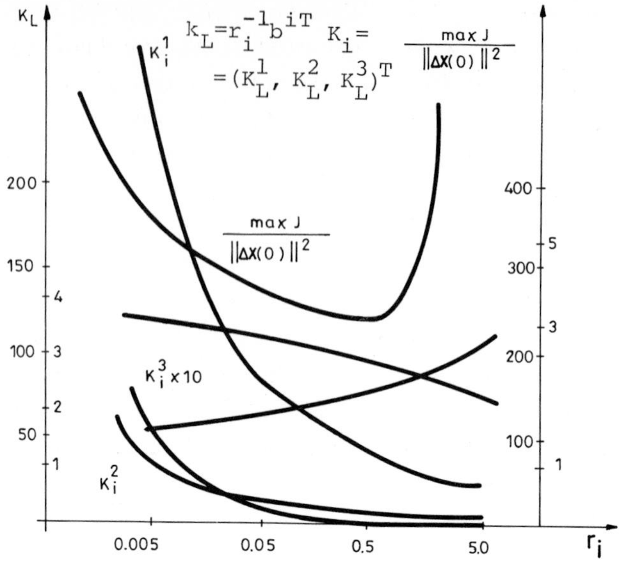

Fig. 3.64. Local control synthesis

$$Q_i = \begin{bmatrix} 10.0 & 0 & 0 \\ 0 & 10.0 & 0 \\ 0 & 0 & 0.1 \end{bmatrix} \quad \begin{matrix} r_i = 0.5; \\ \Pi = 2.5 \end{matrix} \quad K_i = \begin{bmatrix} 184.72 & 8.63 & 0.215 \\ 8.63 & 0.88 & 0.018 \\ 0.215 & 0.018 & 0.00094 \end{bmatrix}$$

$$r_i^{-1} b^{iT} K_i = (43.0, \; 3.59, \; 0.19)^T$$

$$v_o = (7.1, \; 6.5, \; 6.95, \; 5.5, \; 5.8, \; 6.1)^T$$

$$Gv_o = -(22.5, \; 21.5, \; 22.5, \; 16.2, \; 17.1, \; 19.2)^T$$

Table 3.14. Results of local control synthesis and global system stability analysis

global control, i.e., we can find how much the suboptimality of the decentralized control can be improved by introducing the control which takes into account the dynamics of the system.

The second way of partially considering the dynamics of the mechanism in the control synthesis is by varying the distribution of the model to the part associated with subsystems and the part associated with the coupling. That is, in the above control synthesis we have assumed that $\bar{H}_{ii} = 0$. By introducing $\bar{H}_{ii} \neq 0$ in (2.6.5) we can take into account the factor of the mechanism inertia in the synthesis of local control. Fig. 3.66 shows the dependance of $\lambda_M(K_i)$ and of the numbers ξ (estimating coupling in (2.6.23) for $K_i^G = 0$, $\forall i \in I$) on $\bar{H}_{ii}$ for two subsystems. According to this Figure we have taken $\bar{H}_{ii} = 1.5$, $i=1,2,3$ and $\bar{H}_{ii} = 0.5$ for $i=4,5,6$, and the local gains (2.6.13), corresponding to the modified models of the subsystems (2.6.5), have been calculated. The results of this computation are given in Table 3.15. The same Table shows the results of the stability analysis of the global system: the vectors v_o and Gv_o are given. It is obvious that Gv_o satisfies (2.6.24). Even more, this local control satisfies the conditions of practical stability. In order to compare such decentralized control with the decentralized control synthesized for $\bar{H}_{ii} = 0$ when the global control is introduced, the decentralized control from Table 3.14 has been considered but we have also applied global control in form (2.6.34), where $\Delta\bar{P}_i^*$ is to be realized by calculating

$$\Delta\bar{P}_i^* = \vec{e}_i \sum_{j=i}^{n} \vec{r}_{ji} \times \vec{\bar{F}}_j(t,\ q,\ \ddot{q}^i) - \vec{e}_i^{\,o} \sum_{j=i}^{n} \vec{r}_{ji}^{\,o} \times \vec{\bar{F}}_j^{o}(t,\ q^o,\ \ddot{q}^{oi}) \tag{3.3.30}$$

That is, in global control (3.3.30) the term $H_{ii}^o(t,\ \Delta q)\Delta\ddot{q}^i$ in coupling among subsystems is computed. If the estimation of the maximum criterion cost with the local control from Table 3.14 (for $\bar{H}_{ii} = 0$, $\forall i \in I$) and the global control (2.6.34) and (3.3.30) is compared with the estimation of suboptimality of the decentralized control from Table 3.15, it is found that the first control law is less suboptimal. This result is to be expected: when the global control is introduced the better estimation of the term $H_{ii}^o(t,\ \Delta q)$ is taken into account than in the case when it is estimated by $\bar{H}_{ii}$. However, the implementation of only the local control is much simpler and cheaper [20].

These results have been checked by digital simulation of the system S dynamics. The tracking of $x^o(t)$, $u^o(t)$ with various control laws has been simulated. The initial conditions have been chosen at the bound of the region $\bar{X}^I$. Fig. 3.67 shows the results of digital simulation of the manipulator UMS-1 when local control from Table 3.14 with two different global control laws (with $K_i^G = 0.112$, $\forall i \in I$ in both cases) is applied. The calculation of the criterion cost for the simulated tracking (for chosen initial conditions) with various control laws, aggrees closely with the expected relations among the criterion costs.

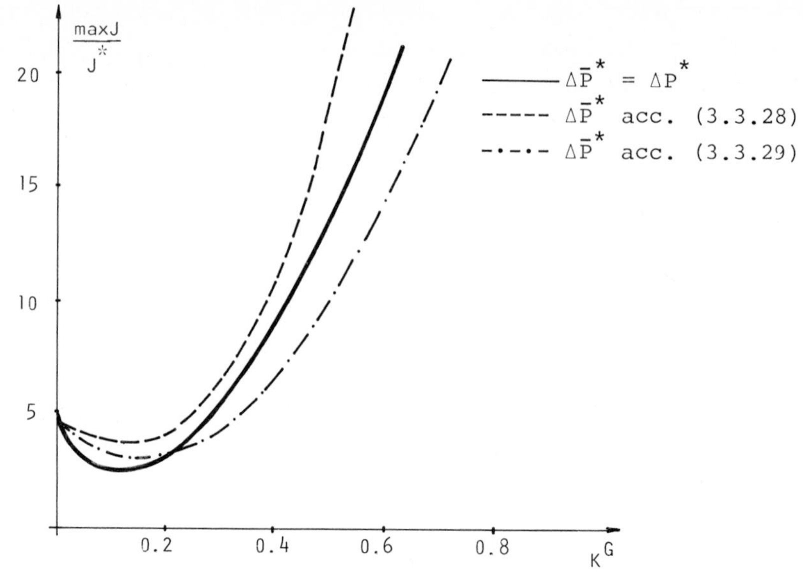

Fig. 3.65. Dependance of suboptimality estimation on global control choice

for $\bar{H}_{ii} = 1.5 \quad i=1,2,3$

$$K_i = \begin{bmatrix} 198.25 & 17.52 & 0.222 \\ 17.52 & 2.56 & 0.029 \\ 0.222 & 0.029 & 0.00086 \end{bmatrix}$$

$$r_i^{-1}\bar{b}^{iT}K_i = (44.5,\ 5.83,\ 0.17)^T$$

for $\bar{H}_{ii} = 0.5 \quad i=4,5,6$

$$K_i = \begin{bmatrix} 188.86 & 11.52 & 0.217 \\ 11.52 & 1.34 & 0.022 \\ 0.217 & 0.022 & 0.00090 \end{bmatrix}$$

$$r_i^{-1}\bar{b}^{iT}K_i = (43.48,\ 4.33,\ 0.18)^T$$

$$v_o = (6.2,\ 5.8,\ 6.05,\ 5.1,\ 5.3,\ 5.7)^T$$

$$Gv_o = -(23.3,\ 22.4,\ 22.9,\ 17.0,\ 17.5,\ 21.1)^T \text{ for } \Delta u^i = \Delta u_i^L$$

Table 3.15. Results of local control synthesis and global system stability analysis

Finally, in order to demonstrate the effect of the global system dynamics on the tracking of the nominal trajectory, we have simulated the case when the nominal programmed control is not calculated on the centralized model of the system. In para. 3.3.2 we presented the simula-

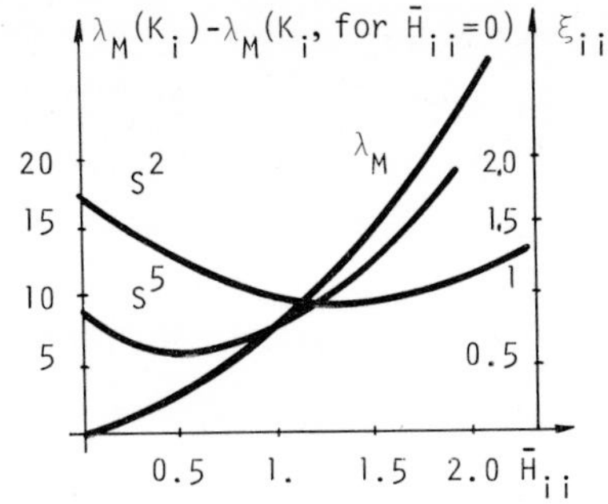

Fig. 3.66. Dependance of local gains and coupling estimation on $\bar{H}_{ii}$

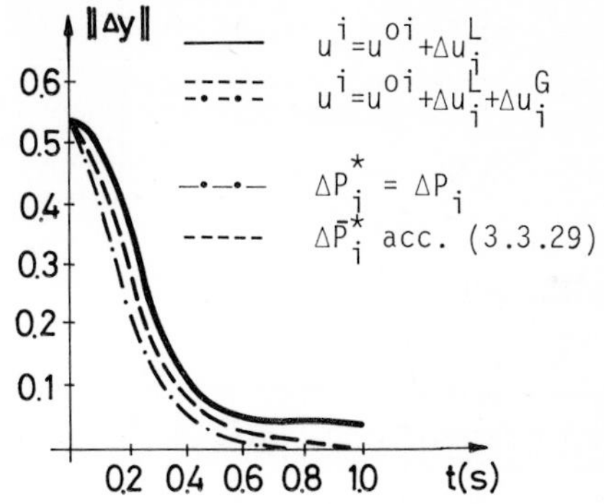

Fig. 3.67. Simulation of tracking of nominal

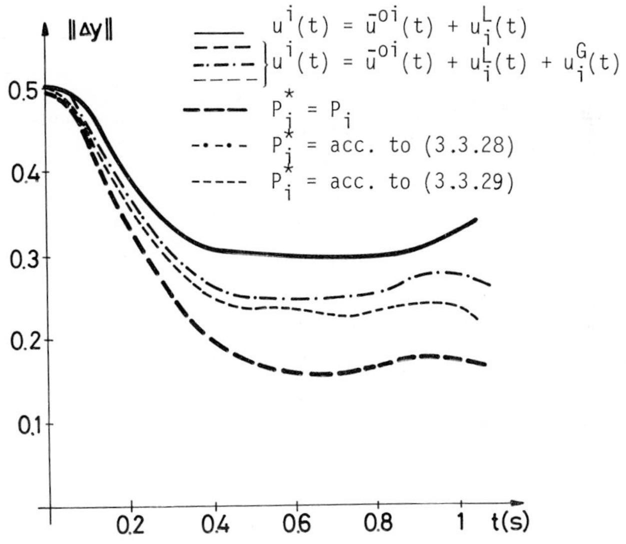

Fig. 3.68. Tracking of nominal with local and global control

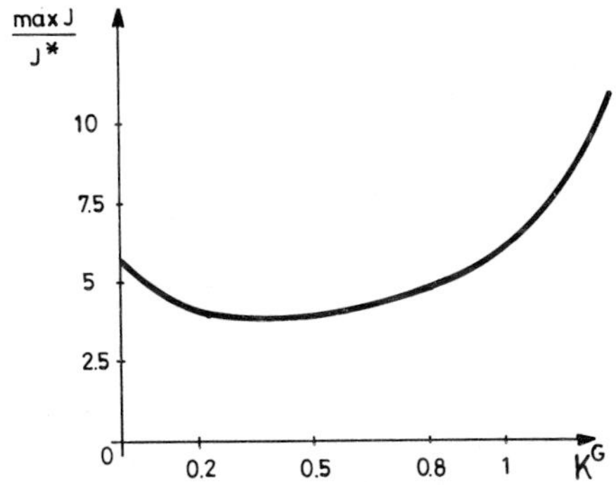

Fig. 3.69. Suboptimality estimation versus K^G (UMS-3B)

$$Q_i = \begin{bmatrix} 100. & 0 & 0 \\ 0 & 0.1 & 0 \\ 0 & 0 & 0.1 \end{bmatrix} \qquad K_{1,2} = \begin{bmatrix} 7965.4 & 87.416 & 5.957 \\ 87.416 & 2.256 & 0.098 \\ 5.957 & 0.098 & 0.0076 \end{bmatrix}$$

$$r_i = 5., \quad \pi_i = 2.5 \quad i\varepsilon I \qquad -r_i^{-1}b^{iT}K_i = -(89.249,\ 1.480,\ 0.115)^T$$

$$i = 1,2$$

$$K_3 = \begin{bmatrix} 7994.8 & 102.19 & 7.969 \\ 102.19 & 2.914 & 0.146 \\ 7.969 & 0.146 & 0.0127 \end{bmatrix} \qquad K_{4,5,6} = \begin{bmatrix} 1353.5 & 0.0388 & 0.229 \\ 0.038 & 0.0029 & 0.0005 \\ 0.229 & 0.0005 & 0.0043 \end{bmatrix}$$

$$-r_3^{-1}b^{3T}K_3 = -(89.414,\ 1.640,\ 0.142)^T, \qquad -r_i b^{iT}K_i = -(36.790,\ 0.092,\ 0.694)^T$$

$$i = 4,5,6$$

Table 3.16. Local optimal regulator for manipulation system UMS-3B

tion of tracking with local output control and without the programmed centralized control. It is obvious that the system performance with such control may be poor, especially in the case of fast motion. However, if we omit the nominal control we neglect not only the coupling among the system in the nominal dynamics but also the dynamics of the actuators, in which case, the trajectory tracking is not guaranteed even if the decoupled subsystems are considered. In order to demonstrate the effect of coupling alone, the case when the nominal programmed control is synthesized on the subsystem level has been considered. This means that we have introduced programmed control $\bar{u}^{oi}(t)$, $\forall t\varepsilon T$, $\forall i\varepsilon I$, satisfying

$$S^i : \dot{x}^{oi}(t) = A^i x^{oi}(t) + b^i \bar{u}^{oi}(t), \tag{3.3.31}$$

i.e. programmed control which ensures the realization of the nominal trajectory under ideal nominal conditions but only if the system is viewed as the set of decoupled subsystems. The local programmed control $\bar{u}^{oi}(t)$ and the local control (2.6.13) ensure tracking of the nominal trajectory $x^{oi}(t)$ for the decoupled subsystem S^i. When this control is applied to the exact model of the considered manipulation system the tracking is unsatisfactory. Fig. 3.68 shows the tracking of the nominal by such local control; the influence of coupling is significant and it is not compensated by the chosen control law. In this case, it is necessary to introduce global control. Fig. 3.68 illustrates the tracking of the nominal by three global control types. It is obvious that global control in this case may significantly improve the tracking of the nominal trajectory. It is also obvious that the control synthesis based on the approximate model of the system is not satisfactory if nominal centralized control is not introduced.

Thus, we have shown the significance of the programmed, centralized control. However, if it is not convenient to introduce nominal centralized control it might be necessary to introduce some global feedbacks, as shown above.

As the second example of decentralized control synthesis using minimization of the local quadratic criteria and synthesis of the global control in the form (2.6.34), we shall consider the stabilization of the manipulation system UMS-3B (Fig. 3.3) around the nominal trajectories from Figs. 3.48 - 3.49. We now assume that the subsystem outputs are $y^i = x^i$, i.e., the whole state vector of the system is measurable. We assume that the regions of practical stability are as follows. $\bar{X}^I = \{\Delta x(0): ||\Delta y^i(0)|| \leq 0.05,\ i=1,2,3,\ ||\Delta y^i(0)|| \leq 0.1,\ i=4,5,6,\ |\Delta p^i(0)| < 9.80 \text{ bar},\ \forall i \varepsilon I\}$, $\bar{X}^t(t) = \{\Delta x(t): ||\Delta y^i(t)|| \leq 0.05 \exp(-1.5t),$ $i=1,2,3,\ ||\Delta y^i(t)|| \leq 0.1 \exp(-1.5t),\ i=4,5,6,\ |\Delta p^i(t)| < 9.80 \text{ bar},$ $\forall i \varepsilon I,\ \forall t \varepsilon T\}$, and $\bar{X}^F(\tau) = \{\Delta x(\tau): ||\Delta y^i(\tau)|| < 0.05 \exp(-1.5\tau),\ i=1,2,3,$ $||\Delta y^i(\tau)|| \leq 0.1 \exp(-1.5\tau),\ i=4,5,6,\ |\Delta p^i(\tau)| < 9.80 \text{ bar},\ \forall i \varepsilon I\}$, for $\tau = \tau_s = 0.65$ sec.

First, as in the previous example, we have analyzed the system performance with the local linear regulators (2.6.13) for various choices of weighting matrices Q_i and r_i and various prescribed degrees of exponential stability Π. On the basis of this analysis the gains of the local regulators have been calculated and presented in Table 3.16. Analysis shows that the system is practically stable with the chosen local regulators. However, as in the previous case, the introduction of global control may reduce the suboptimality of the decentralized control with respect to criterion (2.6.11). If global control in the form (2.6.34) is introduced, where the case (Ga) (the feedback with respect to generalized forces) is considered, the dependance of the estimation of the maximum criterion cost on the global gain K^G is as in Fig. 3.69. On the basis of this function the "optimal" global gains can be chosen. Obviously, the form of this dependance on K^G is similar

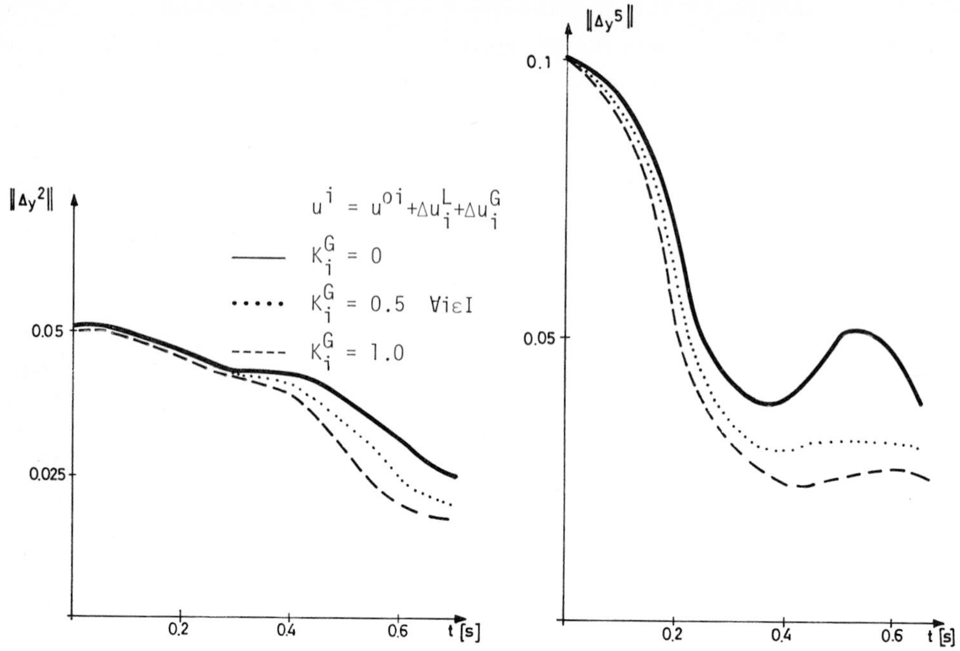

Fig. 3.70. Tracking of nominal with local and global control with various K^G (UMS-3B)

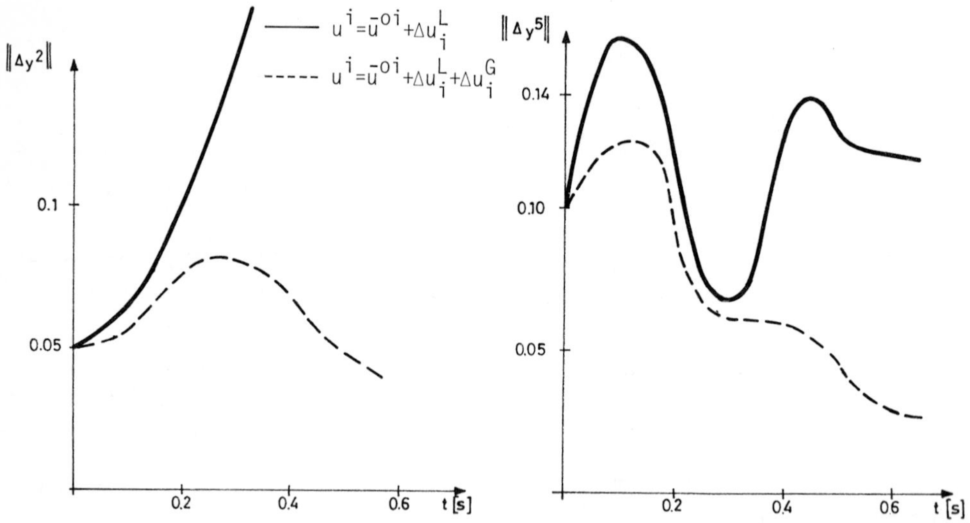

Fig. 3.71. Tracking of nominal with local and global control (UMS-3B)

to same dependance for the manipulator UMS-1 presented in Fig. 3.65 but the influence of the global control on reduction of the criterion cost in the case of the manipulator UMS-3B is much weaker, as a consequence of a weaker effect of the coupling on the manipulation performance.

These results have been checked by digital simulation of the manipulator UMS-3B dynamics. The tracking of $x^O(t)$, $u^O(t)$ by the chosen decentralized optimal regulator (Table 3.16) has been simulated when the global control is not introduced and when global control (2.6.34) (Ga) is introduced with various global gains (Fig. 3.70). The initial conditions are chosen as in Fig. 3.60 in order to compare tracking by various local and global control laws.

As in the case of the manipulator UMS-1, we have simulated the tracking of the nominal $x^O(t)$ by the decentralized regulator and by local, programmed control $\bar{u}^{Oi}(t)$ satisfying (3.3.31). The simulation results (Fig. 3.71) show the effect of the coupling on the stability of the manipulation system UMS-3B. The tracking by this local control (local regulator and local nominal control) and by global control (2.6.34) has been simulated and the results are also presented in Fig. 3.71. These simulation illustrate the significance of the global and centralized, nominal control with the semi-anthropomorphic manipulator UMS-3B.

3.3.4. Synthesis of decentralized observer

In para. 3.3.2. we have considered the case when local control law is synthesized in the form of output feedback. In para. 3.3.3. the local quadratic regulators have been considered, under the assumption that all subsystems state coordinates are measurable. In both cases global control has also been introduced. In this paragraph we shall discuss the case when only subsystems outputs are measurable (as in para. 3.3.2) but local regulators are to be synthesized, as in para. 3.3.3. Since local optimal regulators (2.6.13) demand feedbacks with respect to all subsystems state coordinates, an observer for the estimation of subsystem states should be introduced. As presented in para. 2.7.2. and appendix 2.C, the observer for the estimation of subsystem states should be introduced as a set of local observers. For each subsystem S^i (3.3.6) a local observer (2.C.1) should be introduced. Here, we shall consider the synthesis of local observers and regulators for a particular manipulation system-UMS-3B.

Let us assume that the stabilization of manipulator UMS-3B around nominal trajectories, presented in para. 3.3.1. (Fig. 3.48), should be achieved. In para. 3.3.2, the stabilization of manipulator UMS-3B has been presented, by local output feedback and by global control as a force feedback (Fig. 3.60). It has been shown that by local output feedbacks (when $\Pi_i = 5$ has been adopted) and by global control (3.3.24), the imposed conditions of practical stability are satisfied. Now, let us consider the case when the region of allowed initial conditions $\bar{X}^I$ (around nominal initial condition $x^O(0)$) is wider than in paras. 3.3.2 and 3.3.3. The regions of practical stability are imposed as follows: $\bar{X}_1^I = \{\Delta x^1(0): ||\Delta y^1(0)|| \leq 0.07, |\Delta p^1| < 19.6 \text{ bar}\}$, $\bar{X}_i^I = \{\Delta x^i(0): ||\Delta y^i(0)|| \leq 0.1, |\Delta p^i| \leq 19.6 \text{ bar}\}$, $i=2,3,4,5,6$, $\bar{X}_1^t = \{\Delta x^1(t): ||\Delta y^1(t)|| \leq 0.1 \exp(-2.5t), |\Delta p^1(t)| \leq 19.6 \exp(-2.5t)\}$, $\bar{X}_i^t(t) = \{\Delta x^i(t): ||\Delta y^i(t)|| \leq 0.20 \exp(-2.5t), |\Delta p^i(t)| \leq 19.6 \exp(-2.5t)\}$, $\forall t\varepsilon T$, $i=2,3,4,5,6$ and $\bar{X}_1^F(\tau) = \{\Delta x^1(\tau): ||\Delta y^1(\tau)|| \leq 0.1 \exp(-2.5\tau), |\Delta p^1(\tau)| \leq 19.6 \exp(-2.5\tau)\}$, $\bar{X}_i^F(\tau) = \{\Delta x^i(\tau): ||\Delta y^i(\tau)|| \leq 0.20 \cdot \exp(-2.5\tau), |\Delta p^i(\tau)| \leq 19.6 \exp(-2.5\tau)\}$, $i = 2,3,4,5,6$ for $\tau = \tau_s = 0.65$s (see app. 2.C).

It can be shown that these conditions of practical stability cannot be satisfied by local output feedback and global control (3.3.24) as in para. 3.3.2. We again assume that subsystems outputs are positions, i.e. $y^i = \ell^i$ or $y^i = q^i$. If we want to achieve stabilization of the system but not to increase significantly local gains, we should introduce local regulators. Thus, we should also introduce local observers to reconstruct subsystem state coordinates which are not measurable. Local observers are chosen for all subsystems in a form of Luenberger minimal observers (2.C.1) where the corresponding matrices are chosen as presented in Table 3.17. Matrices D^i are chosen so as to ensure exponential stability of local observers (when considered without the influence of coupling among subsystems). Since the order of subsystem output is $k_i = 1$, the order of the observer is $\ell_i = 2$, $\forall i\varepsilon I$. The degrees of the exponential stability of local observers are also presented in Table 3.17. Then, local quadratic regulators (2.6.13) are synthesized, where the prescribed degree of exponential stability and weighting matrices were given in Table 3.16. The solution of corresponding Riccati equations and local regulators gains were presented in Table 3.16, too. Reconstructed subsystem state $\Delta\hat{x}^i$ (2.C.3) is used to realize local regulator (2.C.4). The ensemble system (system + observer + regulator) has 2n = 12 subsystems and is of the order $N_1 = 30$.

$$D^1 = D^2 = \begin{bmatrix} -11.32 & 47.50 \\ -1133.30 & -66.18 \end{bmatrix}, \quad \tilde{C}^1 = \tilde{C}^2 = \begin{bmatrix} 0 & 1. & 0 \\ 0 & 0 & 1. \end{bmatrix}, \quad h^1 = h^2 (0., 74.9)^T$$

$$D^3 = \begin{bmatrix} -9.77 & 41.04 \\ -850. & -49.63 \end{bmatrix}, \quad \tilde{C}^3 = \begin{bmatrix} 0 & 1 & 0 \\ 0 & 0 & 1 \end{bmatrix}, \quad h^3 = (0., 56.1)^T$$

$$D^i = \begin{bmatrix} -166 & 800. \\ -300. & -80. \end{bmatrix}, \quad \tilde{C}^i = \begin{bmatrix} 0 & 1 & 0 \\ 0 & 0 & 1 \end{bmatrix}, \quad h^i = (0., 80.)^T$$

$$\alpha_1 = \alpha_2 = 32.7, \quad \alpha_3 = 29.7, \quad \alpha_i = 123. \qquad i = 4,5,6$$

$$H_1 = H_2 = \begin{bmatrix} 17.13 & 41.46 \\ 41.46 & 71.79 \end{bmatrix}, \quad H_3 = \begin{bmatrix} 18.24 & 42.77 \\ 42.77 & 88.08 \end{bmatrix}$$

$$H_i = \begin{bmatrix} 47.89 & -6.86 \\ -6.86 & 127.72 \end{bmatrix} \qquad i = 4,5,6$$

Table 3.17. Local observers: matrices, degrees of exponential stability and matrices of Liapunov functions (UMS-3B)

According to the procedure for stability analysis of the system with decentralized regulator and observer, presented in app. 2.C, Liapunov functions for local subsystems are chosen in the form (2.C.8) and Liapunov functions for local observers are chosen as (2.C.12), where matrices H_i are given in Table 3.17. Using these Liapunov functions and constraints upon state coordinates imposed by input constraints and by regions $\bar{X}_i^I$, the regions $\tilde{X}_i$ and $\tilde{\bar{X}}_i$ are estimated according to (2.C.10) and (2.C.14), i.e. the numbers ξ_{ij} and $\bar{\xi}_{ij}$ in (2.C.15) and (2.C.16) are determined using the mathematical model of the system set on a digital computer. Namely, a program has been developed which analyzes the coupling among subsystems and local observers. Thus, the matrix $\bar{G}$ (2.C.20) is in Table 3.18. It can be seen that condition (2.C.19) is not fulfilled in this case. This means that we cannot conclude whether the global system is stable or not. However, simulation of system shows that the system with local regulators and local observers is not practically stable. If we do not want to increase local gains and stability degrees of local observers, we have to introduce global control and global information in the observers.

We have considered the case when global control is introduced in a form of force feedback. Global control is in the form (2.C.25), where $\Delta\bar{P}_i^*$ is measured load acting upon the i-th d.o.f., and K_i^G is calculated in the following way:

$$K_i^G = \bar{K}_i^G \frac{\lambda_M(K_i)}{\lambda_m(K_i)} \frac{||f^i||}{||b^i||}, \qquad \forall i \varepsilon I \tag{3.3.32}$$

where $\bar{K}_i^G \varepsilon R^1$ is constant global gain which will be chosen later and $||b^i||$, $||f^i||$ are the norms of the corresponding vectors. It should be noticed that in (2.C.25) global gain K_i^G has been assumed to be a function of estimated subsystem state $\Delta\hat{x}^i$. In order to simplify global control we choose global gains as constants (3.3.32). We have also introduced this information on coupling in local observers (2.C.26). Now, numbers ξ_{ij}^*, $\tilde{\xi}_{ij}$, $\bar{\xi}_{ij}^*$ and $\tilde{\xi}_{ij}^*$ are determined using the above mentioned program for the analysis of coupling among subsystems. Since $\Delta P_i^* = \Delta P_i$ and K_i^G are constants, the nubmers $\tilde{\xi}_{ij}$, $\bar{\xi}_{ij}^*$ and $\tilde{\xi}_{ij}^*$ are all equal to zero. For $\bar{K}_i^G = 1$, $\forall i \varepsilon I$, the matrix $\bar{G}$ gets the form which is presented in Table 3.19. Obviously, in this case condition (2.C.21) is fulfilled and system is asymptotically stable. According to (2.C.23) the number η is calculated for various values of gains $\bar{K}_i^G$. The results of this analysis are presented in Fig. 3.72. It can be shown that the practical stability of the system can be quaranteed for $\bar{K}_i^G \geq 0.75$, $\forall i \varepsilon I$. This means that by the introduction of local observers and global control and global information on the basis of force measuring, we can ensure the practical stability of the system in a wider region of initial condition $\bar{X}^I$ than in the case when local output feedbacks and global control are applied. However, it should be noticed that by the introduction of force feedbacks we have retreated from the imposed information pattern, i.e. we have assumed that we can measure not only subsystems outputs y^i but coupling ΔP_i as well.

The second possibility to introduce global information is to calculate on-line the coupling among subsystems. We have considered the case

$\bar{G}$=

0.64E+02	0.11E-01	0.15E-01	0.00E+00	0.12E-03	0.22E+00	0.39E+00					
0.63E-02	0.35E+02	0.29E+01	0.13E-02	0.62E+00	0.14E-02		0.90E+00				
0.16E+00	0.12E+01	0.38E+02	0.00E+00	0.12E+00	0.28E-02			0.16E+00			
0.00E+00	0.13E-01	0.00E+00	-0.41E+02	0.12E-02	0.89E-03				0.01E-03		
0.14E-01	0.67E+00	0.36E+00	0.21E-02	-0.40E+02	0.80E-03					0.01E-03	
0.50E+00	0.24E-01	0.23E-01	0.32E-03	0.86E-03	-0.40E+02						0.01E-03
0.14E+04	0.55E-01	0.78E-01	0.34E-03	0.63E-03	0.11E+01	-0.39E+02					
0.31E-01	0.16E+04	0.14E+02	0.65E-02	0.30E+01	0.67E-02		-0.39E+02				
0.19E+01	0.14E+02	0.12E+04	0.62E+03	0.14E+01	0.33E-01			-0.30E+02			
0.00E+00	0.24E+01	0.00E+00	0.14E+04	0.21E+00	0.16E+00				-0.12E+03		
0.25E+01	0.12E+03	0.64E+02	0.37E+00	0.17E+04	0.14E+00					-0.12E+03	
0.89E+02	0.44E+01	0.40E+01	0.57E-01	0.15E+00	0.16E+04						-0.12E+03

Gv_o = (0.83E+02, 0.75E+02, 0.45E+02, -0.40E+02, -0.39E+02, -0.39E+02, -0.60E+03, -0.30E+03, -0.28E+03, -0.14E+04, -0.10E+04, -0.12E+04)^T

Table 3.18. Stability analysis (Π = 2.5, no global control) (UMS-3B)

$$\bar{G} = \left[\begin{array}{cccccc|cccccc}
-0.28E{+}02 & & & & & & 0.39E{+}00 & & & & & \\
 & -0.74E{+}02 & & 0 & & & & 0.90E{+}00 & & & 0 & \\
 & & -0.11E{+}02 & & & & & & 0.16E{+}00 & & & \\
 & & & -0.48E{+}02 & & & & & & 0.01E{-}03 & & \\
 & 0 & & & -0.48E{+}02 & & & 0 & & & 0.01E{-}03 & \\
 & & & & & -0.48E{+}02 & & & & & & 0.01E{-}03 \\
\hline
 & & & & & & -0.39E{+}02 & & & & & \\
 & & & & & & & -0.39E{+}02 & & & 0 & \\
 & & 0 & & & & & & -0.30E{+}02 & & & \\
 & & & & & & & & & -0.12E{+}03 & & \\
 & & & & & & & 0 & & & -0.12E{+}03 & \\
 & & & & & & & & & & & -0.12E{+}03
\end{array}\right]$$

$$\bar{G}v_o = -(0.75E{+}01,\ 0.34E{+}01,\ 0.29E{+}01,\ 0.47E{+}02,\ 0.47E{+}02,\ 0.47E{+}02,\ 0.20E{+}04,\ 0.13E{+}04,\ 0.14E{+}04,\ 0.28E{+}04,\ 0.28E{+}04,\ 0.28E{+}04)^T$$

Table 3.19. Stability analysis ($\Pi = 2.5$, global control-force feedback, $\bar{K}_i^G = 1.,\ \forall i \varepsilon I$)

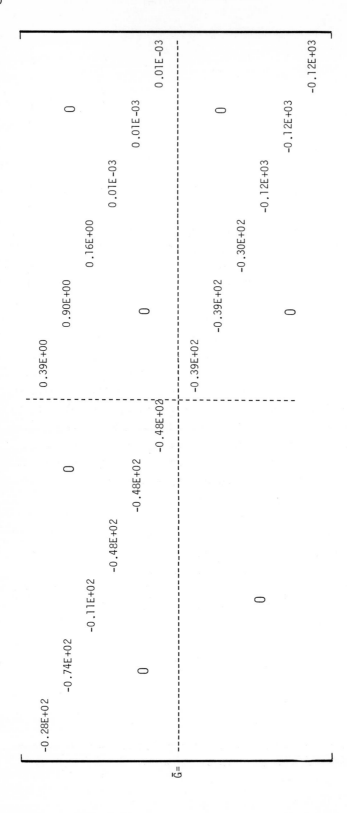

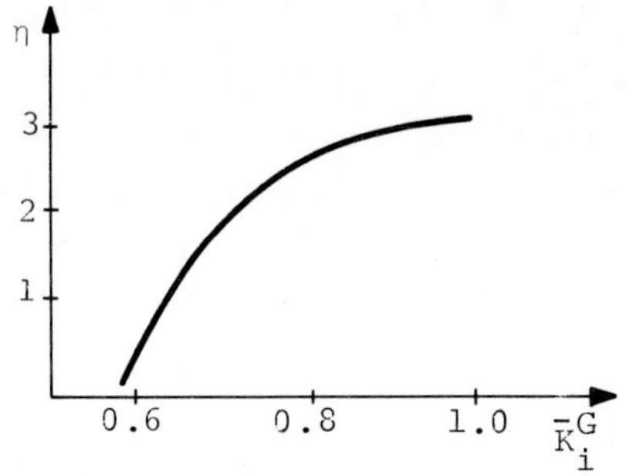

Fig. 3.72. Dependence of stability degree on global gain

when only gravitational moments acting upon each d.o.f. are calculated on-line. However, gravitational moment in the first d.o.f. of this manipulator is obviously equal to zero. On the other hand, from Table 3.18 we can see that with the chosen local gains the stability of the first subsystem cannot be proved (by the applied method for stability analysis). Also, it can be shown that the introduction of gravity forces in (2.C.25) cannot ensure the stabilization of the second and third subsystem. Thus, we should increase stability degrees Π of local subsystems S^i, i = 1,2,3. Table 3.20 shows re-chosen local regulators for these three subsystems. The local regulators for the remaining three subsystems and local observers for all six subsystems are the same as in Tables 3.16 and 3.17. Then, global control is introduced (2.C.25), where global gain is given by (3.3.32) and ΔP_i^* represent on-line calculated gravitational moments. These on-line calculated gravitational moments are also introduced in local observers.

$$Q_i = \begin{bmatrix} 0.1 & 0 & 0 \\ 0 & 100. & 0 \\ 0 & 0 & 0.1 \end{bmatrix} \qquad \begin{aligned} r_i &= 5 \\ \Pi_i &= 10. \end{aligned}$$

$$K_{1,2} = \begin{bmatrix} 25848. & 44.386 & 21.465 \\ 44.386 & 0.837 & 0.079 \\ 21.465 & 0.079 & 0.046 \end{bmatrix}, \quad K_3 = \begin{bmatrix} 26023. & 54.646 & 28.755 \\ 54.646 & 1.143 & 0.118 \\ 28.755 & 0.118 & 0.076 \end{bmatrix}$$

$$-r_1^{-1}b^{iT}K_i = -(321.55, \ 1.187, \ 0.694)^T \qquad i = 1,2$$

$$-r_3^{-1}b^{3T}K_3 = -(322.63, \ 1.326, \ 0.862)^T$$

Table 3.20. Local regulators for the first three subsystems and increased prescribed stability degrees (UMS-3B)

Since gravitational moments are functions only of positions of d.o.f., and since the subsystems outputs are positions of hydraulic pistons, (i.e. ΔP_i^* are functions of Δy^i only), the numbers $\tilde{\xi}_{ij}$ and $\tilde{\xi}_{ij}^*$ in (2.C.29) are all equal to zero. The matrix $\bar{G}$ in this case gets the form presented in Table 3.21 for $\bar{K}_i^G = 1$, $\forall i \varepsilon I$. It can be seen that the system is asymptotically stable, and since $\eta = 2.7$, practical stability of the system also can be guaranteed.

The tracking of nominal trajectories $x^o(t)$, $u^o(t)$ with various forms of global control and information, with local regulators and local observers, has been simulated on a digital computer. The results of simulation are shown in Fig. 3.73. The initial conditions are chosen as follows: $\Delta\ell^1(0) = 0.05$, $\Delta\ell^2(0) = \Delta\ell^3(0) = 0.1$ [m], $\Delta q^4(0) = \Delta q^5(0) = \Delta q^6(0) = 0.1$ [rad], $\Delta\dot{\ell}^i(0) = 0$, $i = 1,2,3$, $\Delta\dot{q}^i(0) = 0$, $i = 4,5,6$, $\Delta p^1(0) = -8.4$ $\Delta p^2(0) = -15.8$, $\Delta p^3(0) = 14.6$, $\Delta p^4(0) = \Delta p^6(0) = 0$, $\Delta p^5(0) = 5.3$ [bar] and initial guess error for local observers: $||w^1|| = -0.3$, $||w^5|| = 1.5$ and $||w^i(0)|| = 0$, for $i = 2,3,4,6$, where $||w^i||$ denotes the norm of observer error $||w^i|| = [(\Delta\dot{\ell}^i - \Delta\hat{\dot{\ell}}^i)^2 + 0.1(\Delta p^i - \Delta\hat{p}^i)^2]^{1/2}$.

The tracking with position feedback has been simulated showing that the system is not practically stable*). When global control (3.3.24) is introduced, the simulation of tracking also shows that system is practically unstable. If we introduce local observers and local regulators presented in Table 3.17 the conditions of practical system stability are not satisfied again. At last, if global control and information are introduced (in the form of force feedback) the tracking of nominal trajectory for these particular initial conditions shows that practical stability is satisfied. Also in the case when local regulators are chosen as in Table 3.20, and global information is introduced as on-line calculated gravitational moments, tracking of nominal trajectory is satisfactory. In Fig. 3.74 the norms of observer errors developing during the above mentioned simulations are presented. It can be easily seen that in the case when force information are introduced observers are completely decoupled and an error in one local observer does not influence other observers. In the other two cases the influence of subsystems upon observers is obvious.

From the results presented above we can see how in this particular

---

*) Local output gains k_i are chosen as in para. 3.3.2.

$\bar{G}=$

-0.96E+02	0.47E-01	0.60E-02	0.16E-03	0.28E-03	0.13E+00	0.28E+01					
0.01E-03	-0.83E+02	0.11E+00	0.00E+00	0.00E+00	0.00E+00		0.31E+01				
0.01E-03	0.01E-03	-0.61E+02	0.00E+00	0.00E+00	0.00E+00			0.01E-03			
0.00E+00	0.01E-03	0.00E+00	-0.41E+02	0.00E+00	0.00E+00				0.01E-03		
0.11E-01	0.47E-00	0.21E+00	0.00E-00	-0.40E+02	0.00E+00					0.01E-03	
0.40E+00	0.01E-01	0.01E-03	0.00E+00	0.00E+00	-0.40E+02						0.01E-03
0.42E+03	0.44E-01	0.38E-01	0.26E-03	0.48E-03	0.95E+00	-0.39E+02					
0.15E+02	0.12E+04	0.33E+02	0.35E+02	0.38E+02	0.35E+02		-0.39E+02				
0.01E-03	0.01E-03	0.82E+03	0.00E+00	0.00E+00	0.00E+00			-0.30E+02			
0.00E+00	0.48E+01	0.00E+00	0.14E+04	0.00E+00	0.00E+00				-0.12E+03		
0.20E+01	0.15E+03	0.19E+03	0.18E+03	0.17E+04	0.19E+03					-0.12E+03	
0.70E+02	0.67E+02	0.11E+03	0.12E+03	0.12E+03	0.16E+04						-0.12E+03

$\bar{G}v_o$ = -(0.55E+01, 0.47E+01, 0.32E+01, 0.41E+02, 0.37E+02, 0.39E+02, 0.14E+04, 0.12E+03, 0.49E+03, 0.13E+04, 0.35E+03, 0.95E+03)^T

Table 3.21. Stability analysis (Π_i = 10 for i = 1, 2, 3, Π_i = 2.5, for i = 4, 5, 6, on-line calculation of gravity forces), (UMS-3B)

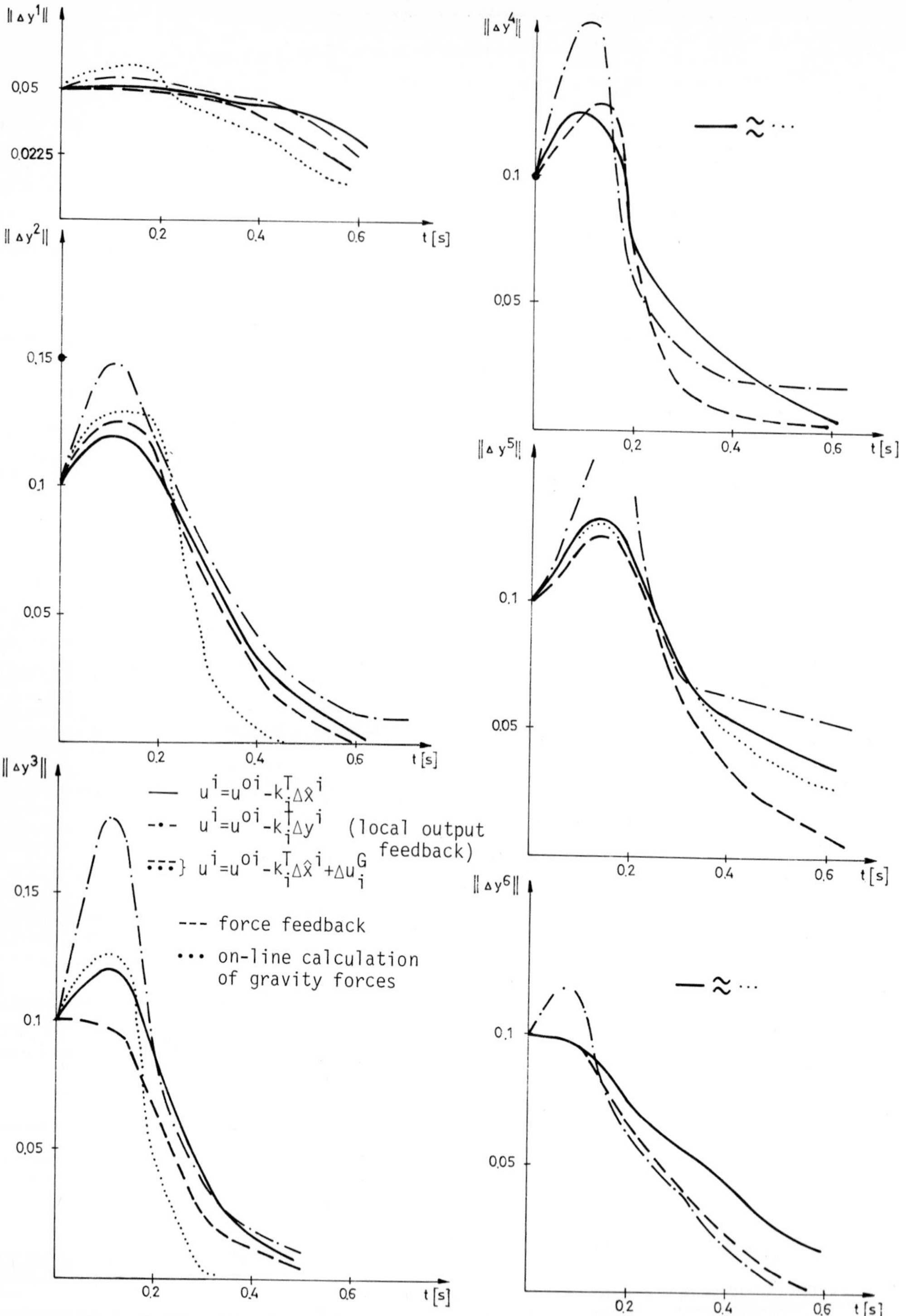

Fig. 3.73. Simulation of tracking of nominal trajectory with local regulators and global control (UMS-3B)

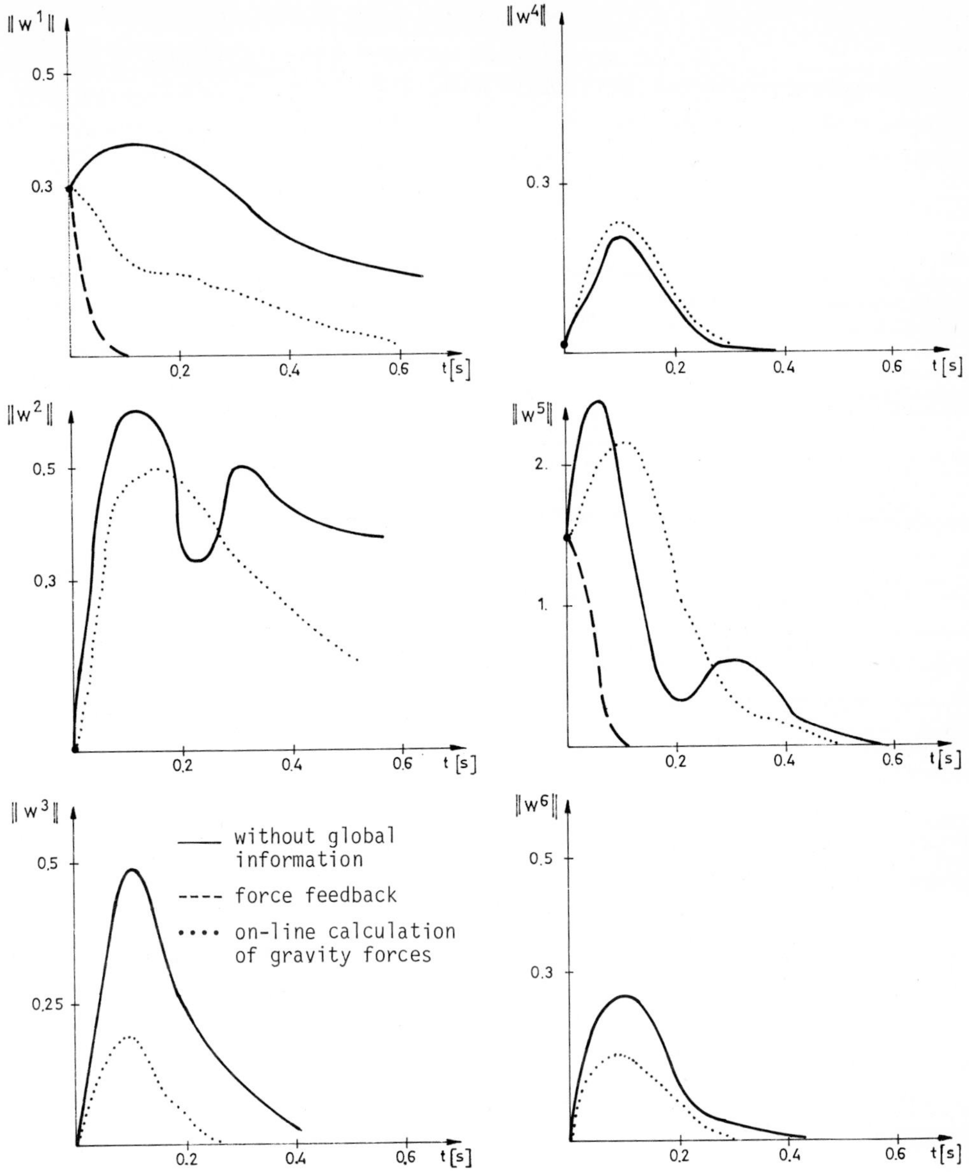

Fig. 3.74. Simulation of tracking of nominal trajectory: observer errors (UMS-3B)

case of semi-anthropomorphic manipulator UMS-3B, the introduction of observer can improve the tracking of desired nominal trajectories. However, a more precise analysis should involve suboptimality of the system with and without observers. From this analysis one should conclude whether it is justifiable to introduce observer keeping in mind that it demands additional computational efforts in on-line implementation of control. The implementation of decentralized observer and regulator by parallel processing is discussed in para. 3.6. It should be stressed once again that with the decentralized observer separation property does not hold, except in the case when force feedbacks are introduced. This fact might be of significance if capability of the system to withstand the white noise is under consideration.

3.4. Control of Orientation of Gripper Using Load-Feedback

In the preceding paragraph various aspects of decentralized and global control were considered when applied to the task of transferring the object along the prescribed trajectory with a prescribed orientation of the object during the motion. In this paragraph we shall consider ways of using force feedback in the synthesis of non-classical control algorithms for solving the same manipulation task.

We should examine ways of applying global control using feedback with respect to some physical variable which need not be direct coupling among subsystems. These methods will be demonstrated with a task met in practice, the so-called "drink test", the task of transferring a vessel containing liquid. This task also belongs to the general class of transfer and orientation of a working object.

If the trajectory is prescribed along which the object should be transferred and if, as in para. 3.3.1, we decouple the manipulator into two functional subsystems $\tilde{S}_o$ and $\tilde{S}_x$ (in the stage of nominal dynamics) then the subsystem $\tilde{S}_o$ (the minimal configuration of the manipulator) should transfer the gripper and the vessel along the prescribed trajectory.

The other subsystem $\tilde{S}_x$ (the gripper) should, in general, satisfy certain dynamic conditions concerning the inertial forces of the object in order to ensure this transfer with the desired orientation of the object in the working space. In order words, it is necessary to satisfy the following constraint on the inertial forces of the object $\vec{F}_o$ and the moment due to inertial forces $\vec{M}_o$

$$f(\vec{F}_o, \vec{M}_o) = 0, \tag{3.4.1}$$

where $\vec{F}_o \varepsilon R^3$, $\vec{M}_o \varepsilon R^3$.

Assuming that no slipping occurs between the object and terminal device, the inertial forces and the moments of inertial forces are determined from the complete manipulator dynamics:

$$\begin{aligned} \vec{F}_o &= \phi_1(q_o^o, \dot{q}_o^o, \ddot{q}_o^o, q_x, \dot{q}_x, \ddot{q}_x) \\ \vec{M}_o &= \phi_2(q_o^o, \dot{q}_o^o, \ddot{q}_o^o, q_x, \dot{q}_x, \ddot{q}_x), \end{aligned} \tag{3.4.2}$$

where: $\phi_i : R^3 \times R^3 \times R^3 \times R^3 \times R^3 \times R^3 \rightarrow R^3$ for $i = 1,2$.

By simultaneously solving the relations (3.4.1) and (3.4.2) we obtain the compensating dynamics of the subsystem $\tilde{S}_x$ (gripper). In the control sense it is possible to satisfy the condition (3.4.1) in a number of ways [23, 24].

One way of calculating the compensating dynamics (the motion of the gripper) is by integrating the differential equations with respect to the compensating coordinates (gripper angles) $q_x(t)$:

$$f(\phi_1(q_o^o, \dot{q}_o^o, \ddot{q}_o^o, q_x, \dot{q}_x, \ddot{q}_x), \ \phi_2(q_o^o, \dot{q}_o^o, \ddot{q}_o^o, q_x, \dot{q}_x, \ddot{q}_x)) = 0 \tag{3.4.3}$$

From the point of view of control implementation, this procedure is inconvenient because real-time calculation of the compensating dynamics (gripper angles) according to the expressions (3.4.3) is a very strong requirement, which may reduce the practical usefulness of the procedure. We therefore propose a procedure for synthesizing compensating movements (gripper movements) based on force feedback. It is obvious that the heart of the dynamic manipulation task is the determination of the inertial force $\vec{F}_o$ and moment $\vec{M}_o$. Thus, the efficiency should be increased by direct measurement and control of the dynamic parameters mentioned. To this end, force transducers are introduced at points of contact between the gripper and object; the forces at these points are thus measured (Fig. 3.75).

If $\vec{R}_\ell$ denotes the force of the ℓ-th transducer on the gripper, and $\vec{r}_{o\ell}$ is the vector from the center of gravity of the object to the ℓ-th transducer, the conditions of kinetostatic equilibrium of the forces and moments acting on the object yield the following expressions (Fig. 3.75)

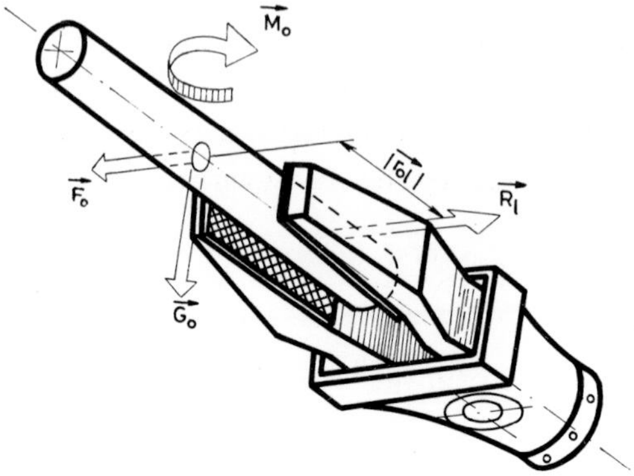

Fig. 3.75. Gripper with force transducers

$$\vec{F}_o + \vec{G}_o + \sum_\ell (-\vec{R}_\ell) = 0,$$
$$\vec{M}_o + \sum_\ell \vec{r}_{o\ell} \times (-\vec{R}_\ell) = 0, \qquad (3.4.4)$$

where $\vec{G}_o \varepsilon R^3$ is the weight of the object and the summation with respect to ℓ covers all transducers on the gripper. $\vec{F}_o$ and $\vec{M}_o$ may be obtained by measuring $\vec{R}_\ell$.

Let $\vec{M}_o^*$ and $\vec{F}_o^*$ be known values which satisfy the dynamic conditions (3.4.1) of the manipulation task under consideration. The deviations $\Delta\vec{F}_o$ and $\Delta\vec{M}_o$ from ideal values may be obtained by measuring $\vec{R}_\ell$. These deviations $\Delta\vec{F}_o$, $\Delta\vec{M}_o$ may be compensated directly by means of the feedback from the transducers to the inputs to the subsystems S^i. That is, from the measurements of $\Delta\vec{F}_o$ and $\Delta\vec{M}_o$ it is possible to determine the deviations of angular velocity and acceleration of the object. Assuming no slipping occurs between the object and the gripper, the angular velocity $\vec{\omega}_o \varepsilon R^3$ and acceleration $\vec{\varepsilon}_o \varepsilon R^3$ of the object will be

$$\vec{\omega}_6 = \vec{\omega}_o; \quad \vec{\varepsilon}_6 = \vec{\varepsilon}_o, \qquad (3.4.5)$$

where $\vec{\omega}_6 \varepsilon R^3$ is the angular velocity of the gripper and $\vec{\varepsilon}_6 \varepsilon R^3$ is the angular acceleration of the gripper.

Applying the basic relations from rigid-body kinematics (1.1.12), the relation between the linear acceleration of the object and that of the gripper may be written as

$$\vec{w}_o = \vec{w}_6 - \vec{\varepsilon}_6 \times \vec{r}^{\,*}_{6o} - \vec{\omega}_6 \times (\vec{\omega}_6 \times \vec{r}^{\,*}_{6o}), \tag{3.4.6}$$

where $\vec{r}^{\,*}_{6o} \varepsilon R^3$ is the vector from the center of gravity of the object to that of the gripper. Using Eqs. (3.4.5) and (3.4.6), it is possible, from the deviations $\Delta\vec{F}_o$ and $\Delta\vec{M}_o$, to determine the deviations of angular velocities and angular accelerations of the object from their ideal values. Thus,

$$\Delta\vec{F}_o = -m_o \Delta\vec{\varepsilon}_o \times \vec{r}_{o6} - m_o \Delta\vec{\omega}_o \times (\Delta\vec{\omega}_o \times \vec{r}_{o6}) = -m_o \Delta\vec{w}_o, \tag{3.4.7}$$

$$\Delta\vec{M}_o = -T_o \cdot \Delta\vec{\varepsilon}_o - \Delta\vec{\omega}_o \times (T_o \cdot \Delta\vec{\omega}_o), \tag{3.4.8}$$

where m_o, T_o are the mass and the inertia matrix of the object and $\vec{r}_{o6} \varepsilon R^3$ is the vector from the gripper joint to the center of gravity of the working object.

From Eqs. (3.4.7) and (3.4.8) it is possible to determine the increments of angular velocities and accelerations so as to satisfy the dynamic conditions (3.4.1) of the manipulation task.

According to (1.1.30), it is possible to calculate the driving torques of the minimal configuration $P_i^o(t)$, $i = 1,2,3$, i.e., of the manipulator with a fixed gripper. Additional driving torques due to the motion of the fourth member and of the working object, as well as to the action of external forces on the object, are calculated using the equations

$$\vec{e}_i [\vec{r}_{6i} \times (\vec{F}_6 + \vec{G}_6) + \vec{M}_6 + \vec{r}_{oi} \times (\vec{F}_o + \vec{G}_o) + \vec{M}_o] = -\tilde{P}_i^o, \ \forall i \varepsilon I \tag{3.4.9}$$

where $\vec{r}_{oi} \varepsilon R^3$ is the vector from the i-th joint to the center of gravity of the object. The nominal moments of the manipulator, when the gripper also moves, are $P_i^o(t) + \tilde{P}_i^o$, for $i = 1,2,3$ and $\tilde{P}_i^o$, for $i=4,5,6$.

If the deviation of the dynamics from the nominal conditions occurs at the level of perturbations, feedback with respect to $\Delta\vec{M}_o$ and $\Delta\vec{F}_o$ should be introduced.

On the basis of (3.4.9), complementary moments in the mechanism joints should be produced to compensate for the deviation from the desired performance. Thus,

$$\vec{e}_i\{\Delta\vec{M}_o - T_6\Delta\vec{\varepsilon}_o - \Delta\vec{\omega}_o \times (T_6\Delta\vec{\omega}_o) - m_6\vec{r}_{6i} \times [\Delta\vec{w}_o + \Delta\vec{\varepsilon}_o \times \vec{r}^{\,*}_{6o} + \\ +\Delta\vec{\omega}_o \times (\Delta\vec{\omega}_o \times \vec{r}^{\,*}_{6o})] + m_o\vec{r}_{oi} \times (-\Delta\vec{w}_o)\} = -\Delta\tilde{P}_i, \ \forall i\varepsilon I \qquad (3.4.10)$$

The force feedback which realizes the gripper dynamics may be established on the basis of (3.4.7), (3.4.8) and (3.4.10). The determination of $\Delta\vec{F}_o$ and $\Delta\vec{M}_o$ from $\vec{R}_\ell$, measured according to the dynamic conditions (3.4.4), and from the solutions of Eqs. (3.4.7), (3.4.8) with respect to $\Delta\vec{\varepsilon}_o$ and $\Delta\vec{\omega}_o$, depends on the particular manipulation task. Thus, using (3.4.7), (3.4.8) (3.4.10), it is possible to establish a force feedback from the transducers to the gripper.

However, it is possible to significantly simplify the control if comparatively slow motion is considered. The kinematic condition concerning the orientation of the vessel (object) must then be satisfied, i.e, the longitudinal axis of the vessel (object) should be kept vertical during the transfer. In this case, the moment of inertial forces $\vec{M}_o$ of the working object should be zero. Certain constraints should be imposed on the values of the inertial forces of the working object so as to prevent jerky motion. In this special case, the dynamic condition (3.4.1) may be written as $\vec{M}_o = 0$, $|\vec{F}_o| < |F_m|$, where $|F_m|$ is the maximum norm of the inertial force. The second condition may be satisfied by ensuring moderate motion of the first three manipulator members. The rest of the dynamic task reduces to the condition requiring the total angular velocity and acceleration of the fourth member (gripper) to be zero. The condition to be satisfied by the reaction forces is

$$\sum_{(\ell)} \vec{r}_{o\ell} \times (-\vec{R}^{o}_{\ell}) = 0, \qquad (3.4.11)$$

where $\vec{R}^o_\ell \varepsilon R^3$ is the ideal value of the reaction force required for the appropriate transfer of the working object. By measuring the reaction forces during the manipulation itself, it is possible to check whether the condition (3.4.11) is satisfied. If this is not the case, it means that certain rotations of the working object have occurred, which in turn means that additional accelerations have occurred. This requires the introduction of correcting driving torques.

We now propose an approximate procedure for an efficient and relatively simple realization of the driving torques necessary to compensate the undesired rotations. Taking into account the nature of the problem, it may be assumed that the changes in angles (i.e. positions) and velocities due to gripper motion are small and negligible compared to the changes in accelerations. We may thus neglect the velocity-related terms on the right-hand side of the equation (3.4.8). In this case, we can directly determine the angular accelerations necessary for compensation purposes as

$$\Delta\vec{\varepsilon}_6 = \Delta\vec{\varepsilon}_o = T_o^{-1}\Delta\vec{M}_o = T_o^{-1}\sum_{(\ell)}(\vec{r}_{o\ell} \times (-\vec{R}_\ell)) \qquad (3.4.12)$$

Thus, the desired motion is produced by superimposing the moments due to the motion of the minimal configuration of the manipulator and the moment due to the complementary motion of the gripper together with the working object.

On the basis of (3.4.10) and the assumption that the rotational velocity of the gripper and object may be neglected, the complementary moments, which must be added to all driving torques in the mechanism joints so as to produce the compensating motion of the gripper and satisfy the dynamic conditions of the manipulation task, have been obtained from the following expressions

$$\vec{e}_i[\Delta\vec{M}_o - T_6T_o^{-1}\Delta\vec{M}_o + m_6\vec{r}_{6i} \times (T_o^{-1}\Delta\vec{M}_o\times\vec{r}^{\,*}_{6o}) +$$
$$+(\vec{r}_{6i}m_6+\vec{r}_{oi}m_o)\times(T_o^{-1}\Delta\vec{M}_o\times\vec{r}_{o6})] = \Delta\tilde{P}_i \quad \forall i\varepsilon I \qquad (3.4.13)$$

It is obvious that these compensating moments prevent rotation of the fourth member and object, i.e., they reduce $\Delta\vec{\varepsilon}_6$ and $\Delta\vec{\varepsilon}_o$ to zero. The last equation enables one to establish the feedback (with respect to $\vec{R}_\ell$) to the inputs to the subsystems S^i necessary for the transfer of the working object, keeping $\vec{M}_o$ equal to zero. The control may thus be simplified: for the first three degrees, we introduce the programmed control $u_o^o(t)$, $\forall t\varepsilon T$, the local control in the form of linear feedback with respect to output Δy^i and the global control in the form of feedback with respect to the moments $\Delta\tilde{P}_i$, which are obtained by measuring the forces $\vec{R}_\ell$. Thus,

$$u_o^i(t) = u_o^{oi}(t) + k_i^T\Delta y^i + K_i^G\Delta\tilde{P}_i(t, \Delta y, \vec{R}_\ell), \text{ for } i=1,2,3 \qquad (3.4.14)$$

For the three degrees of freedom of the gripper we introduce only the global feedback with respect to the deviations of the moment for which the rotational moment of the object is equal to zero (the so-called compensating motion of the gripper).

$$u_x^i(t) = K_i^G \Delta\tilde{P}_i(t, \Delta y, \vec{R}_\ell), \quad \text{for} \quad i = 4,5,6 \qquad (3.4.15)$$

Thus, one can easily determine the control necessary for performing the task of transferring the working object. It is clear, however, that certain approximations have been made and we have introduced (3.4.12) in this global control synthesis. On the other hand, since $\Delta\tilde{P}_i$ in (3.4.13) is a function of, for example, T_6, $\vec{r}_{6i}$ etc., which are functions of the state coordinates, the global feedback in (3.4.14) and (3.4.15) is also a function of time (nominal trajectory) and of the coordinates of the state vector. To simplify the global feedback, it may be assumed that the effect of the deviation of the state vector from the nominal trajectory $x_o^o(t)$ is insignificant and that $\Delta\tilde{P}_i$ is regarded only as a function of time and the instantaneous forces on the transducers, i.e., $\Delta\tilde{P}_i = \Delta\tilde{P}_i(t, \vec{R}_\ell)$. Thus, the control is further simplified and reduced to the feedback with respect to the forces $\vec{R}_\ell$ with time-varying gains. As shown in para. 3.3.2, the influence of gripper dynamics on the degrees of freedom of the minimal configuration of the manipulator (subsystems S^i, $i = 1,2,3$) is very small, so it is not necessary to introduce feedback, with respect to the forces $\vec{R}_\ell$, to the inputs to actuators of the minimal configuration. That is, the programmed control $u_o^{oi}(t)$ and local control suffice to ensure practical stability for the minimal configuration under the conditions of deviation of the gripper from the desired orientation. Hence, we have taken $K_i^G = 0$ for $i = 1,2,3$.

To verify the usefulness of such control synthesis for the process of transferring the liquid, digital simulation of the manipulator dynamics during the transfer of the working object was performed with the control introduced in the form (3.4.13) and (3.4.15) [24]. We simulated the dynamics of the manipulator UMS-1 with six degrees of freedom, whose parameters are given in Table 3.1. (Fig. 3.1.). The mass of the object was $m_o = 0.5$ kg. All degrees of freedom were powered by appropriate actuators: the first three by hydraulic actuators*) and the three degrees of freedom of the gripper by D.C. electro-motors, whose

*) Here, it is assumed that the models of hydraulic actuators are given in the form (3.1.3), where $n_i=3$, with the appropriate connections between q^i, $\dot{q}^i$ and ℓ^i, $\dot{\ell}^i$.

models were of order $n_i = 3$. The order of the system was obviously $\sum_{i=1}^{6} n_i = 18$. We considered the transfer of the working object from the initial position given by $q^o(0) = (0, 1.4, 1.6, 0.15, 0, 0)^T$ [rad], $\dot{q}^o(0) = 0$ to the tip position, the coordinates of which, in the absolute coordinate system, were given by $\vec{S}_B = (0.35, 0.35, 0.35)^T$[m]. The manipulator tip was to be transferred from the initial position (point A) to the point B along a straight line within time $\tau = 2$ sec. The gripper with the vessel containing liquid was to ensure the perpendicularity of the vertical axis of the vessel during the transfer. The manipulation task was divided to two subsystems namely, $\tilde{S}_o$ (minimal configuration) and $\tilde{S}_x$ (gripper). The nominal trajectories of the angles of the minimal configuration $\tilde{S}_o$ which perform the transfer of manipulator tip along a straight line were calculated according to the procedure given in para. 3.2.1. These angle trajectories are shown in Fig. 3.76. The initial position of the manipulator was defined so that the longitudinal axis of the gripper was vertical. What was required in the synthesis of nominal trajectories was that the angles of the gripper remain unchanged with respect to the initial position $q_x^{oi}(t) = q_x^{oi}(0)$ and $\dot{q}_x^{oi}(t) = 0$, $\forall t \varepsilon T$ $i = 4,5,6$, while the maintenance of strict perpendicularity of the vertical axis of the object was not required. With the nominal trajectories $x^o(t)$ defined in this way, we calculated the nominal driving torques of the minimal configuration $P_o^o(t)$, shown in Fig. 3.77. Consequently, in this case, the desired manipulator motion was not produced by the nominal dynamics; the desired orientation was produced by the force feedback.

To produce the nominal dynamics of minimal configuration, local feedbacks for the subsystems S^i, $i = 1,2,3$ were introduced. Only positional feedbacks with respect to the piston strokes of the hydraulic cylinders were introduced. So, the control for the first three degrees of freedom (hydraulic servosystems) was of the form

$$u^i = u^{oi}(t) + k_i(\ell^i - \ell^{oi}(t)), \quad \forall t \varepsilon T, \quad i = 1,2,3, \qquad (3.4.16)$$

where $\ell^i \varepsilon R^1$ are the piston strokes of the hydraulic cylinders, $\ell^{oi}(t)$, $\forall t \varepsilon T$ is the nominal trajectory of the piston corresponding to the nominal trajectory of the angle $q_o^{oi}(t)$, produced by the i-th cylinder. The gains $k_1 = k_2 = 30.00$ $[\frac{mA}{m}]$, $k_3 = 60.00$ $[\frac{mA}{m}]$ were chosen. The motion of the gripper ensuring the perpendicularity of the object during the transfer was produced by means of force feedback (3.4.15). $\Delta\tilde{P}_i$ for $i = 4,5,6$ was determined by measuring the forces between the gripper and working object; thus, the desired motion of the gripper was produced. As stated, force feedback for the minimal configuration was not used in the simulation. We only used force feedback to the inputs to the gripper actuators. The control (3.4.16) was assumed to be sufficient to realize the nominal trajectories $x_o^o(t)$, $\forall t \varepsilon T$ of the minimal configuration with the desired accuracy (depending on the regions X^I, X^t, X^F) even when the angles of the gripper deviate from the constant values $q_x^o(0)$ as a result of force feedback (3.4.15). This follows from the use of powerful actuators for the minimal configuration (hydraulic servosystems).

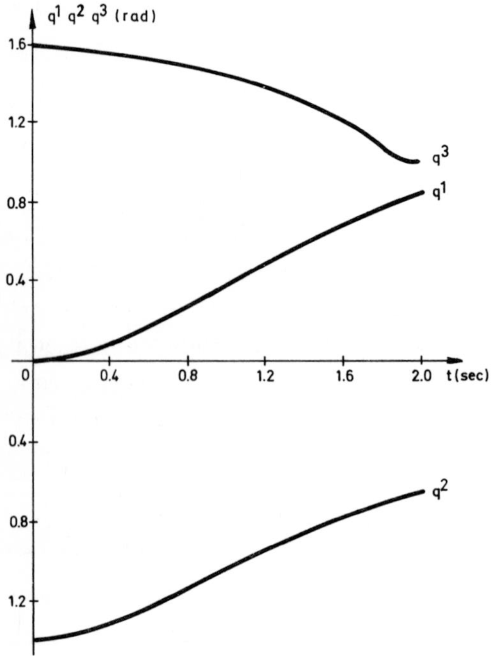

Fig. 3.76. Nominal trajectories of angles of minimal configuration of anthropomorphic manipulator during the transfer of working object along a straight line

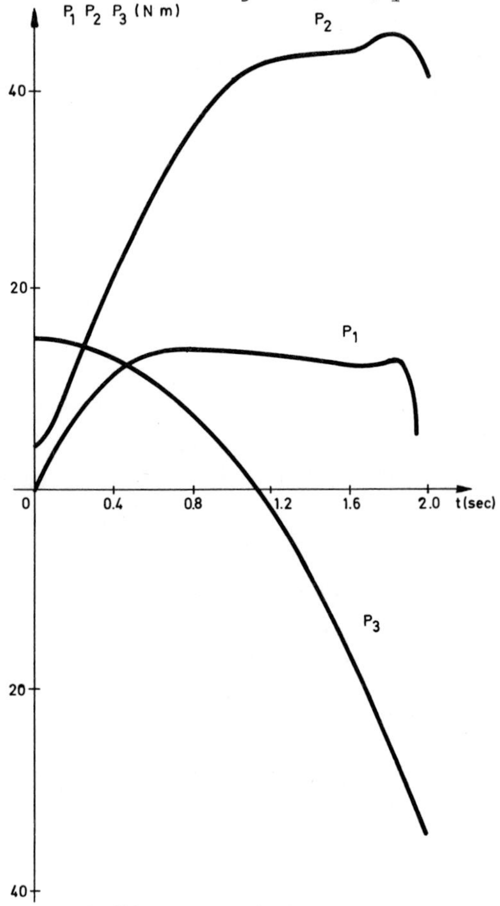

Fig. 3.77. Nominal driving torques of degrees of freedom of minimal configuration of anthropomorphic manipulator during the transfer of working object along a straight line

The analysis of gains K_i^G for i = 4,5,6 shows that they should be as high as possible so as to perform the compensating motion of the gripper in the best possible way. However, taking into account the amplitude constraints on actuator inputs and the technical capabilities, there is no sense in introducing gains which are too high. A compromise solution was therefore accepted and, in this case, the gains of the force feedback were chosen to be $K_i^G = 33$, $i = 4,5,6$. With these gains, the manipulator dynamics were simulated with initial conditions corresponding to the nominal conditions $x^o(0)$. The simulation results are presented in Figs. 3.78 - 3.82. Fig. 3.78 shows the angles of the gripper $q^i(t)$, $(i = 4,5,6)$ which develop as a result of the force feedback. Fig. 3.79 presents the compensating moments of degrees of freedom of the gripper $P_i(t)$, $(i = 4,5,6)$ which develop from the force feedback and which realize the motion of the gripper, annulling the rotational moment of the object and thus ensuring the perpendicularity of the object. These moments are realized by D.C. electro-motors of degrees of freedom of the gripper.

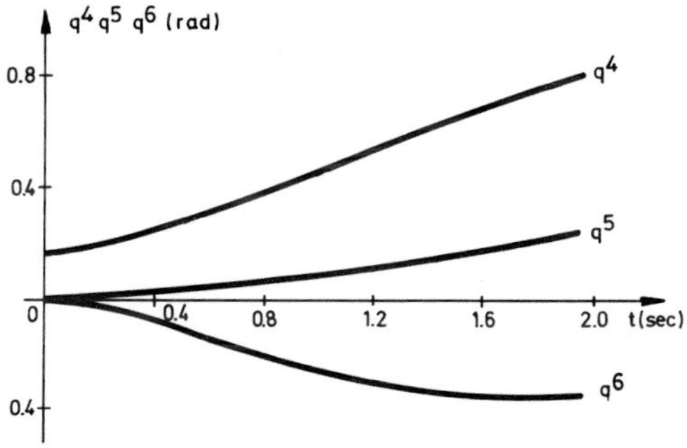

Fig. 3.78. Compensating movements of the manipulator gripper

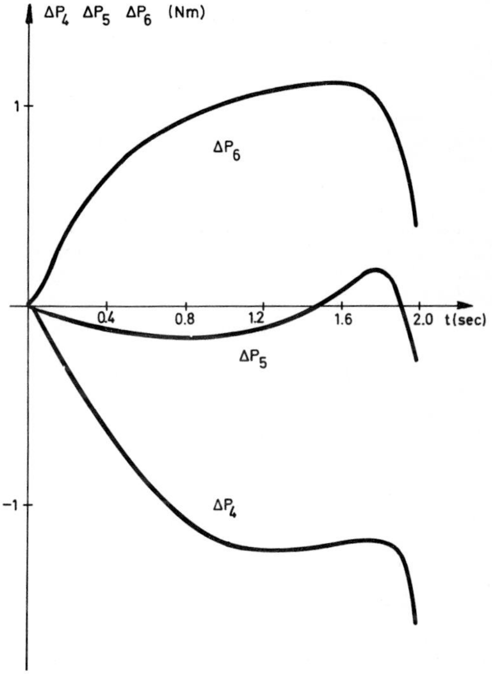

Fig. 3.79. Compensating driving torques of the gripper during the transfer of the working object

Fig. 3.80 presents the components of the force measured by transducers on the gripper, and Fig. 3.81 shows the components of rotational moment of the working object about the axes of the rectangular system, the measurement of which is a basis for on-line calculation of compensating moments of the gripper. Finally, Fig. 3.82 shows the spatial motion of the manipulator in three-quarter projection.

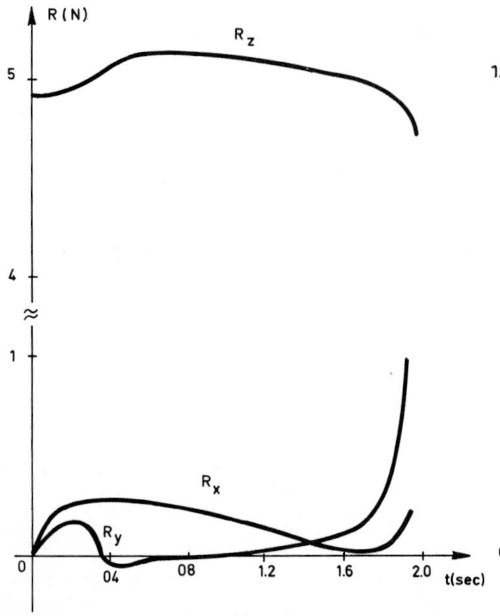

Fig. 3.80. Components of the resultant reaction force along the axes of the Carterian coordinate system during the transfer of the working object

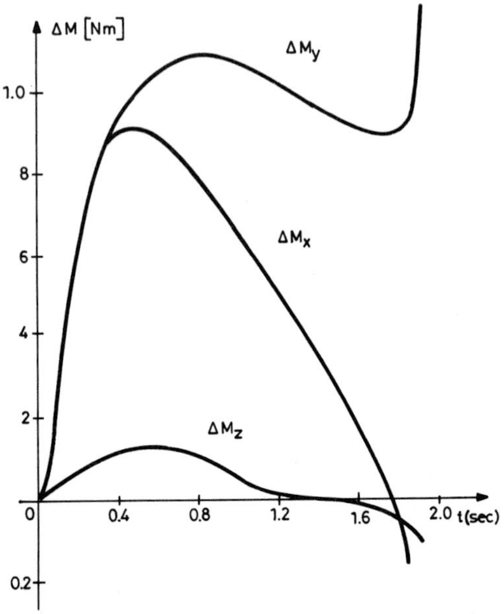

Fig. 3.81. Components of the rotational moment of the working object

The manipulator tip moves along a straight line from the point A to the point B, while the gripper maintains an approximately vertical position (with the allowable limits deviation). The results of this simulation show that the simple control structure introduced may provide a satisfactory system performance and may ensure the performance of the control task imposed. At the same time, this shows the efficiency of global control with respect to forces, since, through force measurement, a physical quantity (in this case, the rotational moment of the object) is measured which directly characterizes the control task [25]. That is, the possibility of introducing global control in the form of force feedback is a very important feature of such systems. In contrast to feedback with respect to the coupling ΔP_i, which produces the nominal dynamics through decoupled control, the feedback

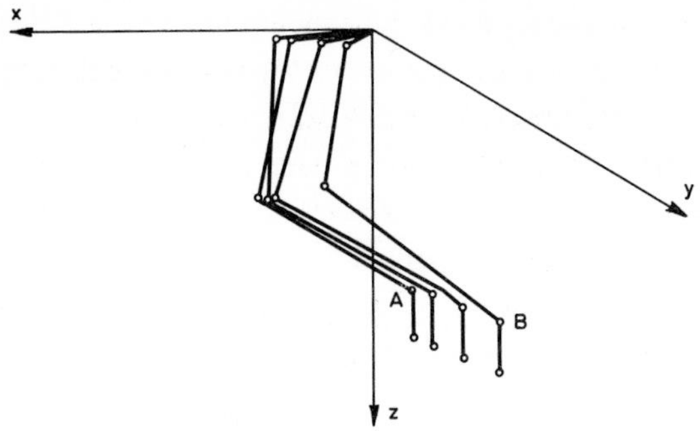

Fig. 3.82. Manipulator motion during the transfer of the working object

with respect to forces produces the desired motion and enables on-line synthesis of the nominal dynamics. So, in this case, it is no longer possible to talk of nominal dynamics in the sense of programmed control and programmed kinematics; the feedback with respect to forces $\vec{R}_\ell$ produces on-line the desired motion of the functional subsystem $\tilde{S}_x$ without calculating separately the nominal dynamics for this subsystem. The nominal dynamics and its realization are synthesized independently for the subsystem of minimal configuration $\tilde{S}_o$, since the control of the subsystem of the gripper $\tilde{S}_x$ is completely left to the force feedback*). This shows the importance of information on forces for such mechanical systems when the aim is to establish comparatively simple control algorithms.

3.5. Assembly Process

One of the most typical and delicate manipulation tasks in industrial practice is the assembly process. Synthesis of industrial robot control in the realization of various assembly processes has been widely studied. Attempts have been made to solve this problem using robots with varying degrees of intelligence depending on the process being considered and the general use of the robot [26 - 28]. Here, as we mentioned in para. 3.1, an assembly process is considered with a mani-

---

*) It is clear that since, in this case, the nominal control is not synthesized for the global system, the analysis of asymptotic system stability of the deviation model (para. 2.5.2) cannot be implemented and that the conditions of practical stability have to be investigated directly.

pulator having no artificial intelligence properties but which is fully autonomous and automated for performing tasks completely defined (in the geometrical and control sense) in advance. We shall consider in detail the assembly phase which is the most delicate task from the point of view of control. In principle, this stage may be reduced to the problem of inserting an object into a fixed hole, where, depending on the process, the working object and the hole have different forms.

In general, the assembly process requires bilateral (or multilateral) manipulation since both the hole and the object may be movable, so the insertion can be carried out with two manipulators. We shall restrict ourselves here to the process of placing the working object into the fixed hole, to which the more general case, when both working parts are movable, may be reduced. The problems of inserting the peg into the fixed hole involve a good many aspects of control synthesis, so that analyzing this task enables us to construct a control strategy for a wide range of industrial assembly processes.

The problem of controlling the manipulator during the assembly process is ideally the problem of good positioning. However, since the insertion of the object into the hole is necessarily accompanied by reaction forces, since the geometry and dimensions of parts are never known precisely enough, and since the manipulator positioning cannot be sufficiently precise, it is necessary to introduce reaction forces feedback [28]. This is in full accordance with the concept of manipulator control (para. 3.4) according to which the problem of manipulator gripper control is solved by feedback with respect to forces between the gripper and the object. In order to synthesize such control with reaction forces feedback, it is necessary to analyze the dynamics of the insertion, i.e., to simulate the manipulator dynamics during the assembly process.

Simulating such a task involves many difficulties, for it is necessary to consider different dynamic configurations of the system, since the mathematical model of the system changes according to the number of contact points between the working object and hole walls, the character of the reaction forces between the peg and hole walls (whether slipping occurs or not) and the like. However, by completely simulating the manipulator dynamics on a digital computer it is possible to recognize all aspects of this task and to synthesize the control and choose the corresponding actuators. It should be said that a static analysis of forces permits an approximate assessment of the reaction forces and force feedback but a precise control synthesis and stability analysis of the system can be obtained only by a complete digital simulation of the manipulator dynamics.

We first analyze the dynamics of the system and the forces developing during the assembly. We then describe the control synthesis and present the simulation results.

Appendix 3.A presents the problems of simulating the assembly dynamics (i.e. it completely describes the algorithm for digital simulation).

It should be noted that in the last few years considerable attention has been paid to the realization of assembly processes by industrial manipulators. At most symposia in the last few years and in journals in the field of industrial robotics, many papers have been concerned with this problem. Two different approaches to the control synthesis have been developed: by means of force feedback, where various dispositions of the force-transducers on the manipulator have been considered [27, 28], and by means of passive aids for realizing fine changes in the motion of the object necessary for inserting it into the hole. Special attention has been devoted to so-called passive compliances, systems of springs whose deflections produce the necessary orientation of the object to be inserted [29]. By such passive devices very satisfactory results have been obtained concerning the allowed clearance between the hole and the object and the relability of such terminal devices. The significant advantage of such aids is their relative simplicity. However, their main disadvantage is that they cannot be easily exchanged. It is difficult to change the task for which they are designed, since each compliance is constructed according to severely defined conditions and dimensions. Hence, they have no universal application to robots, so the main advantage of manipulators and robots over conventional tools (mechanical grippers with fixed programmes) is also lost, i.e. the reprogramability of manipulators is lost. In contrast to these passive aids, force feedback makes it possible to reprogram manipulators in many ways. Thus the manipulators with force feedback control may easily be used in different workplaces. The application of force feedback to assembly tasks is therefore more convenient in spite of the complexity of the dynamics of the processes and of the technical problems arising when force transducers are used.

3.5.1. Analysis of forces in assembly process

Let us consider the process of inserting the object into a hole with a manipulator of configuration shown in Fig. 3.1. The manipulator consists of four members and has six degrees of freedom. The seventh degree of freedom, object gripping, is unimportant for the analysis and simulation. It is also assumed that the object is brought to the hole mouth, coaxially with the hole, which necessitates correct manipulator positioning. Fig. 3.83 shows the initial position of the manipulator and object with respect to the hole under the above assumption.

It is supposed that the object is an orthogonal cylinder of d_m length and $2r_m$ bases diameters; the hole is a hollow cylinder (no upper basis) of D_m length and $2R_m$ basis diameter, so that the clearance upon insertion is $2(R_m-r_m)$, $R_m \geq r_m$. It is also assumed that the centers of gravity of the peg and gripper and the gripper joint are on the same straight line. The next assumption is that the gripper-object contact points are in one plane parallel to the peg cylinder bases (i.e., contact points are on the circle on the object cylindrical surface). Let this contact plane be at the distance d_1 from the joint D and let the distance of the upper cylinder basis from the joint D be denoted by d_2, so that $(d_2-d_1) \leq d_m$. The peg should be placed into the hole up to a depth $h_1 \leq D_m$.

We shall first analyze the forces occurring at the contact points between the object and hole walls during the assembly. The analysis is aimed at studying the effect of friction on the insertion as well as the jamming problem and ways of applying force feedback to avoid undesired effects of hole reaction forces on the insertion. As we have already pointed out, regardless of how well the manipulator has been positioned, contact between the peg and hole may occur, so that reaction forces result.

In force analysis and simulation no aspect of dynamics is neglected. It is assumed that the dynamic coefficient of friction between the peg and the hole surface is the same everywhere and that it equals the static coefficient of friction. This assumption has been introduced for simplicity, but may easily be removed. Another assumption is that both the peg and hole walls are completely rigid, so that, during the simulation, deformation is not considered as this would make the model extremely complex, which is unnecessary for the control synthesis.

Obviously, the given hole and peg geometry ensures that, upon insertion, contact occurs simultaneously at most two points but in some cases involving coaxiality contact occurs at infinitely many points if it occurs along the joint generatrix of the peg cylinder and hole cylinder. One of the two contact points has to be between the hole edge and the peg cylindrical surface, and the other on the edge of the peg cylinder basis and hole cylindrical surface. It is assumed that the peg has already entered the hole, i.e. that jamming has not occurred at the hole edge itself.

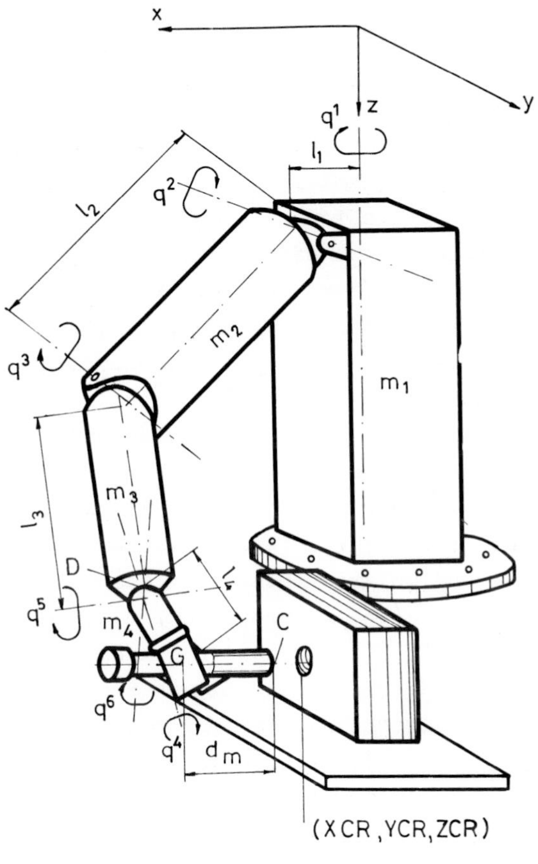

Fig. 3.83. Manipulator UMS-1 in assembly process

In force analysis during insertion four possible situations are considered (under the above assumption that the peg basis has passed through the hole basis) [28, 30].

1. No contact occurs between the hole and the object.

2. One contact point exists.
3. Two contact points exist.
4. Contact occurs along the joint generatrix.

Let us consider each of these situations, respectively.

Case 1, when there is no contact between the hole and peg, represents free manipulator movement without reaction forces from the hole, so this situation does not differ dynamically from the task of transferring the working object in free space along desired trajectory and with a desired orientation. The manipulator dynamics is described by the mathematical model of open kinematic chain dynamics.

Case 2 introduces the problem of the unknown reaction force acting on the manipulator via the object and influencing its dynamics. The problem of determining the reaction force in object-hole contact points arises. Fig. 3.84 illustrates the case of contact between the hole edge and the cylindrical surface of the peg, while Fig. 3.85 shows another possibility of single contact, namely, contact between the edge of the cylindrical base of the peg and the cylindrical surface of the hole. Both figures show the hole and peg section along the plane determined by the contact point and the symmetry axis of the hole cylinder. Reaction forces at contact points are marked in figures. Three components of the reaction force at the point K1 or K2 are to be determined in the direction of the absolute coordinate system axes from Fig. 3.83. In fact, the perpendicular component of the reaction force N_{K1} (perpendicular to the peg cylinder generatrix on which the point K1 is situated, i.e., perpendicular to the hole cylinder generatrix on which K2 is situated) is to be determined as well as the tangential component T_{K1} (or T_{K2}) in the direction of slipping (if it occurs) of the peg along the hole edge (Fig. 3.84) or hole surface (Fig. 3.85).

Two cases are possible. Case 2.a), when friction is such that slipping exists in the peg-hole contact, and 2.b), when no slipping occurs, so that the peg has no tangential linear velocity in the contact point. Assuming that the coefficient of friction μ between the peg material and hole material is known, the following condition should be investigated

$$T_{KI} < \mu N_{KI} \qquad I = 1, 2 \qquad (3.5.1)$$

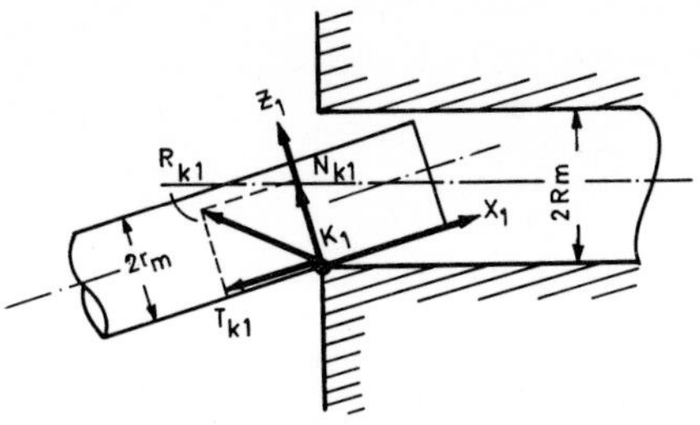

Fig. 3.84. Case of one contact in assembly process-contact between hole edge and peg surface

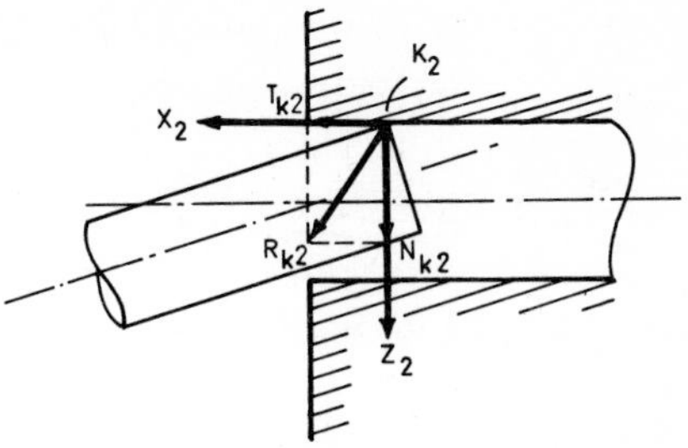

Fig. 3.85. Case of one contact in assembly process-contact between edge of peg cylinder and hole surface

In case 2.a) the peg has a linear velocity at the contact point in the direction of T_{KI} and the reaction force has to satisfy the slipping condition

$$T_{KI} = \mu N_{KI}. \tag{3.5.2}$$

In case 2.b) the reaction force at the momentary contact point has to satisfy the condition (3.5.1). In case 2.a) the contact point changes, and the tangential linear velocity of the peg has to be in the direction of T_{KI}. In case 2.b) the peg can have only rotational velocity around the point KI and the contact point does not change. This imposes certain constraints on the peg acceleration, namely, the peg as a rigid body may have three degrees of freedom of rotation accelerations around the contact point KI, while the three degrees of freedom of translation with respect to the contact point are suppressed, (linear accelerations in case 2.b) at the point KI are equal to zero). It is quite clear that in cases 2.a) and 2.b) the peg´s linear velocity at the point KI can have a tangential component only, while the velocity component in the same direction but in the sense opposite to N_{KI} must equal zero. (If this velocity component has the same direction and sense as N_{KI}, contact is only apparent and no reaction force exists).

In case 2 the peg dynamics are determined by the following forces: the inertia force of the peg itself (according to D´Alembert´s principle) $\vec{F}_o$, the peg gravitational force $\vec{G}_o = m_o\vec{g}$, forces by which the manipulator, i.e., the gripper, acts upon the peg $\vec{R}_\ell$ (ℓ= 1,2,...,L), where L denotes the number of contact points between the peg and gripper at which forces $\vec{R}_\ell$ act on the peg, and the reaction force due to contact at the point KI, $\vec{R}_{KI} \epsilon R^3$. The dynamic equilibrium equation of forces acting on the peg (according to D´Alembert´s principle) is

$$\vec{F}_o + \vec{G}_o + \vec{R}_{KI} + \sum_{\ell=1}^{L} \vec{R}_\ell = 0. \tag{3.5.3}$$

Moments acting on the peg round the center of gravity are: the moment due to inertial forces of the object itself $\vec{M}_o$, the moment due to the manipulator via the gripper, i.e., via forces at gripper-peg contact points, and moment due to reaction forces at the object-hole contact points. Equilibrium of the moments of the peg around the center of gravity (according to D´Alembert´s principle) is

$$\vec{M}_o + \vec{r}_{KIo} \times \vec{R}_{KI} + \sum_{\ell=1}^{L} \vec{r}_{\ell o} \times \vec{R}_\ell = 0, \tag{3.5.4}$$

where $\vec{r}_{KIo}$ is the position vector from the center of gravity o of the peg to the contact point KI, $\vec{r}_{\ell o}$(ℓ = 1,2,...,L) is the position vector from the peg´s center of gravity o to the peg-gripper contact point.

Equations (3.5.3) and (3.5.4) determine the peg dynamics in the case of a single point of contact between the object and hole.

The manipulator dynamics represent the dynamics of a closed kinematic chain. This creates certain problems in constructing the mathematical model of the dynamics (see appendix 3.A).

Case 3, when two points of contact exist, is shown in Fig. 3.86. Contact points K1 and K2 are the same as in Case 2. Depending on the character of the reaction forces at contact points, the following cases may occur. Case 3.a), slipping occurs at both contact points, so the peg will slip into the hole. Case 3.b), no slipping occurs at any contact point - the peg is jammed; 3.c), it may happen that slipping occurs at one contact point and not in the other - in that case the peg rotates round the point without slipping and loses contact in the other point, if this rotation is permitted by the geometry. However, it may also happen that slipping cannot occur at one contact point (e.g. K2) but can occur at the other (K1). The peg would then rotate around K2 but this is not possible on account of the geometry. In this case, the object becomes jammed [28].

Which of these three cases occurs depends upon the nature of the reaction forces at the points K1 and K2, i.e., the following conditions should be investigated

$$T_{K1} < \mu N_{K1}, \qquad T_{K2} < \mu N_{K2} \tag{3.5.5}$$

If condition (3.5.5) is satisfied, there is no slipping at any contact point and case 3.b) occurs. In case 3.a), reaction forces satisfy the following conditions.

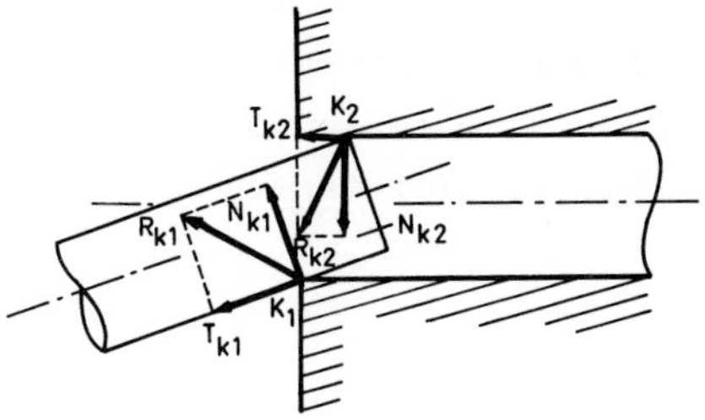

Fig. 3.86. Case of two contact points

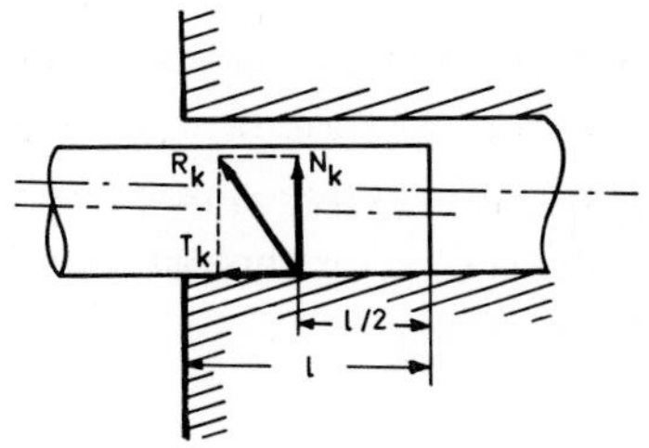

Fig. 3.87. Case of contact along joint generatrix

$$T_{K1} = \mu N_{K1}, \qquad T_{K2} = \mu N_{K2} \tag{3.5.6}$$

In case 3.c) a combination of these conditions occurs, namely,

$$(T_{K1} < \mu N_{K1} \ .\text{AND.}\ T_{K2} = \mu N_{K2})\ .\text{OR.}\ (T_{K1} = \mu N_{K1} \ .\text{AND.}\ T_{K2} < \mu N_{K2}) \tag{3.5.7}$$

However, in this case, it should be checked whether the peg rotation, which would result on account of condition (3.5.7) is permitted by the hole geometry. If it is not, jamming which is dynamically undetermined will occur, since, in that case, a peg deformation may result, which would require the introduction of new degrees of freedom (peg elasticity and the like). The dynamics of peg jamming will not be considered here since elements are assumed to be absolutely rigid bodies and manipulator control should ensure that peg jamming and deformations are avoided. It will suffice to note that no jamming occurs during simulation, i.e., that manipulator control is synthesized in such a way that jamming is avoided. Force feedback must ensure, therefore, that the peg always gets from the manipulator a moment contrary to the external forces moment. Accordingly, case 3.b) should be avoided by applying appropriate control. In case 3.b), since there is no slipping at contact points, both the rotational and the linear peg acceleration are zero, which gives 6 conditions for determining the reaction forces that are to satisfy condition (3.5.5).

In case 3.a) the peg may have only linear velocity at contact points, while linear acceleration and velocity at points KI in the same direction of but opposite sense to N_{KI} is forbidden. Together with conditions (3.5.6) these yield the conditions necessary to determine the reaction forces at points KI.

In case 3.c) conditions (3.5.7) and those requiring no peg acceleration to exist in the same direction of but opposite sense to N_{KI} yield the conditions necessary to determine the reaction forces and peg accelerations. One should check that these are consistent with the geometry.

The equations of dynamic equilibrium of forces and moments acting on the peg in this case are of the following form

$$\vec{F}_o + \vec{G}_o + \vec{R}_{K1} + \vec{R}_{K2} + \sum_{\ell=1}^{L} \vec{R}_\ell = 0 \tag{3.5.8}$$

$$\vec{M}_o + \vec{r}_{K1o} \times \vec{R}_{K1} + \vec{r}_{K2o} \times \vec{R}_{K2} + \sum_{\ell=1}^{L} \vec{r}_{\ell o} \times \vec{R}_\ell = 0 \tag{3.5.9}$$

Case 4 can be reduced to case 2. The reaction force due to the hole acts on the peg along the joint generatrix of the peg´s cylindrical surface and hole cylinder (Fig. 3.87). Since the reaction force acts uniformly at all contact points along the joint generatrix, reaction forces may be substituted by one resultant force $\vec{R}_k$ acting at the mid--point of the line of contact between the peg and hole (Fig. 3.87). This special case may now be treated like the case with one contact point: Eqs. (3.5.1) - (3.5.3) hold, so that slipping and jamming may occur here as well (i.e., the peg cannot slip along the generatrix) but in the last case rotation separating the peg from the hole wall is possible and case 2. occurs (if rotation is permitted by the geometry).

Let us now construct the mathematical model of the manipulator during the assembly process (for the manipulator UMS-1). "The state vector of the mechanical part of the system S^M", $\xi = (q^1, q^2, \ldots, q^6, \dot{q}^1, \dot{q}^2, \ldots$ $\ldots, \dot{q}^6)^T$, $\xi \varepsilon R^{12}$ is introduced. According to Eq. (1.1.30), the mathematical model of the manipulator dynamics during the assembly is given by

$$S^M: P = H'(q)\ddot{q} + h'(q, \dot{q}) + \delta_1 C_{K1} \cdot \vec{R}_{K1} + \delta_2 C_{K2} \vec{R}_{K2}, \tag{3.5.10}$$

where matrix $H':R^6\to R^{6\times6}$ and the vector $h':R^6\times R^6\to R^6$ correspond to manipulator dynamics with the object, while C_{K1}, $C_{K2}\varepsilon R^{6\times3}$.

Obviously, the system S^M is not smooth since, at instants of object--hole impact, discontinuity jumps in the state vector occur, i.e., mapping ξ and P into ξ does not define the continuous map $R^{12}\times R^6\to R^{12}$. However, during time intervals with no impact the system is smooth, so one may write the mathematical model of the manipulator mechanics in the following form [30]

$$S^M:\dot{\xi}=A(\xi)+B(\xi)P+\delta_1\cdot F_1(\xi)\vec{R}_{K1}(\xi,\ \dot{\xi})+\delta_2\cdot F_2(\xi)\vec{R}_{K2}(\xi,\ \dot{\xi}),\ \xi(0)\text{ is given} \tag{3.5.11}$$

where $A(\xi)$ is the vector function $R^{12}\to R^{12}$ of the form

$$A(\xi) = [\dot{q}^T \,\vdots\, (-H'^{-1}h')^T]^T,$$

$B(\xi)$ is the matrix function $R^{12}\to R^{12\times6}$ of the form

$$B(\xi) = [0 \,\vdots\, H'^{-1}]^T,$$

δ_1 and δ_2 are Kronecker deltas such that

$$\delta_1 = \begin{cases} 1 & \text{if there is a contact in point K1,} \\ 0 & \text{if there is no contact,} \end{cases}$$

$$\delta_2 = \begin{cases} 1 & \text{if there is a contact in point K2,} \\ 0 & \text{if there is no contact,} \end{cases}$$

$F_1(\xi)$, $F_2(\xi)$ are matrix functions $R^{12}\to R^{12\times3}$ such that

$$F_I = [0 \,\vdots\, (-H'^{-1}C_{KI})^T]^T \qquad I = 1,\ 2,$$

$\vec{R}_{K1}(\xi,\ \dot{\xi})$, $\vec{R}_{K2}(\xi,\ \dot{\xi})$ are vectors of the reaction forces (the vector function) $R^{12}\times R^{12}\to R^3$. Obviously, the system (3.5.11) is not correctly written in the canonical form, since it is unsolved with respect to $\dot{\xi}$. However, as we have already mentioned, it is difficult to determine the explicit function $\vec{R}_{KI}$ of ξ and $\dot{\xi}$ so we shall accept such a model of the system, which is quite a satisfactory form with respect to the system simulation on a digital computer (as has been shown in appendix

3.A.).

However, model (3.5.11) describes the system dynamics only in time intervals with no impact between the object and hole. Since changes in the values of δ_1 and δ_2 produce step changes on the right-hand sides of the differential equations (3.5.11), the system is not smooth. Impact moments t_i, when step changes in δ_i occur, are to be investigated separately. In moments when δ_i changes from 0 to 1 discontinuity occurs in $\xi(t)$, while in the opposite case (when δ_i changes from 1 to 0) the continuity and smoothness conditions of the system are satisfied. The solution of system (3.5.11) $\xi(t, \xi_o)$ is continuous and uniform with respect to t for $t \in T$, where T is a continuous set of time instants in which the system is observed, except for the discrete times t_i in which the object-hole impact occurs (change in δ_1 and δ_2). In these discrete moments, momentary, step changes in the state vector take place. In order to complete the model for these discrete moments as well, the following sets are introduced. $T^* = (0, \tau)$ denotes the continual time interval, τ is a real positive number, $T_N = \{t_i: i = 1,\dots,N_s\}$ is the set of discrete impact moments, where N_s is the number of impacts in the interval T^*, when either δ_1 or δ_2 change from 0 to 1. The set of moments in which the system S^M is described by the model (3.5.11) and in which the solution $\xi(t)$ of the system (3.5.11) is continuous, is then given by $T = T^* | T_N = \{t: t \in (0, \tau), t \neq t_i, i = 1,2,\dots,N_s\}$.

In order to write the model in the form of a system of differential equations for $t \in T_N$, we introduce

$$S^M: \dot{\xi} = \gamma(t_i) \cdot (\bar{\xi} - \xi(t_i)), \qquad \forall t \in T_N, \tag{3.5.12}$$

where $\xi(t_i) = \lim_{t \to t_i -} (\xi(t))$, $\xi(t)$ is the solution of system (3.5.11), $\gamma(t_i)$ is Dirack's unit function: $\int_{t_i -}^{t_i +} \gamma(t) dt = 1$; $\bar{\xi}$ is the changed vector of the system state, the step change after the impact, where new values of the coordinates are obtained from the impact conditions, depending on friction conditions and the values of δ_1 and δ_2.

The system of equations (3.5.11), (3.5.12) thus represents the complete mathematical model of the manipulator dynamics during the insertion process. The mathematical model is thus written as a set of differential equations whose right-hand sides are not continual functions and which do not satisfy Lipschitz´s conditions with respect to ξ. Nevertheless, this model satisfies for the requirements of simulation and control synthesis. In addition to not being smooth, the system has a variable structure, since depending on δ_1 and δ_2, its structure changes. However, the specificity of the model lies in the fact that the variable part of the system structure is a function of $\xi(t)$, $\dot{\xi}(t)$ and that the functional dependence has not been explicitly introduced, which creates certain problems in control synthesis.

If we suppose that all six d.o.f. are powered by D.C. motors, the models of which are given by (1.2.1), (3.1.1), (3.1.2), then, due to (3.5.11), the model of the global system S can be written in the form

$$S:\dot{x} = \hat{A}(x)+\hat{B}(x)N(u)-\delta_1\hat{F}_1(x)\vec{R}_{K1}(x)-\delta_2\hat{F}_2(x)\vec{R}_{K2}(x), \ \forall t\varepsilon T, \quad (3.5.13)$$

where $\hat{F}_I:R^{18}\to R^{18\times 3}$ for I = 1,2.

The model S defines the system manipulator + actuators $\forall t\varepsilon T$, when impact moments t_i are excluded. At impact moments for $t\varepsilon T_N$, the system is described by the models (3.5.12) and (1.2.1) which need not be combined into a unique model, since it is sufficient to investigate the system at intervals $t\varepsilon T$ and to check whether control synthesized for $t\varepsilon T$ is such as to provide for stability for $t\varepsilon T_N$ as well. During the simulation model for $t\varepsilon T_N$ is being processed as described in appendix 3.A., but it is assumed that in the state coordinates $i_{R2},\ldots i_{R6}$ no changes for $t\varepsilon T_N$ occur. Control for $t\varepsilon T$ is synthesized and it is checked whether jumps in the state vector defined by (3.5.12) cause instability of the system.

It should be stated that system S is of the same form as is S^M and that in (3.5.13) members accompanying δ_1 and δ_2 also appear, changing the structure of the system.

3.5.2. Synthesis of control of manipulator in assembly process

In this paragraph the synthesis of control of manipulator in an assembly process is presented-particularly for the "peg in hole" process, the dynamic analysis of which has been presented in para. 3.5.1. We shall consider the assembly process with 6 d.o.f. manipulator from Fig. 3.83.

The control will be synthesized in two stages according to procedure described in ch. 2, [1, 16]. In the stage of nominal dynamics the nominal trajectories $x^o(t)$, $\forall t \varepsilon T^*$ and programmed control $u^o(t)$, $\forall t \varepsilon T^*$, is synthesized in order to realize the "peg in hole process" in ideal conditions. The synthesis of nominal kinematics is realized by decoupling the manipulator into two functional subsystems, as already explained in para. 3.3. Subsystem $\tilde{S}_o$ - minimal configuration of the manipulator, should produce motion of the gripper joint (tip of the minimal configuration) along the straight line parallel to the axis of the hole cylinder. Subsystem $\tilde{S}_x$ - gripper should ensure the orientation of the axis of the object cylinder in the direction of the hole axis, during the motion. In this, it is assumed that the initial condition $x^o(0)$ is such that the axis of the object cylinder coincides with the hole axis in the initial position, while $\dot{q}^i(0)=0$, $\forall i \varepsilon I$. Calculation of the nominal kinematics $q^o(t)$, $\dot{q}^o(t)$, $\forall t \varepsilon T$ can be performed as described in para. 3.3. The distribution of the tip acceleration is given by (3.2.7), where a_{max} is defined by τ_s - the time interval during which the process should be accomplished. When the nominal trajectory of the state vector $x^o(t)$, $\forall t \varepsilon T^*$ has been calculated, the nominal driving torques $P^o(t)$, $\forall t \varepsilon T^*$ and the programmed control $u^o(t)$, $\forall t \varepsilon T^*$ from model (3.5.13) are calculated, where $\delta_1 = \delta_2 = 0$, $\forall t \varepsilon T^*$, since in nominal conditions it is assumed that no contact between the object and the hole occurs during insertion. Thus, under nominal conditions the reaction forces do not appear, and the manipulator behaves as an open kinematic chain. In this way, the synthesis of the nominal dynamics for the insertion process completely coincides with already considered synthesis of the nominal dynamics in para. 3.3. However, if the assembly process is considered when the initial conditions deviate from the nominal conditions, or if some parameters (geometry, manipulator parameters etc.) deviate from their nominal values, or if errors in realization of the nominal trajectories appear due to unsufficiently precise positioning, noise, friction, clearances, etc., then the contact between the peg and the hole may occur. Thus, the reaction forces appear and system is of variable structure. Due to this, in the stage of perturbed dynamics these specifities of the task should be taken into account. The control synthesis in the second stage partially differs from that presented in para. 3.3.

Let us assume that the control task in the stage of perturbed dynamics can be defined as stated in para. 2.2.: $\forall x(0) \varepsilon X^I$ the control should ensure that $\Delta x(t, \Delta x(0)) \varepsilon \bar{X}^t(t) = X^t - x^o(t)$, $\forall t \varepsilon T^*$ and $\Delta x(t, \Delta x(0)) \varepsilon \bar{X}^F =$

$=x^F - x^o(\tau_s)$, for $t \in T_s = \{t: t=\tau_s\}$, where $\tau=\tau_s$. Regions $\bar{X}^t(t)$ and $\bar{X}^F$ (or X^t and X^F) are bounded by the configuration of the object and hole, and by the constraints upon the velocities and accelerations of the system set in order to prevent hard impacts between the object and hole. It has been assumed that $x^o(0) \in X^I$, $x^o(t) \in X^t$, $\forall t \in T^*$ and $x^o(\tau_s) \in X^F$. We suppose that no external perturbation acts on the system. However the region X^t is so defined that when $x(t, x(0))$ deviates from $x^o(t)$ contact might occur between the object and the hole, and thus the external forces act upon the system. This type of perturbations are taken into account by model (3.5.13), while the perturbations due to the impacts between the object and the hole might be regarded as perturbations of the initial condition type. As already stated in para. 3.5.1, the system can be considered via model (3.5.13), $\forall t \in T$, while in the instants $t \in T_N$ we might regard that the perturbations of the initial conditions type are acting bringing the system state within the region X^t.

The model of deviation of the state from the nominal trajectory $x^o(t)$, $\forall t \in T^*$ may be written on the basis of (3.5.13). Since we want to apply decoupled control of manipulator, we shall consider the model of deviation via model S^M and the actuator models S^i. The model of deviation from the nominal, for the mechanical part of the system, according to (3.5.10), can be obtained in the form

$$S^M: \Delta P = H^{o'}(t, \Delta q)\Delta\ddot{q} + h^{o'}(t, \Delta q, \Delta\dot{q}) + \delta_1 C_{K1}\vec{R}_{K1} + \delta_2 C_{K2}\vec{R}_{K2}$$

$$\Delta q(0),\ \Delta\dot{q}(0) \text{ are given} \tag{3.5.14}$$

and $\Delta P \in R^6$, $\Delta P(t) = P(t) - P^o(t)$, $\Delta q = q - q^o(t)$, $\Delta q \in R^6$, while the definition of the matrix $H^{o'}: T \times R^6 \to R^{6\times 6}$ and of the vector $h^{o'}: T \times R^6 \times R^6 \to R^6$ is obvious. The models of deviations of the actuator states from the nominal S^i ($\forall i \in I$) are given by (2.6.2), with (3.1.2), since the actuators are D.C. motors described by the linear models of order $n_i = 3$.

The control at the stage of perturbed dynamics is chosen in the form of the local controls for the subsystems S^i ($\forall i \in I$) corresponding to actuators. That is, according to the control synthesis described in para. 2.6, the system is viewed as a set of subsystems S^i, and the control is synthesized in the form (2.6.6). The local control should ensure the exponential stability of the free subsystems S^i and it is chosen in the form of incomplete feedback (2.6.9). Here, we shall not repeat the synthesis of the local control and the estimation of the regions in which the free subsystems are locally stable, since this was presented in para. 3.3.

The difference between this and the task considered in para. 3.3, is in the analysis of the global system asymptotic stability, i.e. in the estimation of the coupling among subsystems. We should now examine the coupling of the subsystems in the region $T\times\tilde{X}$, where $\tilde{X}$ is the region $\tilde{X}=\tilde{X}_1\times\tilde{X}_2\times\ldots\times\tilde{X}_6$ and T is the time interval during which the insertion process is observed. The coupling equation given by (3.5.14) may be written in the form

$$\Delta P = \Delta\bar{P} + \delta_1 C_{K1}\vec{R}_{K1} + \delta_2 C_{K2}\vec{R}_{K2}, \tag{3.5.15}$$

where $\Delta\bar{P}$ is the coupling among S^i assuming that there are no reaction forces due to the hole (manipulator motion without contact with the hole).

Considering that $\vec{R}_{K1}$, $\vec{R}_{K2}\to 0$ for $||\Delta x||\to 0$, and $|\vec{R}_{K1}|$, $|\vec{R}_{K2}|\sim||\Delta u||$, and since the control is amplitude-constrained, we get

$$|\vec{R}_{KI}| \leq \bar{R}_{KI}||\Delta x||, \quad \forall\Delta x(t,\ \Delta x(0))\varepsilon\tilde{X}, \quad I = 1,2,\ \forall t\varepsilon T, \tag{3.5.16}$$

($\bar{R}_{KI}$ are the constraints on the forces in the observed region of the state space and control).

As in (2.6.23) we may now determine the constants $\bar{\xi}_{ij}<+\infty$ and the functions ξ_{ij} $(\bar{R}_{K1}, \bar{R}_{K2}, \Delta u_i^G)$, satisfying the conditions

$$\begin{aligned}(\mathrm{grad}v_i)^T[f^i\Delta P_i + b^i\bar{N}(t,\ \Delta u_i^G)] &= (\mathrm{grad}v_i)^T f^i\Delta\bar{P}_i + \\ (\mathrm{grad}v_i)^T[f^i(\delta_1 C_{K1}\vec{R}_{K1} + \delta_2 C_{K2}\vec{R}_{K2}) &+ b^i\bar{N}(t,\ \Delta u_i^G)] \leq \\ \sum_{j=1}^{6}\bar{\xi}_{ij}v_j + \sum_{j=1}^{6}\xi_{ij}(\bar{R}_{K1}, \bar{R}_{K2}, \Delta u_i^G)v_j,& \quad \forall i\varepsilon I,\ \forall(t,\ \Delta x)\varepsilon T\times\tilde{X},\end{aligned} \tag{3.5.17}$$

where $\bar{N}(t,\ \Delta u_i^G)$ is the non-linear function corresponding to amplitude saturation with respect to Δu_i^G, $\bar{\xi}_{ij}>0$, $\forall i,j\varepsilon I$, $\xi_{ij}:R^1\times R^1\times R^1\to R^1$. In (3.5.17) the global control is associated with the part of the coupling acting on the subsystems due to the reaction forces $\vec{R}_{K1}$, $\vec{R}_{K2}$. The coupling among subsystems $\Delta\bar{P}_i$ (3.5.15) is estimated by $\bar{\xi}_{ij}$ and the other part, due to $\vec{R}_{K1}$, $\vec{R}_{K2}$ and Δu_i^G, is estimated by ξ_{ij}. From (2.6.24) the region $\tilde{X}$ in which the system S is asymptotically stable, can be estimated. In this case the elements of the matrix $G \in R^{6\times 6}$ are given by

$$G_{ij} = -\Pi_i \delta_{ij} + \bar{\xi}_{ij} + \xi_{ij}(\bar{R}_{K1}, \bar{R}_{K2}, \Delta u_i^G)$$

Thus, one can estimate the region $\tilde{X}$ in which the asymptotic stability of the system can be guaranteed. Then the region $\tilde{X}(t)$ containing the solution of system S can be also estimated, using (2.6.26), (2.6.27) Finally, by investigating (2.6.28), the conditions for practical stability can be checked.

Let us suppose that the local control is sufficiently "strong" to compensate the influence of coupling $\Delta\bar{P}_i$ originating from the manipulator dynamics if there are no reaction forces, i.e. let us suppose that when local control only is applied the condition $\bar{G}v_o<0$ is satisfied, where the elements of matrix $\bar{G}$ are $\bar{G}_{ij} = -\Pi_i \delta_{ij} + \bar{\xi}_{ij}$. This means that the local control guarantees the asymptotic stability of the system S in the region $\tilde{X}$ if there are no reaction forces present.

In order to satisfy condition (2.6.24), i.e. in order to ensure the asymptotic stability of system S in the region $\tilde{X}$ when the forces $\vec{R}_{K1}$, $\vec{R}_{K2}$ are present, global control should be introduced. In this case, as in para. 3.4, global control is introduced as a feedback with respect to forces measured at contact points between the gripper and object (Fig. 3.75). We mentioned above that the force transducers are introduced at the contact points, $\vec{R}_\ell$ ($\ell=1,2,\ldots,L$). By measuring $\vec{R}_\ell$ one obtains the information on reaction forces between the object and gripper $\vec{R}_{K1}$, $\vec{R}_{K2}$, according to (3.5.8) and (3.5.9).

Global control is chosen in the form [31]

$$\Delta u_i^G = \bar{k}_i^{G'} \sum_{\ell=1}^{L} \vec{R}_\ell + \bar{k}_i^{G''} \sum_{\ell=1}^{L} \vec{r}_{\ell o} \times \vec{R}_\ell \qquad \forall i \in I, \tag{3.5.18}$$

where $\bar{k}_i^{G'}$, $\bar{k}_i^{G''} \in R^3$ are the gain vectors and $\vec{r}_{\ell o}$ has already been defined. Thus choice of global control follows from (3.5.8).

The gains $\bar{k}_i^{G'}$, $\bar{k}_i^{G''}$ should be synthesized so as to satisfy (2.6.24) for as wide a region of initial conditions $\tilde{X}$ as possible and to satisfy condition (2.6.28). Obviously, by Δu_i^G in the form (3.5.18), the influence of the forces $\vec{R}_{K1}$, $\vec{R}_{K2}$ on the system stability should be minimized. Hence, the global gains $k_i^{G'}$, $k_i^{G''}$ are chosen according to the condition [16]

$$\min_{k_i^{G'}, k_i^{G''}} \{(\mathrm{grad}v_i)^T[f^i(C_{K1}\vec{R}_{K1}+C_{K2}\vec{R}_{K2})+b^i\bar{N}(t,\Delta u_i^G)]+C_{1i}k_i^{G'2}+C_{2i}k_i^{G''2}\}, \quad (3.5.19)$$

where C_{1i}, $C_{2i}>0$ are the weighting coefficients.

Thus, we have shown how the conditions for the practical stability of the system S can be checked in the case of the assembly process. However, $\forall t\varepsilon T_N$ a step-wise impulse change in the system state occurs, so one should investigate the conditions under which, for $x(t_1)\varepsilon X^t$, $x(t, t_1, x(t_1))\varepsilon X^t$, $\forall t\varepsilon(t_1, \tau)$ and $x(t, t_1, x(t_1))\varepsilon X^F$ for $t = \tau_s$. As already mentioned this analysis reduces to analyzing the practical stability when the impulse perturbations act upon the system S, bringing the state within the region X^t.

Finally, it should be noted that the global control (3.5.18) permits an exact tracking of the nominal trajectory when $x(0) \neq x^o(0)$ but $x(0)\varepsilon X^I$. However, it has been assumed that there is no error in the response of the model relative to the response of the real system, i.e., that the programmed control realizes the insertion process if the initial condition coincides with the ideal nominal condition. Thus, $\Delta\dot{x} = 0$, for $\Delta x = 0$, i.e., the model of deviation is in equilibrium on the nominal trajectory $x^o(t)$, $\forall t\varepsilon T^*$. This is why, in this case, practical stability can be examined by analyzing asymptotic stability around the nominal trajectory. However, if the parameters of the system and the dimensions of the hole and object are not identified precisely enough (tolerances in dimensions, friction etc.) then the computed nominal programmed control might not be able to realize the insertion process even under ideal conditions. It would then be necessary to adapt system to the different task conditions. Obviously, global control in the form of force feedback permits this. If there is contact between the object and the hole because of "incorrect" nominal control, the force feedback "produces" additional driving torques. This produces deviation of the state from the nominal so the system can adapt to the new situation. Obviously, in this case, the above analysis of asymptotic stability is useless since the system is no longer in equilibrium for $\Delta x = 0$. The practical stability should therefore be investigated directly.

For the sake of investigating the efficiency of the proposed control we simulate the insertion of the peg into the hole using a 6 d.o.f. manipulator (Fig. 3.83). The parameters of this manipulator are given in Table 3.1 [31].

The mass and moments of inertia of the peg (an object in the form of an orthogonal cylinder) are included in the mass and the moments of inertia of the gripper (fourth member) (m_o = 0.3[kg], J_{xo} = 0.0003 [kgm^2], J_{yo} = 0.0010[kgm^2], J_{zo} = 0.0010[kgm^2]).

The geometrical parameters are as follows. The coordinates of the center of the circular base of the hole are XCR = -0.008[m], YCR=0.32920 [m], ZCR = 0.305[m]. The radius of the base of the hole is R_m=0.01003 [m]. The radius of the base of the object is r_m = 0.01[m]. The clearance between the hole and object is 0.00003[m], i.e., 0.3%. The distances d_1 and d_2 are d_1 = 0.15[m], d_2 = 0.25[m]. The object is a cylinder of length $d_m \geq 0.1$[m]. The axis of the hole is parallel to the x-axis of the absolute coordinate system. The coefficient of friction is $\mu = 0.1$.

As actuators we have used D.C. electro-motors, a product of Globe Industrial Division of TRW INC, of type 102A200-8 for the first three degrees of freedom of the manipulator and of type 5A540-2 for the degrees of freedom of the gripper.

Since the object is to be inserted into the hole up to a depth D_m = 0.15[m], the coordinates of the manipulator tip and the tip of the gripper are $x_D(\tau_s)=d_2-D_m+XCR$, $y_D(\tau_s)=YCR$, $z_D(\tau_s)=ZCR$, $x_G(\tau_s)=d_m-D_m+XCR$, $Y_G(\tau_s)=YCR$, $z_G(\tau_s)=ZCR$, where τ_s is the interval for the assembly process. Nominal trajectories of the manipulator angles are calculated according to the procedure described in para. 3.3.

The programmed control $u^o(t)$ and nominal trajectories $x^o(t)$ are calculated to realize the assembly when the initial conditions are so chosen that $\dot{q}^o(0) = 0$ and the position of the tip is $x_D^o(0)=d_2+XCR$, $y_D^o(0)=YCR$, $z_D^o(0)=ZCR$ and the center of the base of the object is $x_C^o(0)=XCR$, $y_C^o(0)=YCR$, $z_C^o(0)=ZCR$. The nominal trajectory has to be realized by local control (3.3.7). Thus, local gains $k_i = (50,3)^T$ for i=1,2,3 and $k_i = (30,2)^T$ for i=4,5,6 are chosen for the local stabilization of each free subsystem S^i. The decentralized control (3.3.7) together with programmed input $u^o(t)$ is implemented in the assembly process.

Several cases have been simulated. The initial manipulator position is such that the center of the base of the object has the coordinates

$x_C = -0.00745$ [m], $y_C = 0.3296$ [m] $=$ YCR $+ 0.0004$ [m], $z_C = 0.3046$ [m] $=$ ZCR $- 0.0004$ [m], so the object has already entered the hole. The manipulator tip coordinates are $x_D = 0.2408$ [m], $y_D = 0.3004$ [m], $z_D = 0.3001$ [m], so the slope between the object axis and hole axis is 0.117 [rad].

The initial conditions for the angles and velocities are $q^1 = -0.2364$ [rad] and $q^2 = 1.4353$ [rad], $q^3 = 1.6641$ [rad], $q^4 = q^5 = q^6 = 0$; $\dot{q}^i = 0$, $\forall i \varepsilon I$.

First, for the purpose of checking the accuracy of the manipulator dynamics simulation when reaction forces due to the hole are present, we simulated the case when the programmed kinematics is incorrect, which produces object jamming. The simulation results are presented in Fig. 3.88. Fig. 3.88. shows the components of the reaction forces at the contact point with respect to the absolute axes.

We then simulated the case when the programmed kinematics is correctly prescribed, but force feedback was not introduced. The simulation results are presented in Fig. 3.89.

Finally, the previous case was simulated, but with the feedback introduced with respect to the working object moment. All feedback gains are indicated in figures. The results of this simulation are given in Fig. 3.90g. The figures illustrate only the beginning of the simulation, the first 0.5 sec., since this suffices to recognize the dynamic phenomena and feedback effect.

Figs. 3.90c,d show the nominal and realized trajectories of the manipulator for the last case (when global feedback is introduced). Fig. 3.90b shows the change of slope between the hole axis and the object axis during assembly. Fig. 3.90a shows the motion of the point on the object whose x-coordinate is largest during assembly. That is, the initial conditions are so chosen that contact between the peg and the hole occurs for up to $t \geq 0.5$ sec. However, as can be seen from Figs. 3.90b,c,d, the slope between the peg axis and the hole axis is reduced during insertion and the trajectories of the angles return to the nominal trajectories. Figs. 3.90e,f show the driving torques in the d.o.f. of the manipulator, which develop during assembly.

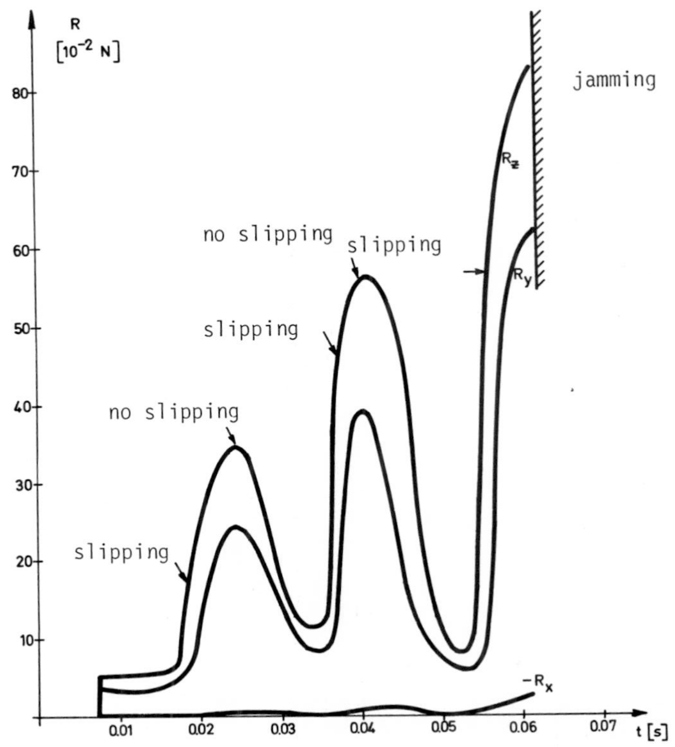

Fig. 3.88. Reaction forces at contact point for the case of force feedback and wrong kinematics ($\bar{k}_i^{G'} = 0, \bar{k}_i^{G''} = (100, 100, 100)^T$)

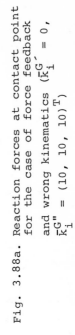

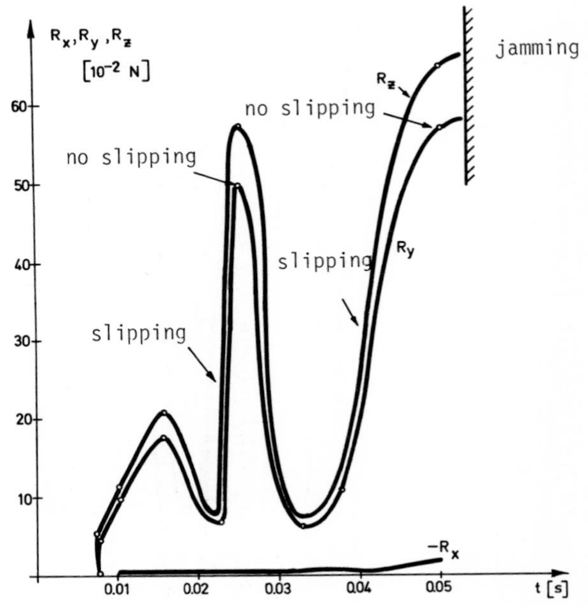

Fig. 3.88a. Reaction forces at contact point for the case of force feedback and wrong kinematics ($\bar{k}_i^{G'} = 0$, $\bar{k}_i^{G''} = (10, 10, 10)^T$)

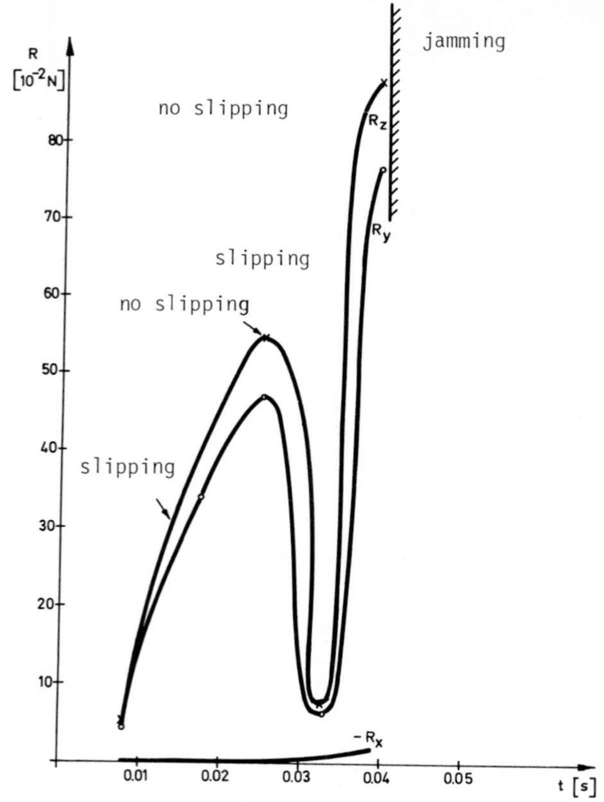

Fig. 3.88b. Reaction forces at contact point without force feedback and wrong kinematics

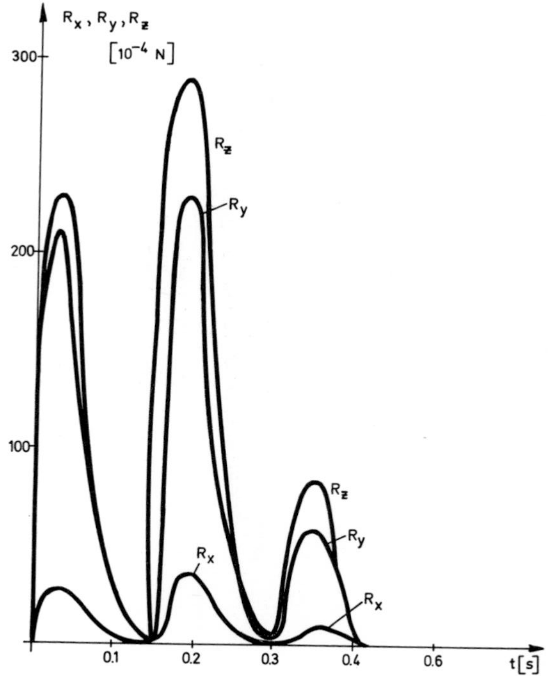

Fig. 3.89. Simulation of "peg in hole" process without force feedback and with correct kinematics: reaction forces at contact point

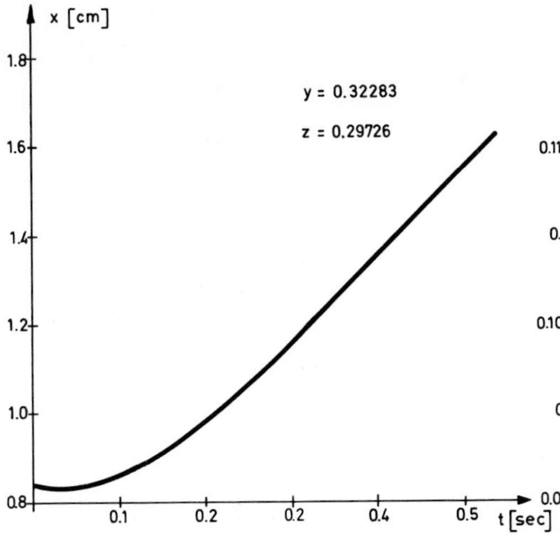

Fig. 3.90a. Simulation of "peg in hole" process with exact nominal kinematics and force feedback: motion of point of contact between object and hole

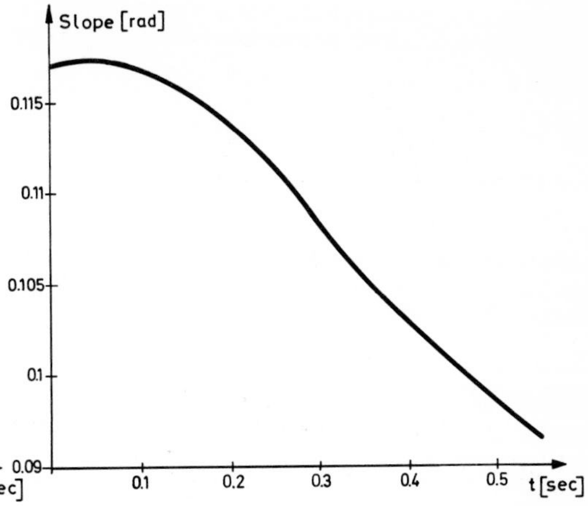

Fig. 3.90b. Change of slope of object axis with respect to hole axis

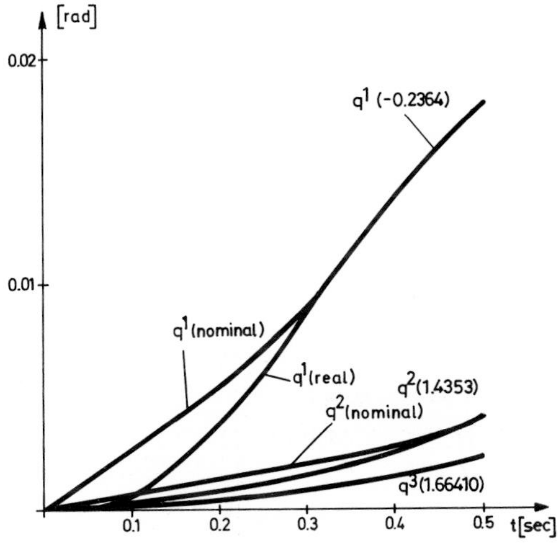

Fig. 3.90c. Kinematics of minimal configuration in "peg in hole" process: nominal and realized trajectories

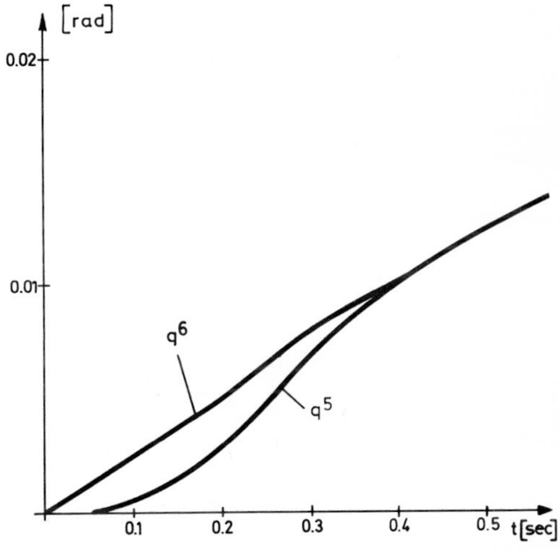

Fig. 3.90d. Kinematics of gripper in "peg in hole" process

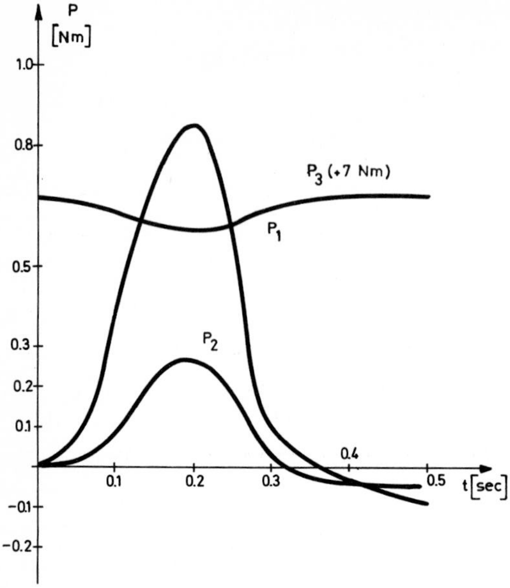

Fig. 3.90e. Driving torques of minimal configuration

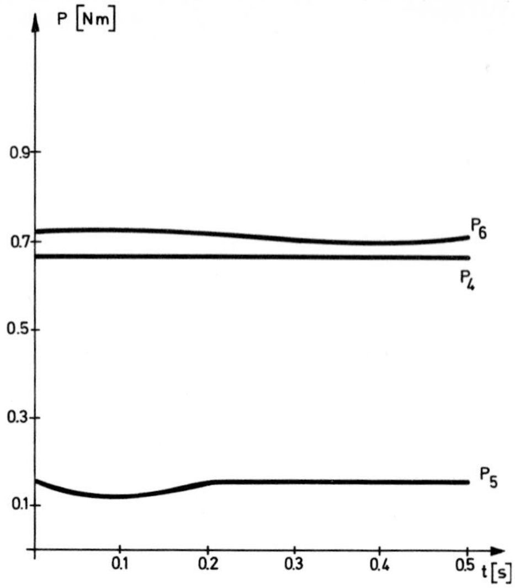

Fig. 3.90f. Driving torques of gripper

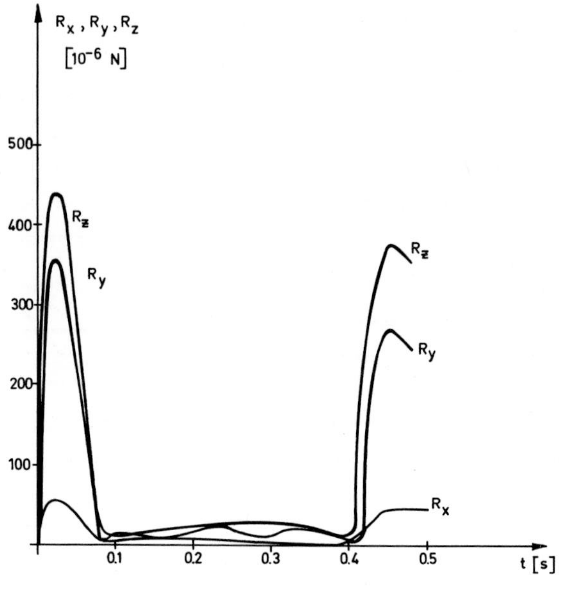

Fig. 3.90g. Reaction forces at point of contact of object and hole with force feedback applied ($\bar{k}_i^{G'} = 0$, $\bar{k}_i^{G''} = (100, 100, 100)^T$)

3.6. Multiprocessor Realization of Decentralized and Global Control

Although in this book the problems of implementing the robot control are not of prime interest, we shall briefly consider some ways of implementing the control laws considered. We mentioned in para. 2.7, that due to advances in technology it is convenient to realize various control laws by multiprocessor systems. To realize control by digital computer it is necessary to synthesize the corresponding control laws in a discrete-time domain. Para. 2.7.1 described the synthesis of a linear discrete centralized regulator and paras. 3.2.2 and 3.2.4 described this synthesis in the case of the manipulation system UMS-2. Para. 2.7.2 described the synthesis of a decentralized, discrete regulator and observer and the synthesis of global control. Here, we shall briefly consider multiprocessor implementation of decentralized control and we shall describe the synthesis of a discrete, decentralized regulator and observer for one very simple manipulator but we shall not consider other ways of implementation of such control (analogue, hybrid etc.).

We briefly consider the realization of the control (2.7.28), (2.7.25), (2.7.30) and of the decentralized observer. We consider the case when programmed control is synthesized at the level of subsystems, i.e. when it satisfies (2.7.29), so that it is also necessary to introduce global control (2.7.30). If a manipulation system or robot with a greater number of degrees of freedom is in question, the realization of the decentralized regulator and observer, although much simpler than the realization of the centralized observer and regulator with respect to the number of operations which have to be performed on one sampling interval in order to calculate the next system state and the control on the following interval, can require the use of a powerful computer for the realization of a sufficiently short sampling period. Current technology permits the use of many cheap microprocessors in parallel operation. The structure of decentralized control and observer makes them very suitable for realization by parallel processing. Various versions of multi-processor realization of both decentralized ang global control (2.7.30) and observer are possible. The choice of the multi-processor structure depends on the computing speed of the microprocessors, the number of operations needed to calculate the estimated state and control, as well as the periods of discrete time satisfying the problem considered. For the calculation of control by (2.7.15), for a system with n subsystems of order n_i, $\forall i \in I$, it is necessary to perform $\sum_{i=1}^{n} n_i$ multiplications and $\sum_{i=1}^{n} 2n_i$ - additions, where the number of additions can be reduced to $\sum_{i=1}^{n} n_i$ if the terms in expression (2.7.25) which depend on the nominal trajectories $x^{io}(k)$ are calculated in advance and memorized along with the programmed control $u^o(k)$. To estimate the state with the decentralized observer it is, in general, necessary to perform on each sampling interval $\sum_{i=1}^{n} n_i^2+\ell_i^2+\ell_i(k_i+1)$ multiplications and $\sum_{i=1}^{n} n_i(n_i-1) + \ell_i(\ell_i-1) + \ell_i(k_i+1)$ additions, where this number of operations can be reduced by a suitable choice of matrix D^i, as well by certain savings in the program. Concerning the number of operations which have to be performed on one sampling interval, it is not possible to give a general analysis: calculation of global control by on-line computation of the coupling among the subsystems depends on the dynamical characteristics of the system considered and on its complexity, i.e. the complexity of the approximate relations used to estimate the coupling. In para. 2.6 it was shown that the more exact the estimation of the coupling, the less suboptimal is the applied decentralized and global control. If a linear approximation of the coupling (2.6.4) is used to calculate the coupling on one sampling interval, it is necessary to perform $n\times n+N$ multiplications and $n\times(n-2)+N$ additions. However, if the coupling is calculated by the more exact estimation (2.6.29), which also takes into account the second--order term with respect to the system state, the number of operations which have to be performed is, in general, much greater namely, $N+n\times n+(n+1)n^2/2$ multiplications and $N+n(n-3+(n+1)n/2)$ additions. Apart from that, in order to calculate global control in the form (2.6.34) it is also necessary to perform $\sum_{i=1}^{n} 2n_i^2$ multiplications and $\sum_{i=1}^{n} 2n_i(n_i-1)$ additions in order to calculate the "global gain" for each subsystem.

However, it should be remembered that these operations in the calculation of global control are valid only when it is necessary to introduce all terms in the expressions for estimating the coupling (2.6.43) and (2.6.29), which is usually not the case. The proposed control synthesis is aimed at minimizing the number of operations which have to be performed to calculate the control, i.e. to choose the right calculations, necessary to achieve satisfactory tracking of the prescribed nominal trajectory. Thus, in the proposed approximations for the coupling (2.6.43) and (2.6.29) one should only consider those terms, which are necessary to achieve the appropriate system performance. Para. 3.3

described some methods of approximating the coupling for a particular manipulation system. The only approximations used were some components of the generalized force acting on the coupling among the subsystems (e.g. gravity forces, inertial forces of members, etc.). Consequently, one only uses those components of the driving torques in the mechanism joints which contribute essentially to the coupling among subsystems. With such an analysis it is possible to choose of a control which does not require too many operations. If many operations are necessary to estimate the coupling, i.e. to estimate (2.6.43) and (2.6.29), the advantages of decentralized control over the centralized regulator would be lost. The linear optimal regulator, synthesized on the basis of the centralized linearized system model, in general requires $n \times N$ multiplications and $n \times (N-1)$ additions on each sampling interval. Compared with the decentralized regulator and by introducing global control, the number of operations required by the optimal regulator may be smaller. However, if the coupling is not estimated using all the terms in expression (2.6.34), the number of operations, in this case, is smaller than in the case of the centralized regulator. If the coupling is estimated by (2.6.29), the number of operations can be greater but control with such an estimation of the coupling can also be less suboptimal than in the linearized case. The second advantage of decentralized and global control over the centralized regulator is that the coupling among subsystems changes more slowly than the subsystem dynamics, i.e. the dynamics of mechanical part S^M of the system is slower than the dynamics of the subsystem S^i, which usually means that global control can be realized over a longer sampling interval than is necessary for the realization of local control. Thus, the coupling can often be calculated on more basic sampling intervals than the number of intervals on which local control and local observers are calculated. This results in a substantial reduction in the number of operations needed to be performed on one basic sampling interval. When one has thus estimated the number of operations, needed on one sampling interval to calculate local and global control and observer, and when one has estimated the longest sampling interval satisfying the control [32], one can then determine the number of microprocessors necessary for parallel processing. It is of course necessary to consider the processing speed of the microprocessors as well as the required computing precision (indicating the number of bytes necessary for the calculations, whether calculations use fixed- or floating-point arithmetic, the accuracy of the A/D and D/A convertors, the method of performing the arithmetic operations, etc). Depending on the

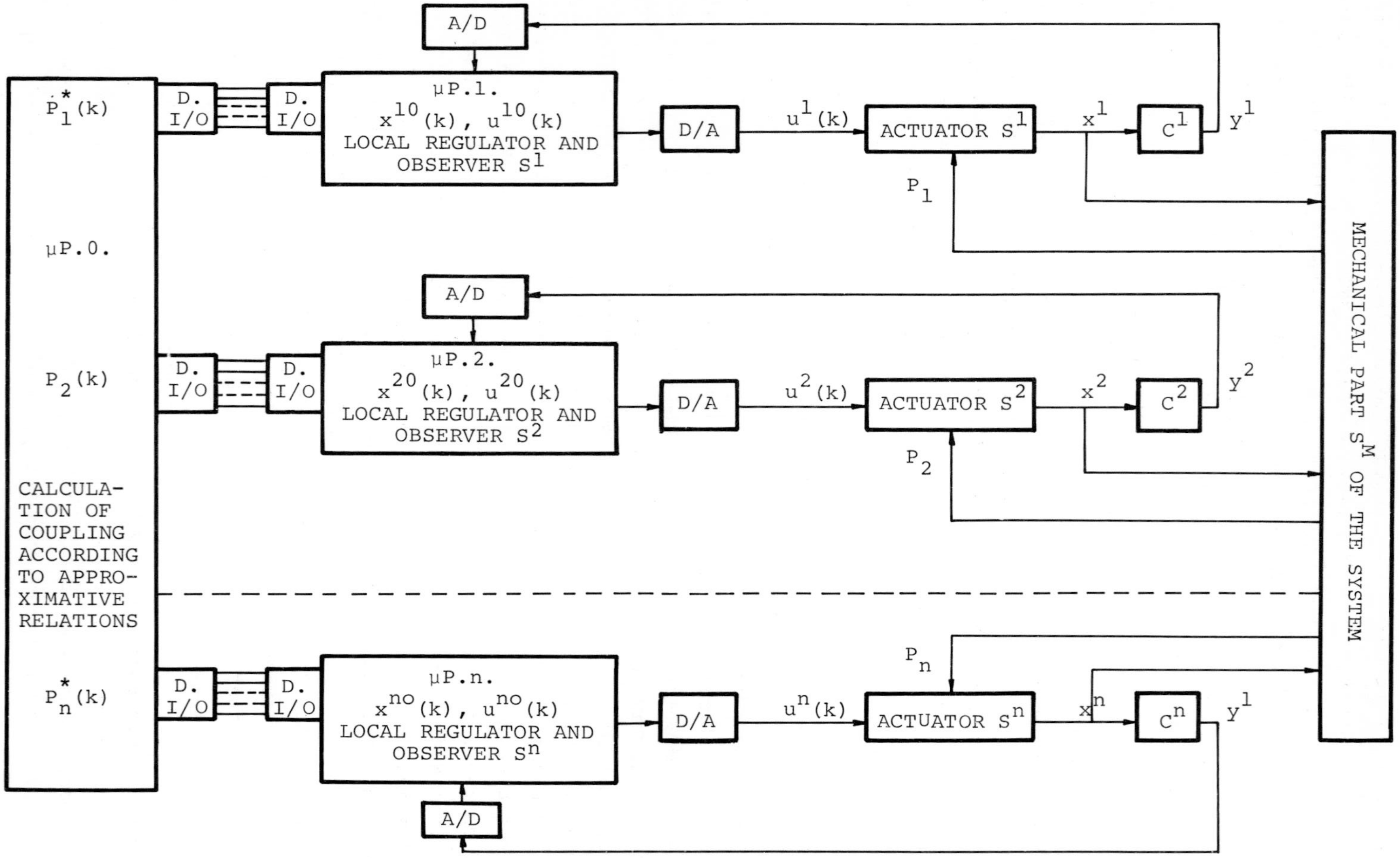

Fig. 3.91. Block-scheme of one version of multi-processor robot control

number of microprocessors required, one distributes the calculations among the various microprocessors. Decentralized control and decentralized observer, from that aspect, extremely are appropriate because it is very easy to share calculations among the microprocessors without unnecessary interconnections. This considerably increases the reliability of such a control system. Each microprocessor computes the local control and local observer for one or more subsystems, while the central processor synchronizes the work of several such microprocessors and calculates the coupling among subsystems.

Here, we briefly describe one possible version of the multiprocessor solution of the observer and control [23]. Fig. 3.91 shows the block-scheme of such a solution. In this case, to each degree of freedom of the manipulation system or robot, a microprocessor is adjoined, which realizes the local observer and local regulator for the subsystem S^i. Microprocessor No. 0 synchronizes the work of all microprocessors and dictates to the other microprocessors the nominal trajectories $x^{io}(k)$ and the programmed control $u^{io}(k)$ for the task to be performed by the system. This microprocessor communicates with the higher control levels which decide which movements have to be made by the mechanical system, i.e., which prescribe the control task in terms of the regions X^I, X^F and X^t, the settling time τ_s and the time τ in which the task is to be performed. Since this book does not consider the higher levels of control we shall also not consider the implementation of these levels. In principle, microprocessor No. 0 can be used to realize control at these levels. Alternatively, new microprocessors can be added, depending on the complexity of the decisions to be made at these levels, on the amount of informations to be processed, etc. Here we suppose that microprocessor No. 0 is used. This microprocessor generates the nominal trajectories $x^o(k)$ and nominal control $u^o(k)$. This calculation can be realized off-line with the supposed nominal initial conditions $x^o(0)$, and the calculated nominal trajectories and control are memorized in this computer. The second possibility is to generate the nominal trajectories $x^{io}(k)$ on-line in this microprocessor, to generate on-line the nominal control $\bar{u}^{io}(k)$ at the level of subsystems, in microprocessors adjoined to the individual subsystems S^i, i.e. the d.o.f. of the mechanism. We suppose that the nominal trajectories $x^{io}(k)$ are prescribed thus and that it is possible to determine the control $\bar{u}^{io}(k)$ realizing $x^{io}(t)$ when neither coupling nor perturbations act on the subsystems (i.e. $\bar{u}^{io}(k)$ satisfies (2.7.29)) [34].

The n microprocessors adjoined to the subsystems S^i, determine the local observers (2.7.22), (2.7.24) and the local regulators (2.7.25) for tracking the nominal trajectory $x^{io}(k)$, which is obtained from microprocessor No. 0. Each local microprocessor is equipped with A/D converters, to which the subsystem outputs y^i are connected. On the basis of these the subsystem state is reconstructed by means of the local observer on each sampling interval, and the control (being fed to the D/A converters) is calculated.

If it also is necessary to determine global control, the "leading microprocessor" (No. 0), gathers information about the instantaneous estimated states $x^{io}(k)$ of the subsystems and calculates the coupling P_i^* using the approximation coupling model, as explained earlier. The calculated coupling P_i^* is sent by the "leading microprocessor" to the other microprocessors, which use it to calculate global control (2.7.30). The global control, together with the programmed, nominal control and the local control is fed to the D/A converters of the microprocessors adjoined to the individual subsystems. As we said earlier, coupling often changes more slowly than the actuator dynamics, so P_i^* can be calculated over a period longer than the basic sampling period T_D. The microprocessors can communicate through the shared memory or digital input-outputs (D.I/O) and the interrupts, as shown in Fig. 3.91 [33].

The basic advantage of such a microprocessor structure is the relatively little communication among the microprocessors, i.e. little exchange of information among the microprocessors on one sampling interval. All the links among microprocessors are via the "leading microprocessor", which synchronizes the work of all microprocessors, determines the sampling time, sends data about the nominal trajectories $x^{io}(k)$ and the programmed nominal control, and eventually receives the data about the estimated subsystem states $\hat{x}^i$ and calculates the approximative values of coupling P_i^*. Between the microprocessors themselves, which determine the local observers and regulators as well the global control (if necessary), there is no exchange of information. If the nominal control is synthesized off-line, then it is possible to prevent the nominal trajectories of the subsystem state vector trajectories x^{io} and the nominal control u^{io} being transferred during the on--line system operation. Instead, they can be memorized in the subsystem microprocessor memories. The exchange of information between the microprocessors is thus reduced even further which is a very significant factor in being able to achieve sufficient operational speed of the multi-processor system. It is known that one of the most significant factors limiting the operational speed of multi-processor systems is the time needed for information exchange between the individual microprocessors. If, by increasing the number of microprocessors, the number of data which have to be exchanged on one sampling interval is significantly increased, then "saturation" of the system´s operational speed can occur, when the increase in the number of microprocessors does not shorten the sampling interval. Clearly, the multi-processor stem proposed alows relatively small exchange of information among sy- the microprocessors. Moreover, as we have often said, the proposed method of control synthesis, based on knowing the behaviour of the system S, allows the introduction of only those feedback loops, necessary to satisfactory system performance. This ensures that only necessary exchanges of data occur among the microprocessors.

Since the "leading microprocessor" is charged with most of the data exchange, it is possible to divide the task performed by this microprocessor among several microprocessors, i.e., to introduce supplementary microprocessors, which would together synthesize the nominal trajectories (if performed on-line), send the data on the nominal states to the other microprocessors and calculate the coupling. In this case it is very important that the algorithms for synthesizing the nominal trajectories, i.e. the coupling, can be efficiently divided for parallel processing and that minimal data exchange occurs. The proposed procedure of decoupling the system S into two functional subsystems during the synthesis of nominal dynamics (para. 2.2), thus offers significant advantages.

As shown with the manipulation system in para 3.3.2, synthesis of the nominal states using two functional subsystems $\tilde{S}_o$ and $\tilde{S}_x$ permits parallel synthesis of the nominal states for the subsystem of the minimal configuration and the gripper subsystem. This synthesis can be suitably implemented by two microprocessors. The same is true for the calculation of the coupling: this can be performed in parallel for the minimal configuration subsystem and the gripper subsystem. This can ensure that the "leading microprocessors" work fast enough not to prolong the sampling period.

To demonstrate multi-processor implementation of the proposed control form, a control has been determined for a simple manipulator with two degrees of freedom, using parallel processing with two microprocessors.

We stress that this manipulator is of no greater practical significance and that two microprocessors are not needed for the control of such a system. The control was intended to demonstrate how decentralized and global control, as well the decentralized observer, can be implemented by parallel processing [33].

The scheme of the manipulator for which the control has been implemented is shown in Fig. 3.92. For actuators we used D.C. servomotors PRINTED MOTORS LTD., type G9M4T, the model of which is given by (1.2.1), where $n_i = 3$, $x^i = (q^i, \dot{q}^i, i_R^i)^T$, so that $N = 2\times 3 = 6$. The manipulator parameters are shown in Table 3.22. The actuator models are given in Table 3.22. In the system S we measured only the manipulator angles q^1, q^2. The other state vector coordinates (angular velocities $\dot{q}^1$, $\dot{q}^2$ and the rotor currents i_R^1, i_R^2) were not accessible to measurement. Thus, the subsystem outputs are given by $y^i = q^i$, $k^i = 1$, $i = 1,2$. To implement the control we used INTEL 8080 microprocessors.

The task to be performed by this manipulator is as in para. 2.7.2 or para 2.2. It is necessary to transfer the system from the region X^I to the region X^F in a finite time τ and the system state during transfer should belong to a bounded region X^t. It is supposed that the control process includes the operator who, by means of a terminal can set conditions to be met by the manipulator.

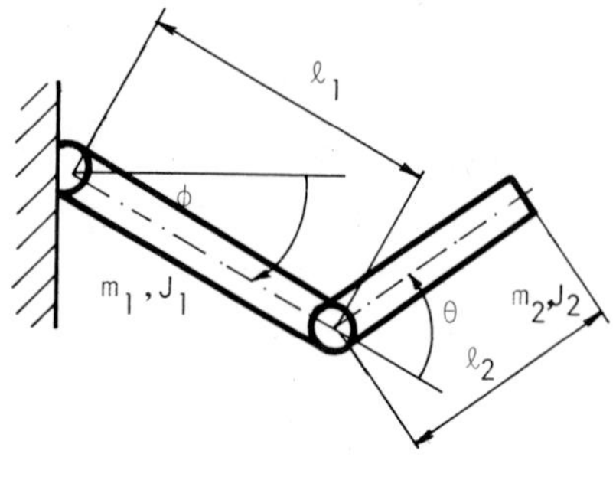

Fig. 3.92. Manipulator with two degrees of freedom

In the case in question the operator sets the state $x(\tau)$ to which the system is to be transferred in the time interval τ. It is supposed, that the region X^I of the initial conditions, the region X^F around the final state and the region X^t, to which the state should belong during

$m_1 = 1.5[kg]$

$m_2 = 1[kg]$

$\ell_1 = 0.25[m]$

$\ell_2 = 0.2[m]$

$$A = \begin{bmatrix} 0. & 1. & 0. \\ 0. & -0.1425 & 30.7 \\ -1800. & -80. & -100. \end{bmatrix}$$

$$b^i = (0, \quad 0, \quad 1800.)^T$$

$$f^i = (0, \quad -25., \quad 0)^T$$

Table 3.22. Parameters of the actuator model (with analogous feedback with respect to the output)

the transfer are known. The microprocessor system should synthesize the nominal trajectories, which transfer the system state from the supposed nominal initial state into the defined final state in a given time interval τ. The synthesis of the nominal trajectory and the nominal local programmed control is performed off-line.

In order to track the nominal trajectories when the manipulator state deviates from the supposed initial state, it is necessary to determine the decentralized regulator (2.7.21). As the whole state vector is not being measured, it is necessary to introduce an observer to reconstruct the system state. A decentralized observer (2.7.22) - (2.7.24) was therefore determined. This shows, that to achieve a sufficiently good tracking of the nominal trajectories and a sufficiently accurate estimation of the system state it is appropriate to introduce global control as well which should reduce the effect on the system stability of the coupling between two subsystems. It also shows, that to satisfactorily estimate the system state and calculate nominal states, local control and global control, it is necessary to use floating point arithmetic (12-bit A/D converters and 10-bit D/A converters were used).

A package with "floating point" arithmetic was available, operating with 24-bit numbers (2-8-bit locations for the mantissa and 1-8-bit for the exponent). This provided satisfactory computing accuracy. However, since available microprocessors do not dispense with a hardware multiplyer, the floating point operations are relatively slow (multiplication about 1.2 msec, addition about 0.6 msec, etc.). To calculate the inputs for each regulator (2.7.25) it is necessary to perform $n_i=3$ multiplications and $2n_i - 1 = 5$ additions on each sampling interval. To determine the local observer it is necessary to perform 8 multiplications and 6 additions on each sampling interval, if matrix D^i is chosen so that matrix C^i is a unit matrix (see Table 3.22). Consequently, the total number of operations to be performed in one sampling interval for both subsystems is 22 multiplications and 22 additions, which requires 40 msec. When the other operations, which have to be performed are also taken into account, as well the computations to determine global control, it is estimated that the sampling interval can be executed in 60 msec. Since the shortest time constant of the mechanical system part is about 0.5 sec and the mechanical time constant of the actuator is about 50 msec., it is evident that the sampling period which can be achieved with one microprocessor is too long. Hence, we adopted the solution with microprocessors, described above. However, since higher control levels are not implemented the leading microprocessor is connected directly to the operator, and since the nominal trajectories are synthesized off-line, the leading microprocessor (microprocessor No. 0 in Fig. 3.91) is free of any real-time calculations. Hence, control is implemented by only two microprocessors, not by three as would be the case if the scheme in Fig. 3.91 were strictly observed. The local regulator and local observer are located in the microprocessor No. 0 for the first subsystem S^1 and the second microprocessor computes the local observer and local regulator for the second subsystem S^2. The first microprocessor synchronizes the operations of the system, determines the sampling interval using the program clock and inerrupts the system at each given sampling interval T_D. By using two microprocessors a sampling period of 30 msec was achieved, which proved satisfactory.

The leading microprocessor receives via the operator´s terminal the data about the final state $x^o(\tau)$ into which the system should be transferred, data about the transfer time τ and data about the regions X^I, X^t and X^F required during the transfer. Based on these data, the microprocessor generates the nominal trajectories and nominal local control for both subsystems and stores the data about the trajectory in its memory. Since the nominal control is computed off-line, in principle it can be also stored in the memory of the second microprocessor (for the second subsystem). In this case, however, it was decided that the first microprocessor should send to the second the nominal trajectories of the state coordinates and the nominal programmed control of the second subsystem in real time and that this should be done in order to examine possibilities of data exchange between the microprocessors. This would be necessary if the nominal trajectories were on-line. Since a very simple manipulator is in question, the generation of the nominal trajectories is trivial. Defining the region X^t, in which the manipulator state should be during the transfer from $x^o(0)$ to $x^o(\tau)$, is reduced to prescribing the velocity profiles for both degrees of freedom (i.e. the acceleration law during the movement) which can be produced by the given actuator. Alternatively, it should be ensured that it is possible to determine the nominal local control satisfying (2.7.29).

Fig. 3.93 shows the nominal trajectories, synthesized in such a way that they satisfy the triangular velocity profile.

The local regulators and local observers were than synthesized. Based on actuator models given in Table 3.22, the actuator models were determined in the discrete-time domain (matrix A_D^i and vectors b_D^i, f_D^i) given in Table 3.23 (for sampling period T_D = 30 msec)*). With the chosen weighting matrix Q_i and element r_i (also given in Table 3.23), the solution of the Riccatti difference equation ($K_i(0)$ in Table 3.23) was determined and thus the gains of the local discrete regulator obtained. Table 3.23 also shows the choice of matrix D^i for the local minimal observer (2.7.22), the corresponding vectors E^i and h^i and matrix C^i. In order to check whether the minimal observer was chosen exactly (so that its asymptotic stability is ensured), the behaviour of the subsystem with the discrete regulator was simulated and the subsystem state was estimated by the chosen observer, whereby an error was made in estimating the initial state. Fig. 3.94. shows the error of the estimated state with respect to the real system state as a function of time. It is evident, that the estimation error was reduced with time, showing that the observer was asymptotically stable. In order to check whether the subsystem model was exactly set and whether it was satisfied by the local regulator, a digital simulation was made of the free subsystem (without the influence of coupling) with the synthesized regulator, and the subsystem state, estimated by the local observer, was used. Stabilization around $x^2 = 0$ for perturbations of the initial conditions type was simulated. The local observer and local regulator were determined on the microprocessor and the behaviour of only one degree of freedom was examined with this control (other

---

*) Matrix A_D^i and vectors f_D^i, b_D^i are determined by Taylor series expansions of $\exp(A_D^i T_D)$, etc.

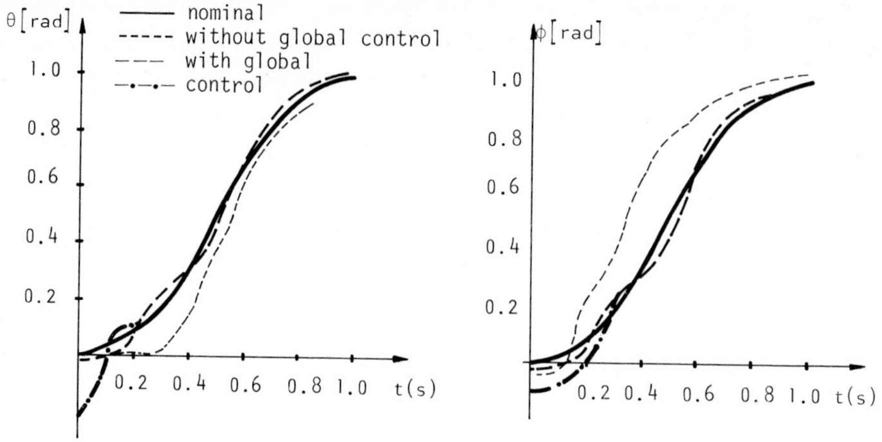

Fig. 3.93. Trajectory realization with and without global control

$$A_D^i = \begin{bmatrix} 0.877 & 0.023 & 0.005 \\ -9.548 & 0.449 & 0.183 \\ -11.097 & -0.802 & -0.164 \end{bmatrix} \qquad b_D^i = \begin{bmatrix} 0.122 \\ 9.548 \\ 11.097 \end{bmatrix}$$

$$Q_i = \begin{bmatrix} 3000. & 0. & 0. \\ 0. & 10. & 0. \\ 0. & 0. & 0.3 \end{bmatrix} \qquad r_i = 0.01$$

$$K_i(0) = \begin{bmatrix} 1.510 & 117.341 & 136.377 \\ 117.341 & 9116.660 & 10595.650 \\ 136.377 & 10595.650 & 12314.550 \end{bmatrix}$$

$$-r_i^{-1} b_D^{iT} (A_D^{iT})^{-1} (K_i(0) - Q_i) = -(0.334,\ 0.062,\ 0.021)^T$$

$$D^i = \begin{bmatrix} 0.449 & 0.188 \\ -0.802 & -0.164 \end{bmatrix} \quad E^i = \begin{bmatrix} -9.548 \\ -11.097 \end{bmatrix} \quad h^i = \begin{bmatrix} 9.548 \\ 11.097 \end{bmatrix} \quad C^i = \begin{bmatrix} 0 & 1 & 0 \\ 0 & 0 & 1 \end{bmatrix}$$

Table 3.23. The parameters of the actuator discrete model, weighting matrices of the quadratic criterion, parameters of the discrete local regulator and parameters of the discrete observer.

degree of freedom was fixed). The results of simulation and of the real system behaviour are presented in Fig. 3.95. These show a good correlation between the model and the real system (actuator); a certain difference in behaviour is the consequence of the gravity force which was not taken into account during subsystem modelling. It should be mentioned, that the data taken from the catalogue for the given motor were not quite accurate. This made it necessary to define the subsystem model in several iterations. Fig. 3.95 shows the behaviour of the subsystem (of the second degree of freedom) in reacting to perturbations of the initial conditions, with a regulator and observer synthesized and the actuator model parameters taken from the catalogue; it shows that the stabilization is weak and that the subsystem behaviour differs significantly from the simulated subsystem model. With improved actuator parameters, a good correspondence between the model and the real subsystem was achieved, as well as satisfactory stabilization.

Local regulators and observers synthesized in this way (given in Table 3.23) were used to track the nominal trajectories in Fig. 3.93. The local regulators and observers were implemented by two microprocessors, as explained above. The results of simultaneous tracking for both subsystems with local regulator and local programmed control are also presented in Fig. 3.93. Fig. 3.96 illustrates the tracking of the nominal trajectory only for the second degree of freedom, when the first is immobile (so that, effectively, coupling has no effect on subsystem S^2). It can be seen, that there is significant difference in tracking the free subsystem S^2 and tracking when the subsystems are coupled. In order to compensate the effect of coupling, global control is introduced in the form of feedback with respect to the coupling, which is computed on-line in the leading microprocessor. In the case in question, the determination of the coupling among subsystems only considered the gravitational forces acting on the manipulator members because they contribute significantly to the coupling among subsystems, and their calculation is relatively simple. The on-line calculation of the gravitational moments, acting on both members reduces to 3 multiplications and 2 additions and calculating the sines of three manipulator angles. Since the time change of the gravitational moments is much slower than the actuator dynamics (i.e. than the chosen sampling speed), the number of operations per sampling period was reduced by dividing the calculation of the gravitational moments between two sampling intervals: in one sampling interval sines of manipulator angles

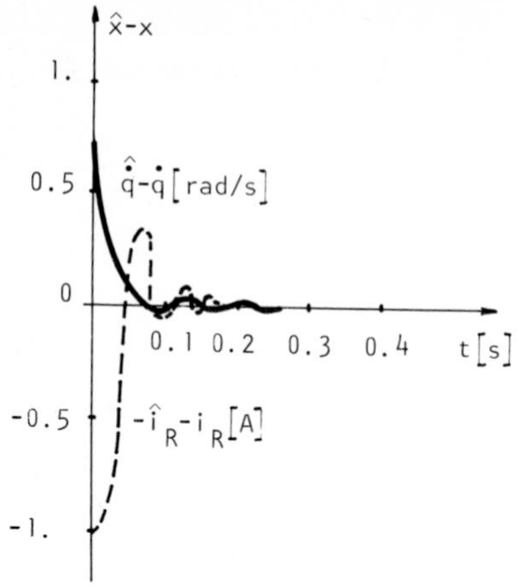

Fig. 3.94. Estimation error with chosen observer

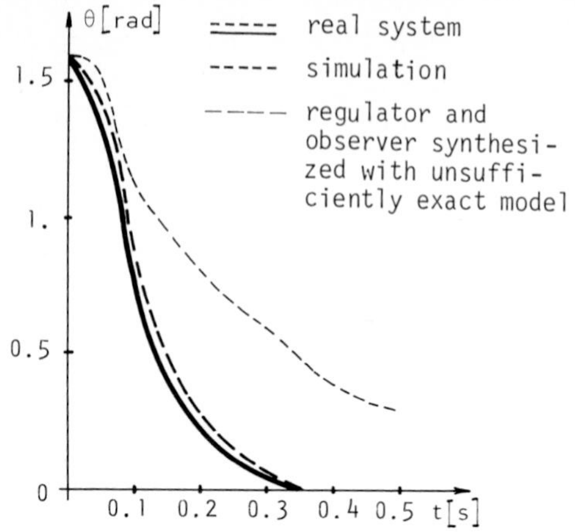

Fig. 3.95. Comparison of the real subsystem regulation and simulation

are calculated and in the other the multiplications and additions are performed. In order to accelerate the calculation of the sines of the angles table look-up method was used. On each sampling interval the first microprocessor sends to the second one data about the nominal trajectories of the state vector coordinates of the second subsystem $S^2x^{o2}(k)$, which have to be realized on the k-th sampling interval, and the data about the nominal programmed control for the second subsystem $u^{o2}(k)$. It also sends the data about the calculated gravitational

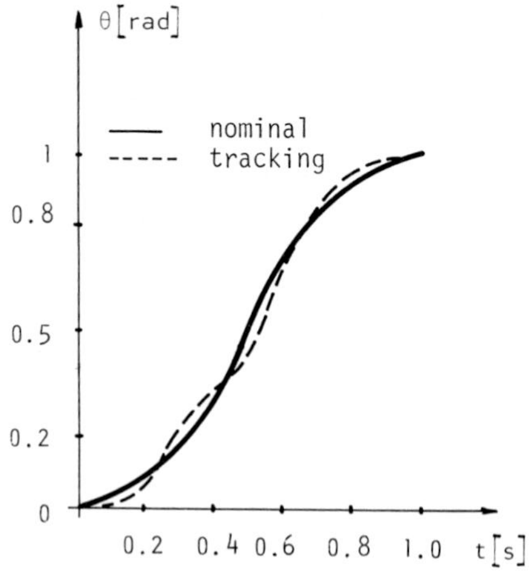

Fig. 3.96. Tracking of the nominal states of the decoupled second degree of freedom

moment acting on the second member $G^2(k)$. Data is transferred by the digital inputs and outputs and the interrupts. The second microprocessor sends to the first one data about the value of the instantaneous angle of the second member $q^2(k)$, which is obtained from the A/D converter. Each microprocessor estimates the state on the next interval using the local observers $\hat{x}^i(k)$, calculates the local control (2.7.25) from the memorized regulator gains, estimated states and memorized nominal trajectories $x^{oi}(k)$. With the calculated gravitational moments each microprocessor calculates global control in the form (2.7.30), where $K_i^G(x^i(k))$ was chosen as K_i^G sgn $(\Delta u_i^L(k))$ (see para. 3.3) and the global gain K_i^G was chosen as $(b_D^{iT} b_D^i)^{-1} b_D^{iT} f_D^i$, where $(b_D^{iT} b_D^i)^{-1}$ b_D^{iT} is the minimal inversion of b_D^i. Tracking of the nominal, when both local and global control are applied, is illustrated in Fig. 3.93. It can be seen that introducing global control significantly improves the tracking. It is evident, that the calculated gravitational moments are also introduced into the observers for the sake of state estimation.

Further improvement of the tracking is possible by a more exact calculation of the coupling among the subsystems, which would also take into account the inertial forces. However, in the case in question, even with such a simple approximation of moments satisfactory results were achieved.

The above example was intended to demonstrate the possibility of using parallel processing to implement the decentralized and global control and observer. By analogy, n microprocessors can clearly be connected in parallel for controlling a manipulator with n degrees of freedom. Such microprocessor structures can thus be used to control manipulators having a much more complex mechanical configuration and more degrees of freedom. In this case the use of more microprocessors would have made much more sense than in the above example. With a more complex manipulation system where inertial and other dynamical forces are more prominent than in the above case the calculation of the coupling among subsystem would need to consider the inertial forces. Global control would therefore be more complex and would require a more complex distribution of the coupling among the microprocessors.

3.7. Conclusion

This chapter has considered the synthesis of various forms of centralized and decentralized control in terms of various concrete examples. We also considered two ways of synthesizing nominal trajectories (paras. 3.2.1 and 3.3.1) for some manipulation tasks commonly met in practice. We considered the synthesis of a centralized optimal regulator in continuous and discrete-time domains (para. 3.2.2) together with the synthesis of an asymptotic regulator. We concentrated on the synthesis of decentralized control. We only considered two particular forms of linear local control and discussed only previously adopted forms of global control. Obviously, it is possible to investigate other local control laws, for example nonlinear local control and the like [35]. Global control might also be introduced in various forms, as in para. 3.3.2 where instead of (2.6.34) a simpler control law (3.3.24) was used for global control. We have also shown how by such global control can obviate the need to synthesize the nominal control on the centralized model of the system.

Para. 3.3.3 gave a more rigorous synthesis of local and global control using suboptimality analysis. The relation between local and global control was investigated in various cases. Para. 3.3.2 presented some simpler forms of local and global control which are easier to determine in practice. In this case we considered only the practical stability of the global system. It is obvious that the suboptimality analysis in para. 3.3.3, can be applied to analyze the suboptimality of the simplified laws of local and global control. The complete local state feedback may be realized by introducing observers for local subsystems and by estimating the subsystem states.

Paragraph 3.3.2 showed how the dynamics of the mechanism influences the system performance with decentralized control and how increase in the prescribed speed of the manipulator influences global stability. However, the manipulator was of an anthropomorphic type with strong coupling among d.o.f. With a number of industrial manipulators the coupling among d.o.f. is much weaker, so the application of local control may suffice. The same paragraph compared the behaviour of the manipulator UMS-1 with the behaviour of two other industrial manipulator UMS-2 and UMS-3B (with linear d.o.f.). With these two manipulator types the introduction of global control is not necessary, except for very fast motion, when the dynamic factor is more important. Thus, for each particular manipulator system it is necessary to perform the analysis demonstrated in paras. 3.3.2 and 3.3.3 and to determine whether decentralized control is sufficient or whether it is desirable to introduce global control. In so doing, one should determine which form of global control is the most convenient from the point of view of the quality of the tracking, the suboptimality of the control applied and the simplicity and cost of the control system.

We also considered the use of force feedback, when these forces are not couplings among subsystems. Two characteristic manipulator tasks were considered. We considered the possibility of using force feedback to solve the control task of transferring an object along a prescribed trajectory under certain dynamic conditions (para. 3.5). In contrast to the control considered in paras. 3.2 - 3.3, in this case, force feedback ensured on-line generation of the necessary motion of the gripper [1, 24, 36]. The purpose of this discussion was to explore various aspects of force feedback application. Force feedback enables one to directly determine the compensating motion satisfying the particular dynamic conditions of the task considered.

Finally, we considered the implementation of control at the lowest hierarhical levels. We discussed the implementation by microcomputer of local and global control and the state observers for manipulation systems. We have not considered various possible analogous or hybrid implementation of robot control or various ways of implementing control by microcomputer, (various centralized regulators, adaptive control, nonlinear control etc.). We did, however, consider the possibility of implementing the proposed control structure (ch. 2) by parallel processing and we proposed a multiprocessor configuration with this control structure.

References

[1] Vukobratović, K.M., D.M.Stokić, D.S.Hristić, "Dynamic Control of Anthropomorphic Manipulators", Proc. of IV International Symp. on Industrial Robots, Tokyo, 1974.

[2] Vukobratović K.M., D.M.Stokić, D.S.Hristić, "Algorithmic Control of Anthropomorphic Manipulators", (in Russian), Izvestiya AN SSSR, Teknicheskaya kibernetika, No. 3, 1976.

[3] Kahn, M.E., B.Roth, "The Near Minimum Time Control of Open Loop Articulated Kinematic Chains", Trans. of the ASME, Journal of Dynamic Systems, Measurement and Control, September, 164-172, 1971.

[4] Cvetković, S.V., M.K.Vukobratović, "Towards Real - Time Dynamic Control of Manipulators", Preprints of III IFToMM Symp. on Theory and Practice of Robots and Manipulators, Udine 1978, also in IFToMM Journal of Mechanism and Machine Theory, Vol. 15, No. 6, 1980.

[5] Vukobratović, K.M., D.Juričić, "One Way of Moving the Artificial Upper Extremities", IEEE Trans. on Bio-Medical Eng., Vol. BME-10, No. 2, 1969.

[6] Vukobratović, K.M., D.M.Stokić, D.S.Hristić, "A New Control Concept of Anthropomorphic Manipulators", Proc. of Second Conference of Remotely Manned Systems, Los Angeles, June, 1975.

[7] Vukobratović, K.M., D.M.Stokić, "Engineering Approach to Dynamic Control of Industrial Manipulators, Part I: Synthesis of Dynamic Nominal Regimes", (in Russian), Mashinovedeniya AN SSSR, No. 3, 1981.

[8] Vukobratović, K.M., D.S.Hristić, D.M.Stokić, "Dynamic Control Synthesis of Industrial Manipulation Systems", Proc. of 10th International Symp. on Industrial Manipulation Systems, Milano, 1980.

[9] Kirćanski, M., "Contribution to Synthesis of Nominal Trajectories of Manipulators via Dynamic Programming", Proc. of International Conference on Systems Engineering, Coventry, 114-122, 1980.

[10] Athans, M., P.L.Falb, Optimal Control, McGraw-Hill, 1966.

[11] Vukobratović, K.M., D.S.Hristić, D.M.Stokić, "Synthesis and Design of Anthropomorphic Manipulator for Industrial Application", The Industrial Robot, Vol. 5, No 2, 1978.

[12] Kirćanski N., Vukobratović M., "Control of Manipulators by Means of Discrete Regulators", Trans. of the Institute of Measurement and Control, No. 1, 1981.

[13] Vukobratović, K.M., D.S.Hristić, D.M.Stokić, "Algorithmic Control of Anthropomorphic Manipulators", Proc. of V Intern. Symp on Industrial Robots, Chicago, Illinois, Sept. 1975.

[14] Vukobratović, K.M., V.Potkonjak, "Contribution to Computer-Aided Design of Industrial Manipulators Using Their Dynamic Properties", Journal of IFToMM Mechanism and Machine Theory, Vol. 17, No. 1, 1982.

[15] Vukobratović, K.M., et al., "New Method of Artificial Motion Synthesis and its Application to Locomotion Robots and Manipulators", Proc. of 7th IFAC Symp. on Automatic Control in Space, Minhen, 1976.

[16] Vukobratović, K.M., D.M.Stokić, "Contribution to the Decoupled Control of Large-Scale Mechanical Systems", Automatica, Jan., No. 1, 1980.

[17] Šiljak, D.D., M.Vukčević, "Decentralization, Stabilization and Estimation of Large-Scale Linear Systems", Proc. of I Symp. on Large-Scale Systems, Udine, 1976.

[18] Šiljak, D.D., Large-Scale Dynamic Systems: Stability and Structure, North-Holland, New York, 1978.

[19] Vukobratović, K.M., D.M.Stokić, "One Engineering Concept of Dynamic Control of Manipulators", Journal of Dynamic Systems, Measurement and Control, Trans. of the ASME, Vol. 103, No. 2, 1981.

[20] Vukobratović, K.M., D.M.Stokić, "Choice of Decoupled Control Law of Large Scale Mechanical Systems", Proc. of Second Symp. on Large-Scale Systems Theory and Applications, Toulouse, 1980, also in Large-Scale Systems-Theory and Application, June, 1981.

[21] Tomović R., M.K.Vukobratović, General Sensitivity Theory, Elsevier Publ. Co., 1972.

[22] Vukobratović, K.M., D.M.Stokić, M.V.Kirćanski, "A Procedure for Interactive Dynamic Control Synthesis of Manipulators", Prepr. of IV IFToMM Symp. on Theory and Practice of Robots and Manipulators, Warsaw, 1981.

[23] Vukobratović, K.M., D.S.Hristić, D.M.Stokić, "Dynamic Control of Industrial Manipulators and its Application", Journal of Industrial Robots, No 1, 1981.

[24] Vukobratović, K.M., D.M.Stokić, "Dynamic Control of Manipulators via Load-Feedback", Journal Mechanism and Machine Theory, Vol. 17, No. 2, pp. 107-118, 1982.

[25] Vukobratović, K.M., D.M.Stokić, "Significance of Force-Feedback in Controlling Artificial Locomotion-Manipulation Systems", Transaction on Biomedical Engineering, December, 1980.

[26] Nevins, J., D.Whithey et al., Exploratory Research in Industrial Modular Assembly, Report R-996, The Charles Stark Draper Laboratory, Cambridge, Mass., 1976.

[27] Nevins, J.I., D.E.Whitney, "Research Issues for Automatic Assembly", Proc. I IFAC/IFIP Symp. on Information Control Problems in Manufacturing Technology, Tokyo, 1977.

[28] Simunović, S., "Force Information in Assembly Processes", V Internat. Symp. on Industrial Robots, Chicago, 1975.

[29] Drake, S.H., P.C.Watson, S.Simunović, "High Speed Robot Assembly of Precision Parts Using Compliance Instead of Sensory Feedback", Proc. of I IFAC/IFIP Symp. on Information Control Problems in Manufacturing Technology, Tokyo, 1977.

[30] Stokić, M.D., M.K.Vukobratović, "Simulation and Control Synthesis of Manipulator in Assembling Mechanical Elements", Third Symposium IFToMM on Theory and Practice of Robots and Manipulators, Udine, 1978, also in IFToMM Journal of Mechanism and Machine Theory, Vol. 16, pp. 71-79, 1981.

[31] Stokić, M.D., M.K.Vukobratović, "Simulation and Control Synthesis of Manipulator in Assembling Technical Parts", Trans. of the ASME, Journal of Dynamic Systems, Measurement and Control, December, 1979.

[32] Levis, H.A., A.R.Schlueter, M.Athans, "On the Behaviour of Optimal Linear Sampled-Data Regulators", Int. J. of Control, Vol. 13, No. 2, 343-363, 1971.

[33] Stokić, M.D., M.K.Vukobratović, "Further Development of Dynamic Algorithms for Industrial Robots Control", International Conference on Systems Engineering, Coventry, England, 1980.

[34] Vukobratović, K.M., D.M.Stokić, M.V.Kirćanski, "Contribution to Dynamic Control of Industrial Manipulators", Proc. of XI ISIR, Tokyo, 1981.

[35] Cvetković S.V., "Synthesis of Suboptimal Control of Manipulators Using the Liapunov's Function", (in Serbian), Proc. I Yugoslav Symp. on Industrial Robotics and Artificial Intelligence, Dubrovnik, 167-178, 1979.

[36] Vukobratović, K.M., Legged Locomotion Robots and Anthropomorphic Mechanisms, Monograph, Institute "M.Pupin", (in English) Beograd, 1975, also published by Mir, Moscow, (in Russian), 1976.

Appendix 3. A.
Simulation Algorithm for Assembly

We consider problems arising when using a digital computer to simulate the manipulator dynamics of inserting the peg into the hole. The reaction forces arising during the insertion create difficulties in determining the mathematical model of the manipulator dynamics. In the case of no contact between the peg and hole walls, the mathematical model of the manipulator is given by differential equations of open kinematic chain dynamics. The differential equations of open kinematic chain dynamics may be determined by one of the procedures for forming the mathematical model of open kinematic chains. Here, we used the method for automatic setting of differential equations on a digital computer based on D'Alembert's principle. This was presented in detail in ch. 1. In this case, differential equations of manipulator dynamics with 6 degrees of freedom (Fig. 3.1) are obtained in the following from

$$P_i = \vec{e}_i \cdot [\sum_{j=1}^{n_s} (\vec{M}_j + \vec{r}_{ji} \times \vec{\Gamma}_j)] = \sum_{k=1}^{6} H_{ik}\ddot{q}^k + h_i, \qquad \forall i \varepsilon I \tag{3.A.1}$$

where the $P_i \varepsilon R^1$ denote driving torques in manipulator joints, $\vec{e}_i \varepsilon R^3$ are unit vectors of joint rotation axes, $\vec{M}_j \varepsilon R^3$ are moments of inertial forces due to rotation of the member j, $\vec{r}_{ji} \varepsilon R^3$ is the position vector from the center of the i-th joint to the center of gravity of the j-th member, $\vec{\Gamma}_j = \vec{F}_j + \vec{G}_j$, $\vec{F}_j \varepsilon R^3$ is the inertial force of the j-th member, $\vec{G}_j \varepsilon R^3$ is the vector of the gravitational force acting on the j-th member, and q^i are the internal manipulator angles, as shown in Fig. 3.1. It is supposed that all six degrees of freedom are powered by corresponding actuators with P_i driving torques. The manipulator obviously has 4 members (n_s=4), the fourth of which (the gripper) has three rotations relative to the preceding member. According to (1.1.25) the following notation may be introduced

$$\vec{M}_j = \sum_{k=1}^{6} \vec{b}_{jk} \cdot \ddot{q}^k + \vec{b}_j^o, \tag{3.A.2}$$

$$\vec{\Gamma}_j = \sum_{k=1}^{6} \vec{a}_{jk} \cdot \ddot{q}^k + \vec{a}_j^o + \vec{G}_j, \tag{3.A.3}$$

So, from (3.A.1) we get

$$H_{ik} = \vec{e}_i\Big[\sum_{j=i}^{4}(\vec{b}_{jk}+\vec{r}_{ji}\times\vec{a}_{jk})\Big] \qquad i,\ k\varepsilon I, \tag{3.A.4}$$

$$h_i = \vec{e}_i\Big[\sum_{j=1}^{4}(\vec{r}_{ji}\times\vec{a}_j^{\,o}+\vec{r}_{ji}\times\vec{G}_j+\vec{b}_j^{\,o})\Big]. \tag{3.A.5}$$

The introduction of vector q, vector $P = (P_1, P_2,\ldots,P_6)^T$, matrix $H = [H_{ik}]$ and vector $h = (h_1, h_2,\ldots,h_6)^T$ enables the equations (3.A.1) to be written in matrix form (see ch. 1). Thus,

$$P = H\ddot{q} + h \tag{3.A.6}$$

If we consider the manipulator dynamics with a peg in the gripper, the peg dynamics should be incorporated into the equations (3.A.1), (3.A.4), (3.A.5). According to (1.1.11), the linear (translational) acceleration of the gripper is

$$\vec{w}_6 = \vec{w}^6 + \vec{\varepsilon}_6 \times \vec{r}_{6,6} + \vec{\omega}_6 \times (\vec{\omega}_6\times\vec{r}_{6,6}), \tag{3.A.7}$$

where $\vec{w}_6\varepsilon R^3$ denotes the linear acceleration of the 4-th member, $\vec{w}^6\varepsilon R^3$ is the linear acceleration of the center of joint 4 (point D in Fig. 3.83) having three degrees of freedom, $\vec{\varepsilon}_6\varepsilon R^3$ is the rotational acceleration of the 4-th member, $\vec{r}_{66}\varepsilon R^3$ is the vector from the 4-th joint to the center of gravity of the 4-th member, and $\vec{\omega}_6\varepsilon R^3$ is the angular velocity of the 4-th member. The linear acceleration $\vec{w}_P$ of the peg has the form

$$\vec{w}_P = \vec{w}^6 + \vec{\varepsilon}_6 \times \vec{r}_{P6} + \vec{\omega}_6 \times (\vec{\omega}_6\times\vec{r}_{P6}), \tag{3.A.8}$$

where $\vec{r}_{P6}\varepsilon R^3$ denotes the vector from the 4-th joint to the center of gravity of the peg.

Since the vectors $\vec{r}_{P6}$ and $\vec{r}_{6,6}$ have the same direction, $\vec{r}_{P6} = \frac{||\vec{r}_{P6}||}{||\vec{r}_{6,6}||}\vec{r}_{6,6}$ holds, so we get

$$\vec{w}_P = \vec{w}^6 + \frac{||\vec{r}_{P6}||}{||\vec{r}_{6,6}||}(\vec{w}_6-\vec{w}^6). \tag{3.A.9}$$

The inertial force of the peg has the form

$$\vec{F}_P = m_P\vec{w}_P = m_P[\vec{w}^6 + \frac{||\vec{r}_{P6}||}{||\vec{r}_{6,6}||}(\vec{w}_6 - \vec{w}^6)] = \sum_{j=1}^{6} \vec{a}_{Pj}\ddot{q}_j + \vec{a}_P^o. \quad (3.A.10)$$

The moment of inertial forces of the **peg** is obtained in the form

$$\vec{M}_P = T_P \cdot \vec{\varepsilon}_6 + \vec{\lambda}_P = \sum_{j=1}^{6} \vec{b}_{Pj}\ddot{q}^j + \vec{b}_P^o, \quad (3.A.11)$$

where elements of the matrix $T_P \varepsilon R^{3\times 3}$ are given by

$$T_P^{jk} = \sum_{\ell=1}^{3} q_{6\ell}^k q_{6\ell}^j J_{P\ell} \quad (3.A.12)$$

and the vector $\vec{\lambda}_P$ by

$$\vec{\lambda}_P = \begin{bmatrix} (J_{P2} - J_{P3})(\vec{\omega}_6\vec{q}_{62})(\vec{\omega}_6\vec{q}_{63}) \\ (J_{P3} - J_{P1})(\vec{\omega}_6\vec{q}_{63})(\vec{\omega}_6\vec{q}_{61}) \\ (J_{P1} - J_{P2})(\vec{\omega}_6\vec{q}_{61})(\vec{\omega}_6\vec{q}_{62}) \end{bmatrix}, \quad (3.A.13)$$

where J_{Pi} $(i = 1,2,3)$ denotes the moments of the peg's inertia about the main axes of inertia, $q_6^{k\ell} = A_6^{k\ell}$ are elements of the transformation matrix $A_6 = [A_6^{k\ell}]$, which provides the transformation from the internal coordinate system of the 4-th member to the absolute system, $\vec{q}_{6j}$ are the column vectors of the matrix $A_6 = [\vec{q}_{61}, \vec{q}_{62}, \vec{q}_{63}]$. The absolute system is shown in Fig. 3.83 and the internal system of the 4-th member has its axes directed along the main inertia axes so that in position $q^i = 0$, $\forall i \varepsilon I$, the directions of axes of the internal and absolute system coincide.

According to (3.A.1), (3.A.10) and (3.A.11) the dynamics equations of the manipulator with a peg are obtained in the following form

$$P_i = \sum_{k=1}^{6} H_{ik}\ddot{q}^k + h_i + \vec{e}_i[\vec{r}_{Pi}\times(\vec{F}_P + \vec{G}_P) + \vec{M}_P], \quad \forall i \varepsilon I, \quad (3.A.14)$$

where $\vec{G}_P \varepsilon R^3$ denotes the vector of the gravitational force acting on the peg. Equations (3.A.1) may be written in the form

$$P = H'\ddot{q} + h', \quad (3.A.15)$$

where matrix $H' = [H'_{ik}]$ and vector $h' = (h'_i)$. These elements, according to (3.A.14), are given by

$$H'_{ik} = H_{ik} + \vec{e}_i(\vec{r}_{Pi}\times\vec{a}_{Pk} + \vec{b}_{Pk}), \quad \forall i, k \varepsilon I, \quad (3.A.16)$$

$$h_i' = h_i + \vec{e}_i(\vec{r}_{Pi}\times\vec{a}_P^o+\vec{b}_P^o+\vec{r}_{Pi}\times\vec{G}_P). \qquad (3.A.17)$$

We thus obtain the differential equations of the manipulator dynamics when there is no contact between the object and hole.

If there is one point of contact, the differential equations take the following form

$$P_i = \sum_{k=1}^{6} H_{ik}'\ddot{q}^k + h_i' + \vec{e}_i(\vec{r}_{KIi}\times\vec{R}_{KI}), \qquad \forall i\varepsilon I, \quad I = 1,2, \qquad (3.A.18)$$

where $\vec{r}_{KIi}$ denotes the vector from the center of joint i to the contact point KI of the peg and hole wall. The reaction force $\vec{R}_{KI}$ is a function of q, $\dot{q}$, $\ddot{q}$ and, if this dependence were to be determined, differential equations in the form (3.A.15) would be obtained. This refers to the case of a closed kinematic chain. In order to avoid direct determination of the dependence of $\vec{R}_{KI}$ on $\ddot{q}$, we shall use another procedure.

Let us consider the construction of mathematical models of the manipulator (the right-hand sides of differential equations of motion) on one integration interval. The first thing to be determined is whether the peg and hole are in contact. First, one determines whether there is contact between cylindrical surface of the object and the edge of the base of the hole. According to Fig. 3.83 the equation of the circle of the hole in the absolute system is

$$(y-y_R)^2 + (z-z_R)^2 = R_m^2, \qquad x = 0, \qquad (3.A.19)$$

where $(0, y_R, z_R)$ are the coordinates of the center. The equation of the cylindrical surface of the peg in the internal coordinate system is

$$\tilde{y}^2 + \tilde{z}^2 = r_m^2, \qquad d_1 \le \tilde{x} \le d_2. \qquad (3.A.20)$$

The relation between the coordinates of the internal and absolute system is given by

$$A_6\cdot\tilde{\vec{r}} + \vec{r}_{4,1} = \vec{r}, \qquad (3.A.21)$$

where $\vec{r}_{4,1}$ is the vector from joint 1 (where the origin of the absolute system is placed) to joint 4 (the coordinate origin of the internal system). From (3.A.21) we get

$$\tilde{\vec{r}} = A_6^{-1}(\vec{r}-\vec{r}_{4,1}) = Q(\vec{r}-\vec{r}_{4,1}), \qquad (3.A.22)$$

so equation (3.A.20) in the absolute system takes the form

$$[q_{21}(x-x_{4,1}) + q_{22}(y-y_{4,1}) + q_{23}(z-z_{4,1})]^2 +$$

$$+ [q_{31}(x-x_{4,1}) + q_{32}(y-y_{4,1}) + q_{33}(z-z_{4,1})]^2 = r_m^2 \qquad (3.A.23)$$

$$d_1 \leq [q_{11}(x-x_{4,1}) + q_{12}(y-y_{4,1}) + q_{13}(z-z_{4,1})] \leq d_2.$$

The solution of equations (3.A.19) and (3.A.23) gives the section between the peg surface and hole edge. This system of two algebraic second-order equations is solved on a digital computer by applying the Fletcher-Powel numerical procedure for finding the local minimum of a function of several variables. The question to be answered is whether the section between the cylindrical surface of the peg and hole edge is a single point. This is done by calculating the tangent to the circumference of the base of the hole and the tangent to the ellipse on the cylindrical surface of the peg for $x = 0$. If these two tangents coincide the section consists of only one point, which is a satisfactory solution. If a section exists and the tangents do not coincide, there are two sections between the cylindrical surface of the peg and hole edge but this is mechanically impossible. If this happens, the integration interval is too long and an impact between the peg and hole must have occurred at the preceding interval, which must have been "swallowed" by the integration interval. It is therefore necessary to return the simulation of the dynamics to the preceding interval and to reduce the integration interval (e.g. by bisecting the interval). With the new integration interval the new object position is determined and solution of the system of equations (3.A.19) and (3.A.23) is checked. The integration interval is thus reduced until only one solution of the system (3.A.19), (3.A.23) exists, in which case, the impact of peg and hole occurs at the moment of computing the model.

The second possible contact point (according to para. 3.5.1) is between the edge of the top of the peg cylinder and hole wall. The equation of the hole wall in the absolute system is

$$(y-y_R)^2 + (z-z_R)^2 = R_m^2, \qquad 0 \leq x \leq D_m. \qquad (3.A.24)$$

The equation of the top of the peg cylinder in the internal coordinate system is

$$\tilde{y}^2 + \tilde{z}^2 = r_m^2 , \quad \tilde{x} = d_2. \tag{3.A.25}$$

By applying (3.A.22), equation (3.A.25) in the absolute system reads as follows

$$[q_{21}(x-x_{4,1}) + q_{22}(y-y_{4,1}) + q_{23}(z-z_{4,1})]^2 +$$
$$+ [q_{31}(x-x_{4,1}) + q_{32}(y-y_{4,1}) + q_{33}(z-z_{4,1})]^2 = r_m^2 \tag{3.A.26}$$
$$q_{11}(x-x_{4,1}) + q_{12}(y-y_{4,1}) + q_{13}(z-z_{4,1}) = d_2$$

The solution to equations (3.A.24) and (3.A.26) on a digital computer is sought using the numerical procedure mentioned above. If the solution does exist, then by considering tangents as above, one determines whether or not it is consistent with the geometrical conditions. If only one solution exists, then all is well; if not, the above method is used.

One thus determines whether one, two or no points of contact exist between the peg and the hole. The special case (when contact is along the joint generatrix) is studied by examining the coaxiality of the axes of the peg and hole. One of the two sets of equations (3.A.19), (3.A.23) and (3.A.24), (3.A.26) provides the solution.

Let us consider the case when a solution to one of the two sets of equations does exist. At the contact point of the peg and hole the linear velocity of the peg must be examined. This must not have a component normal to the generatrix of the cylindrical surface on which contact exists (Figs. 3.84, 3.85). This is why a coordinate system is placed at the contact point K1 with one axis (the z1-axis) directed through the center of the circle passing through K1 the cylindrical surface of the peg, the x1-axis in the direction of that generatrix of the peg´s surface which contains K1, and the y1-axis perpendicular to these two. If contact occurs at the point K2, the axes of the coordinate system are arranged as follows. The z2-axis is taken through the center of the circle on the hole wall which passes through K2, the x2--axis is taken in the direction of the x-axis of the absolute system, and the y2-axis is perpendicular to those two. Let the matrix $E_I (3\times3)$

be the matrix which transforms these coordinate systems into the absolute system. Let the linear velocity of the peg at the point KI be $\vec{v}_{PI} = (v^x_{PI}, v^y_{PI}, v^z_{PI})^T$ with projections in the absolute system. The projections of this velocity onto the coordinate axes of the internal coordinate systems are given by

$$\vec{v}^I_{PI} = (v^{xI}_{PI}, v^{yI}_{PI}, v^{zI}_{PI})^T = E_I \cdot \vec{v}_{PI} \qquad (I = 1,2). \qquad (3.A.27)$$

The velocity component v^{zI}_{PI} in the direction of the zI-axis has to be

$$v^{zI}_{PI} \geq 0 \qquad (3.A.28)$$

If this is the case, then there is no contact so no reaction force is present. If $v^{zI}_{PI} < 0$ the velocity must be zero in the direction of zI--axis.

This presents a problem when simulating the impact between the object and the hole. At the instant of contact the state vector of the manipulator model is instantaneously changed, since velocity changes instantaneously upon impact. The impact of the manipulator and hole may be regarded as plastic if it is assumed that the manipulator control is robust enough and that impact between two big masses (the one being infinite and the other big) occurs at low speed. Therefore, we may say that at the moment of impact the velocity of the object, i.e, the linear velocity of the manipulator at the point of contact in the direction of the impact instantaneously decreases to zero. This is why the state vector must change instantaneously during the simulation.

If friction at the point KI is such that no slipping occurs, $\vec{v}^I_{PI}$ must equal zero. However, this can be verified only by checking the reaction forces, so they ought to be investigated first. Reactions are determined iteratively during the simulation. It is supposed that there is no slipping at the point KI, i.e., that $\vec{v}^I_{PI} = 0$. Since this velocity is a function of q, $\dot{q}$ an instantaneous change in the state vector is assumed on the integration interval observed in order that $\vec{v}_{PI} = 0$. Since it may be assumed that control for the first three degrees of freedom of the manipulator are sufficiently robust to ensure that the state coordinates (q^i, $\dot{q}^i$, i = 1,2,3) corresponding to these degrees of freedom do not change instantaneously as a result of the manipulator-hole impact, the whole change may be ascribed to the gripper degrees of freedom (q^i, $\dot{q}^i$, i = 4,5,6). In order to calculate $\dot{q}^i$ (i =

4,5,6) so that $\vec{v}^{I}_{PI} = 0$, we may write

$$\vec{v}_{PI} = \sum_{k=1}^{6} \vec{d}_{Pk}\dot{q}^{k} = D_P\dot{q} = D_{P1}(\dot{q}^1,\ \dot{q}^2,\ \dot{q}^3)^T + D_{P2}(\dot{q}^4,\ \dot{q}^5,\ \dot{q}^6)^T, \tag{3.A.29}$$

where $D_P = [D_{P1} \vdots D_{P2}]$ is a 3×6 matrix. From the condition $\vec{v}_{PI} = 0$ it is possible to compute the new angular velocities of the gripper assuming that angular velocities of the minimal configuration remain unchanged). Thus,

$$(\dot{q}^4,\ \dot{q}^5,\ \dot{q}^6)^T = D_{P2}^{-1}D_{P1}(\dot{q}^1,\ \dot{q}^2,\ \dot{q}^3)^T. \tag{3.A.30}$$

When the new velocities have been thus calculated we shall calculate the reaction forces under the asumption that there is no slipping.

The peg acceleration at the contact point KI may be expressed in a matrix form in terms of $\ddot{q}$ as follows.

$$\vec{w}_{PI} = GAM\ddot{q} + G\vec{AM}1, \tag{3.A.31}$$

where GAM is a 3×6 matrix and GAM1 is a 3×1 vector. Equation (3.A.18) may be written in the matrix form

$$P = H'\ddot{q} + h' + C_{KI}\cdot\vec{R}_{KI}, \qquad I = 1,2 \tag{3.A.32}$$

where C_{KI} is a 6×3 matrix with rows

$$C_{KI} = [r^z_{KIi}\cdot e^y_i - r^y_{KIi}\cdot e^z_i,\ \ r^x_{KIi}\cdot e^z_i - r^z_{KIi}\cdot e^x_i,\ \ r^y_{KIi}\cdot e^x_i - r^x_{KIi}\cdot e^y_i] \qquad i\varepsilon I, \tag{3.A.33}$$

where $\vec{r}_{KIi}$ is the vector from the joint i to the contact point KI, $\vec{e}_i = (e^x_i,\ e^y_i,\ e^z_i)^T$.

From (3.A.32) we obtain

$$\ddot{q} = H'^{-1}(P - h' - C_{KI}\cdot\vec{R}_{KI}), \qquad I = 1,2 \tag{3.A.34}$$

Assuming no slipping at the point KI, from (3.A.31) and (3.A.34) we obtain

$$GAM\ H'^{-1}(P - h' - C_{KI}\cdot\vec{R}_{KI}) + G\vec{AM}1 = 0, \tag{3.A.35}$$

which gives the reaction forces in the form

$$\vec{R}_{KI} = (GAMH'^{-1} \cdot C_{KI})^{-1} [GAMH'^{-1}(P-h') + GA\vec{M}1] \qquad (3.A.36)$$

if the matrix GAM $H'^{-1} \cdot C_{KI}(3\times 3)$ is nonsingular. Having computed $\vec{R}_{KI}$, we now calculate the components of this force in the direction of the coordinate system axes at the point KI.

$$\vec{R}^{I}_{KI} = E_I \vec{R}_{KI} = (R^{xI}_{KI}, R^{yI}_{KI}, R^{zI}_{KI})^T \qquad (3.A.37)$$

With these components we check to see whether the condition (3.5.1) for no slipping is satisfied. If it is satisfied, the assumption of no slipping holds, so from (3.A.34) we can calculate the accelerations $\ddot{q}$, i.e., the right-hand sides of the differential equations of motion on the integration interval observed.

If condition (3.5.1) is not satisfied, the assumption that no slipping occurs at the point KI does not hold and the reaction force $\vec{R}_{KI}$, calculated from (3.A.36), does not correspond to the physical conditions.

In this case we assume that slipping does occur at the point KI. This means that the components v^{xI}_{PI} and v^{yI}_{PI} of the linear velocity are non--zero, while the third component v^{zI}_{PI} equals zero. Since this third component is zero, the momentary change in the system state vector must also be introduced in this case: angular velocities of the manipulator have to be changed so that the condition $v^{zI}_{PI} = 0$ is satisfied (if it is not already satisfied by the momentary angular velocities of the manipulator). Since we are considering the impact between the object and hole walls, the problem of the momentary change in the velocity of the manipulator is unsolved. We shall treat the problem as follows. We are considering v_{PI} as a function of $\dot{q}$. From (3.A.29) and (3.A.27) we may write

$$\vec{v}^{I}_{PI} = E_I D_P \dot{q} = \bar{D}_P \dot{q} = \bar{D}_{P1}(\dot{q}^1, \dot{q}^2, \dot{q}^3)^T + \bar{D}_{P2}(\dot{q}^4, \dot{q}^5, \dot{q}^6)^T. \qquad (3.A.38)$$

The condition $v^{zI}_{PI} = 0$ gives

$$\sum_{j=1}^{3} \bar{D}^{3j}_{P1}\dot{q}^j + \sum_{j=1}^{3} \bar{D}^{3j}_{P2}\dot{q}^{j+3} = 0, \qquad (3.A.39)$$

where $\bar{D}^{kj}_{P1}$ and $\bar{D}^{kj}_{P2}$ are corresponding elements of the matrices $\bar{D}_{P1}$ and $\bar{D}_{P2}$. As in the previous case, we shall assume that the control for the first three d.o.f. of the manipulator is robust enough to prevent momentary changes in angular velocities due to impact. Accordingly, the complete change in angular velocities due to impact is ascribed to the d.o.f. of the gripper. We shall assume that the velocity change due to impact is governed by that d.o.f. of the gripper which mostly affects the linear velocity of the object at the point KI in the direction perpendicular to the contact. Let the degree j(j = 4,5,6) be that for which $\bar{D}_{P2}^{3j-3} = \max_k \bar{D}_{P2}^{3k}$ holds. From (3.A.39) we calculate

$$\dot{\bar{q}}^j = - \frac{\sum_{k=1}^{3} \bar{D}_{P1}^{3k} \dot{q}^k + \sum_{\substack{k=1 \\ k \neq j-3}}^{3} \bar{D}_{P2}^{3k} \dot{q}^{k+3}}{\bar{D}_{P2}^{3j-3}}, \tag{3.A.40}$$

where $\dot{\bar{q}}^j$ is the new angular velocity in the j-th degree of freedom of the gripper, which satisfies the condition $v_{PI}^{zI} = 0$. Having thus computed $\dot{q}^i$, we now calculate the direction of the tangential velocity of the peg at the point KI:

$$\delta = \frac{v_{PI}^{yI}}{v_{PI}^{xI}} = \frac{\sum_{k=1}^{6} \bar{D}_P^{2k} \dot{q}^k}{\sum_{k=1}^{6} \bar{D}_P^{1k} \dot{q}^k}. \tag{3.A.41}$$

This is calculated with respect to the xI-axis of the coordinate system placed at the contact point.

The reaction force $\vec{R}_{KI}$ can now be computed if slipping occurs in the point KI. As we have shown the tangential component of the reaction force at the point KI must be in the direction of slipping. In addition, the reaction force must satisfy the condition (3.5.2). If the components of the reaction force $\vec{R}_{KI}^{I}$ are considered in relation to the internal coordinate system connected to KI, then from (3.5.2) and (3.A.41) we may write

$$R_{KI}^{yI} = \delta R_{KI}^{xI}, \quad T_{KI}^{I} = \sqrt{R_{KI}^{xI\,2} + R_{KI}^{yI\,2}} = \mu R_{KI}^{zI},$$

$$R_{KI}^{xI} = \frac{\mu R_{KI}^{zI}}{\sqrt{1+\delta^2}}, \quad R_{KI}^{yI} = \frac{\delta \mu R_{KI}^{zI}}{\sqrt{1+\delta^2}}.$$

In addition, we must require the acceleration component at the point KI in the direction perpendicular to contact to be zero. From the transformation matrix E_I (I = 1,2) and equation (3.A.31) we find that components of the vector of the linear acceleration of the object at the point KI in relation to the coordinate system at KI are given by

$$\vec{w}_{PI}^{I} = E_I \vec{w}_{PI} = \overline{GAM} \cdot \ddot{q} + \overrightarrow{\overline{GAM1}} \qquad (3.A.42)$$

where the matrix $\overline{GAM} = E_I \cdot GAM$ has dimensions 3×6 and the vector $\overrightarrow{\overline{GAM1}} = E_I \cdot \overrightarrow{GAM1}$ has dimensions 3×1. From (3.A.37) we get

$$\vec{R}_{KI} = E_I^{-1} \vec{R}_{KI}^{I} \qquad (3.A.43)$$

and by substitution in (3.A.34) we obtain

$$\ddot{q} = H'^{-1}(P - h' - C_{KI} E_I^{-1} \vec{R}_{KI}^{I}) . \qquad (3.A.44)$$

Since $\vec{R}_{KI}^{I}$ is given by

$$\vec{R}_{KI}^{I} = \left(\frac{\mu}{\sqrt{1+\delta^2}} , \frac{\mu\delta}{\sqrt{1+\delta^2}} , 1\right)^T \cdot R_{KI}^{zI} = \vec{g}_\delta R_{KI}^{zI}, \qquad (3.A.45)$$

where $\vec{g}_\delta$ is a 3×1 vector, by substituting in (3.A.44) and (3.A.45) we get

$$\vec{w}_{PI}^{I} = \overline{GAM}[H'^{-1}(P - h' - C_{KI} E_I^{-1} \vec{g}_\delta \cdot R_{KI}^{zI})] + \overrightarrow{\overline{GAM1}} . \qquad (3.A.46)$$

From the condition $w_{PI}^{zI} = 0$ we obtain

$$R_{KI}^{zI} = \frac{\{\overline{GAM}[H'^{-1}(P-h')]\}^3 + \{\overrightarrow{\overline{GAM1}}\}^3}{\{\overline{GAM}\ H'^{-1} \cdot C_{KI} E_I^{-1} \vec{g}_\delta\}^3} , \qquad (3.A.47)$$

where $\{\}^3$ refers to the third member of the corresponding 3×1 vector. Equation (3.A.47) yields the perpendicular component of the reaction force at the point KI, assuming that the denominator is non-zero. The remaining two components of the reaction force are calculated using R_{KI}^{zI}, computed as above, and (3.A.45). The accelerations $\ddot{q}$ are calculated from (3.A.44).

Thus, if slipping occurs at the point KI, the accelerations $\ddot{q}$ can be computed, i.e., the right-hand sides of the differential equations of the manipulator dynamics can be determined.

We have thus investigated the structure of the differential equations of the manipulator dynamics on one integration interval in the case of a single point of contact between the working object and the hole. Let us briefly recall the setting of these differential equations for this case on one integration interval (Fig. 3.A.1).

First, by examining the solutions of the system of algebraic equations (3.A.19), (3.A.23) and (3.A.24), (3.A.26) we determine whether object--hole contact has occurred and, if it has, at which point. In the case of a single point of contact, we examine the velocity at this point. If $v_{PI}^{zI} > 0$, then no contact occurs, since no reaction forces act on the object. In this case, $\ddot{q}$ is calculated from (3.A.15) as

$$\ddot{q} = H'^{-1}(P - h'), \tag{3.A.48}$$

which is equivalent to the situation No 1. If $v_{PI}^{zI} = 0$, we check to see whether the components v_{PI}^{xI} and v_{PI}^{yI} are non-zero. If one or both is non-zero, it is possible that slipping occurs at the contact point. From (3.A.41) the slope δ of the velocity is calculated and the reaction force is found from (3.A.45), (3.A.47).

However, if no slipping occurs, the manipulator velocities must be changed instantaneously according to (3.A.30) in order to reflect the lack of slipping at KI. In this case, the reaction force is calculated from (3.A.36), (3.A.37) and the condition (3.A.71) is checked.

If $v_{PI}^{zI} < 0$, the velocities of the gripper degrees of freedom have to be changed instantaneously. We first investigate the possibility of no slipping at the contact point and then calculate the new angular velocities of the manipulator from equation (3.A.30) and the condition $\vec{v}_{PI} = 0$. The reaction force is calculated from (3.A.36), (3.A.37) and the condition (3.5.1) checked. If this condition is not satisfied, we pass on to the other assumption - that slipping does occur at the point KI. The change in angular velocity is calculated from (3.A.40) and the reaction force is found from (3.A.45), (3.A.47).

In fact, during the simulation the following effects may be observed. By changing the integration interval we ensure that impact occurs at the discrete moment when the accelerations $\ddot{q}$ are calculated (the right--hand sides of the system state equations). At the moment of impact the vector of the system state has to be changed instantaneously, either to make the velocity at the point KI equal to zero or to permit slipping. On the next integration interval the velocity is such that the object either slips or is jammed as a result of the momentary velocity, which need not be changed any further. When a velocity component $v_{PI}^{zI} > 0$ occurs at the contact point the object is separated from the contact point. Of course, regardless of v_{PI}^{zI}, one should always calculate $\ddot{q}$ from (3.A.48) and check whether (3.A.42) yields $w_{PI}^{zI} > 0$, which means that although at the point KI the velocity $v_{PI}^{zI} \leq 0$, the reaction force not longer acts on the manipulator.

If there are two simultaneous contacts between the object and the hole, an analogous method is used. By studying the solutions to the algebraic systems (3.A.19), (3.A.23) and (3.A.24), (3.A.26) we see that two simultaneous contacts occur. If two sections exist for $x = 0$ and $\tilde{x} = d_2$, the integration interval should be decreased in order to make the moment at which the object-hole contact occurs coincide with the discrete moment at which the right-hand sides of the differential equations of the manipulator mathematical model are determined.

If two contacts occur, we should calculate the linear velocities of the object at the contact points K1 and K2 and check whether the perpendicular components are $v_{P1}^{z1} > 0$ and $v_{P2}^{z2} > 0$. In this case simulation is reduced to one of the two previous cases. If $v_{P1}^{z1} < 0$ and $v_{P2}^{z2} < 0$, the vector of the system state has to be changed so that $v_{P1}^{z1} = 0$ and $v_{P2}^{z2} = 0$. As we said in para. 3.5.1, three situations are possible.

Let us first suppose that there is no slipping at either point of contact. This means that the following condition has to be satisfied

$$\vec{v}_{P1} = 0 \text{ and } \vec{v}_{P2} = 0 \tag{3.A.49}$$

It is quite clear from (3.A.49) and from (3.A.29) that the angular velocities of the manipulator after impact have to equal zero.

The equations of moments around the axes of the manipulator joints may, according to (3.A.32), be written in the form

$$P = H'\ddot{q} + h' + C_{K1}\cdot\vec{R}_{K1} + C_{K2}\cdot\vec{R}_{K2} \tag{3.A.50}$$

From (3.A.50) the acceleration $\ddot{q}$ is calculated in the form

$$\ddot{q} = H'^{-1}(P - h' - [C_{K1} \mid C_{K2}]\cdot[R_{K1}^T \mid R_{K2}^T]^T). \tag{3.A.51}$$

If slipping occurs at neither contact point, the accelerations of all six degrees of freedom of the manipulator must also equal zero: $\ddot{q}^i = 0$, $\forall i \varepsilon I$. From (3.A.51) the reaction forces at both contact points can be computed:

$$[\vec{R}^T_{K1} \vdots \vec{R}^T_{K2}]^T = [C_{K1} \vdots C_{K2}]^{-1}(P - h'), \qquad (3.A.52)$$

assuming the matrix $[C_{K1} \vdots C_{K2}]$ to be nonsingular. From (3.A.37) we can therefore find the components of the reaction force with respect to the internal coordinate systems at the contact points KI (I = 1,2) and check whether the condition (3.5.5) is satisfied. If it is, the assumption of no slipping holds and the peg is jammed. The values of the reaction calculated from (3.A.52) are correct and the accelerations $\ddot{q}$ are calculated from equation (3.A.51).

If condition (3.A.75) is not satisfied, we have to make the other assumption, i.e., that slipping occurs at one point but not at the other. At the point where slipping occurs (e.g. I=1) the velocity component v^{z1}_{P1} must be zero ($v^{z1}_{P1} = 0$) and at the other point the linear velocity must equal zero ($\vec{v}_{P2} = 0$). The new angular velocities of the manipulator are to be calculated so as to satisfy these conditions. The same procedure as in the case of a single contact is applied: the dependence of the velocities $\vec{v}^I_{PI}$ on the angular velocities $\dot{q}$ is considered in the form (3.A.38) and those 4 degrees of freedom are determined which most affect linear velocities, i.e., those degrees of freedom for which the elements of the matrix $\bar{D}_{PI}$ for I = 1,2 have their maximum value. One determines the four angular velocities which most influence the linear velocities of the object. Let us denote the two remaining angular velocities by $\dot{\tilde{q}}$. Then, from (3.A.38) we may write

$$[v^{z1}_{P1}, v^{x2}_{P2}, v^{y2}_{P2}, v^{z2}_{P2}]^T = \bar{D}\cdot\dot{\bar{q}} + \tilde{D}\dot{\tilde{q}}, \qquad (3.A.53)$$

where $\bar{D}$, $\tilde{D}$ are matrices with dimensions 4×4, 4×2, respectively, into which the corresponding elements of the matrices D_P have been grouped, according to the procedure described above. From the condition that the left-hand side of (3.A.53) equals zero, changes (due to impact) in angular velocities are calculated as

$$\dot{\bar{q}} = -\bar{D}^{-1}\tilde{D}\dot{\tilde{q}}, \qquad (3.A.54)$$

assuming that the matrix $\bar{D}$ is nonsingular. From (3.A.41) we then calculate the direction of the tangential velocity at the point K1 where slipping occurs. The force $\vec{R}^1_{P1}$ at the point K1 may be expressed in terms of R^{z1}_{P1} by applying equation (3.A.45). From (3.A.42) the linear accelerations at K1 and K2 may be expressed in the form

$$[\vec{w}^{1T}_{P1} \mid \vec{w}^{2T}_{P2}]^T = [\overline{GAM}_1 \mid \overline{GAM}_2]\ddot{q} + [\overline{\vec{GAM1}}^T_1 \mid \overline{\vec{GAM1}}^T_2]^T, \qquad (3.A.55)$$

where $\overline{GAM}_I$, $\overline{\vec{GAM1}}_I$, I = 1,2, are matrices and vectors for the corresponding contact points, according to (3.A.47). According to the assumption that slipping does not occur at K2 and does occur at K1, the conditions

$$w^{z1}_{P1} = 0, \quad \vec{w}^2_{P2} = 0 \qquad (3.A.56)$$

must be satisfied.

From (3.A.51) and (3.A.45) we may write

$$\ddot{q} = H'^{-1}(P - h' - [C_{K1} \mid C_{K2}] \cdot [E_1^{-1}\vec{g}_\delta \mid E_2^{-1}] \cdot [R^{z1}_{P1} \mid \vec{R}^{2T}_{P2}]^T), \qquad (3.A.57)$$

where E_1, E_2 are the corresponding matrices. From (3.A.55), (3.A.56), (3.A.57) the reaction forces may be computed as

$$[R^{z1}_{P1} \mid \vec{R}^{2T}_{P2}]^T = ([\{\overline{GAM}_1\}^3 \mid \overline{GAM}_2]\; H'^{-1} \cdot [C_{K1} \mid C_{K2}]\,[E_1^{-1}\vec{g}_\delta \mid E_2^{-1}])^{-1}$$

$$\cdot\; ([\{\overline{GAM}_1\}^3 \mid \overline{GAM}_2] \cdot H'^{-1}(P - h') - [\{\overline{\vec{GAM1}}_1\}^3 \mid \overline{\vec{GAM1}}^T_2]^T),$$

(3.A.58)

assuming the corresponding matrix to be nonsingular. $\{\}^3$ here refers to the third row of the matrix $\overline{GAM}_1$, i.e., the third element of the vector $\overline{\vec{GAM1}}_1$. When $\vec{R}^2_{P2}$ is calculated in this way, one should check whether the condition (3.5.7) is satisfied. One should also check whether slipping at the point K1 is consistent with the geometrical conditions. If it is not, jamming will occur but we have not incorporated this in the simulation.

Finally, if condition (3.5.7) is not satisfied, the third possibility is considered, namely, that slipping occurs at both contact points. In this case, the components of the linear velocities at K1 and K2 in the direction perpendicular to contact must be zero, i.e.,

$$v_{P1}^{z1} = 0, \quad v_{P2}^{z2} = 0 \tag{3.A.59}$$

In order to satisfy conditions (3.A.59), changes in angular velocities due to impact, are calculated by analogy with the single-contact case. From the three degrees of freedom of the gripper those two are chosen which most affect the linear velocity components (according to the elements of the matrices D_P). Angular velocities in these degrees of freedom are determined so as to satisfy conditions (3.A.59). From (3.A.41) one then finds the directions of tangential velocity at the contact points (which coincide for K1 and K2). The reaction forces R_{K1}^{1} and R_{K2}^{2} may then, according to (3.A.45), be expressed in terms of R_{K1}^{z1} and R_{K2}^{z2} as

$$\vec{R}_{K1}^{1} = \vec{g}_{\delta}^{1} R_{K1}^{z1}, \quad \vec{R}_{K2}^{2} = \vec{g}_{\delta}^{2} R_{K2}^{z2} \,. \tag{3.A.60}$$

If slipping occurs at both points, the following conditions have to be satisfied.

$$w_{P1}^{z1} = 0, \quad w_{P2}^{z2} = 0. \tag{3.A.61}$$

From (3.A.51) and (3.A.60) we may write

$$\ddot{q} = H'^{-1}(P - h' - [C_{K1} \,\vdots\, C_{K2}] \cdot [E_1^{-1}\vec{g}_{\delta}^{1} \,\vdots\, E_2^{-1}\vec{g}_{\delta}^{2}] \cdot (R_{K1}^{z1}, \; R_{K2}^{z2})^T) \tag{3.A.62}$$

From (3.A.53), (3.A.61) and (3.A.62) the reaction forces are given by

$$(R_{P1}^{z1}, R_{P2}^{z2})^T = ([\{\overline{GAM}_1\}^3 \,\vdots\, \{\overline{GAM}_2\}^3] \cdot H'^{-1} \cdot [C_{K1} \,\vdots\, C_{K2}] [E_1^{-1}\vec{g}_{\delta}^{1} \,\vdots\, E_2^{-1}\vec{g}_{\delta}^{2}])^{-1} \cdot$$

$$\cdot \; ([\{\overline{GAM}_1\}^3 \,\vdots\, \{\overline{GAM}_2\}^3] \cdot H'^{-1}(P - h') - [\{\overrightarrow{\overline{GAM1}}_1\}^3 \{\overrightarrow{\overline{GAM1}}_2\}^3]^T)$$

$$\tag{3.A.63}$$

The accelerations $\ddot{q}$ are obtained from (3.A.62).

Thus, the reaction forces and accelerations are calculated on one integration interval in the case of two simultaneous contacts between the object and the hole.

Contact is along the joint generatrix if solutions to the system of algebraic equations (3.A.19), (3.A.23) exist and if, at the same time, the main axis of symmetry of the object cylinder is parallel to the axis of hole cylinder. In this case contact occurs along the joint generatrix of the cylinders. The reaction forces distributed along the line of contact may be substituted by a single resultant reaction force having its point of application in the middle of the line of contact. This situation is thus reduced to situation No 2 with respect to simulation.

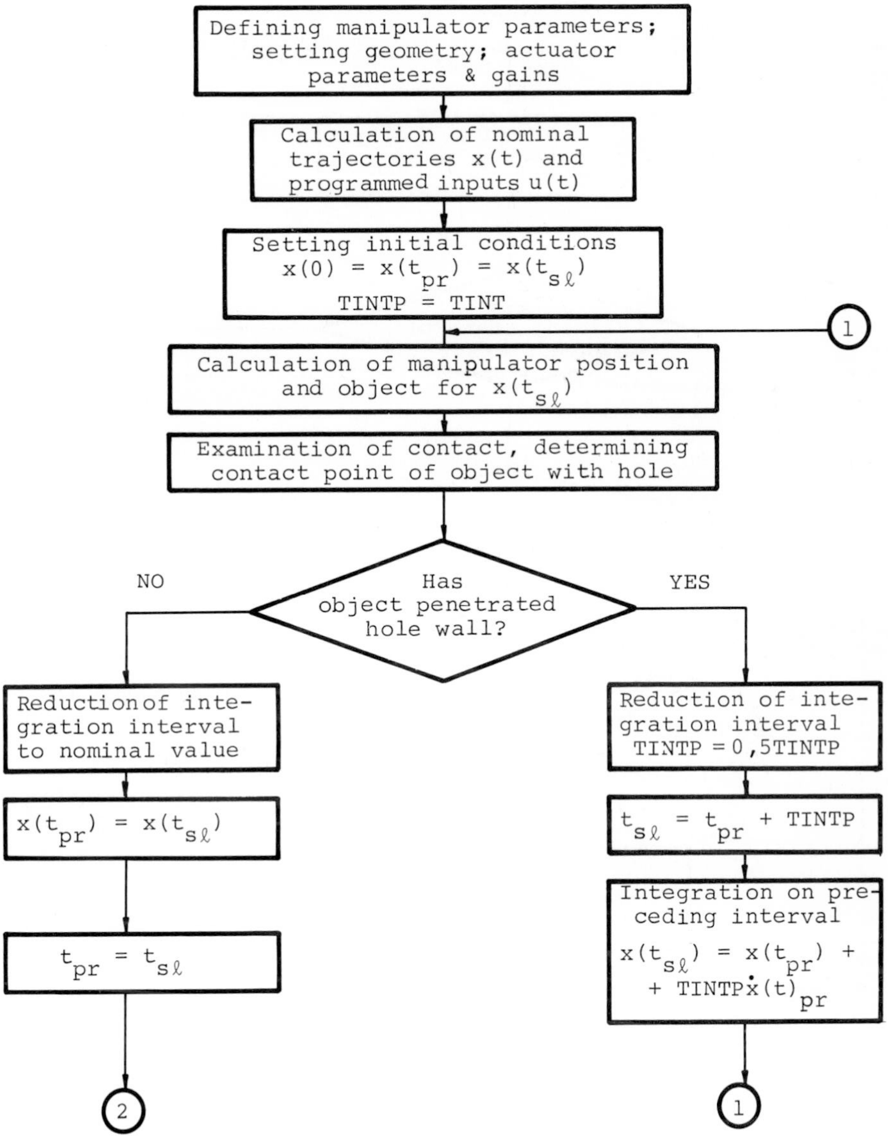

Fig. 3.A.1. Flow-chart for simulation of "peg in hole" process

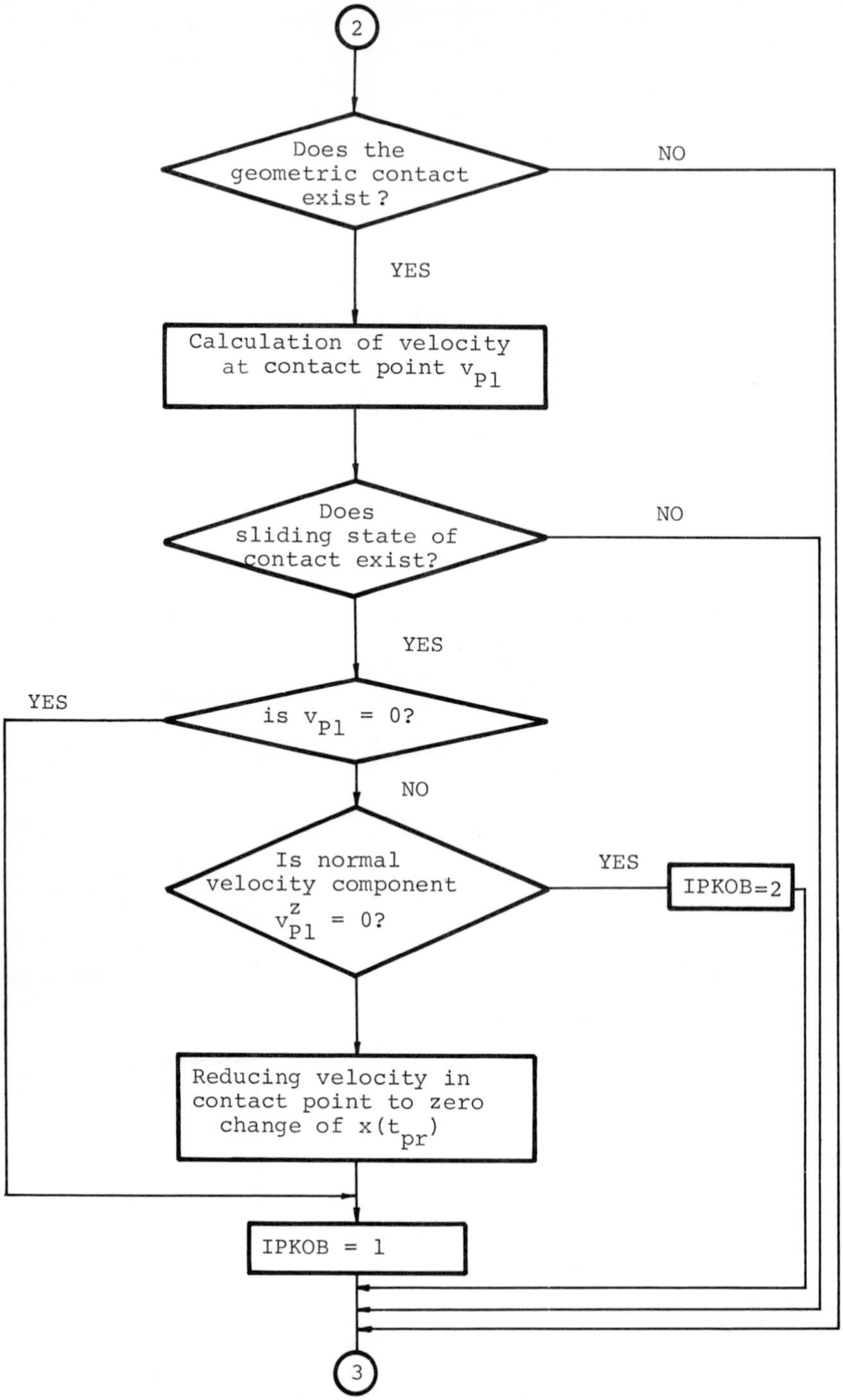

Fig. 3.A.1.1. (contd.)

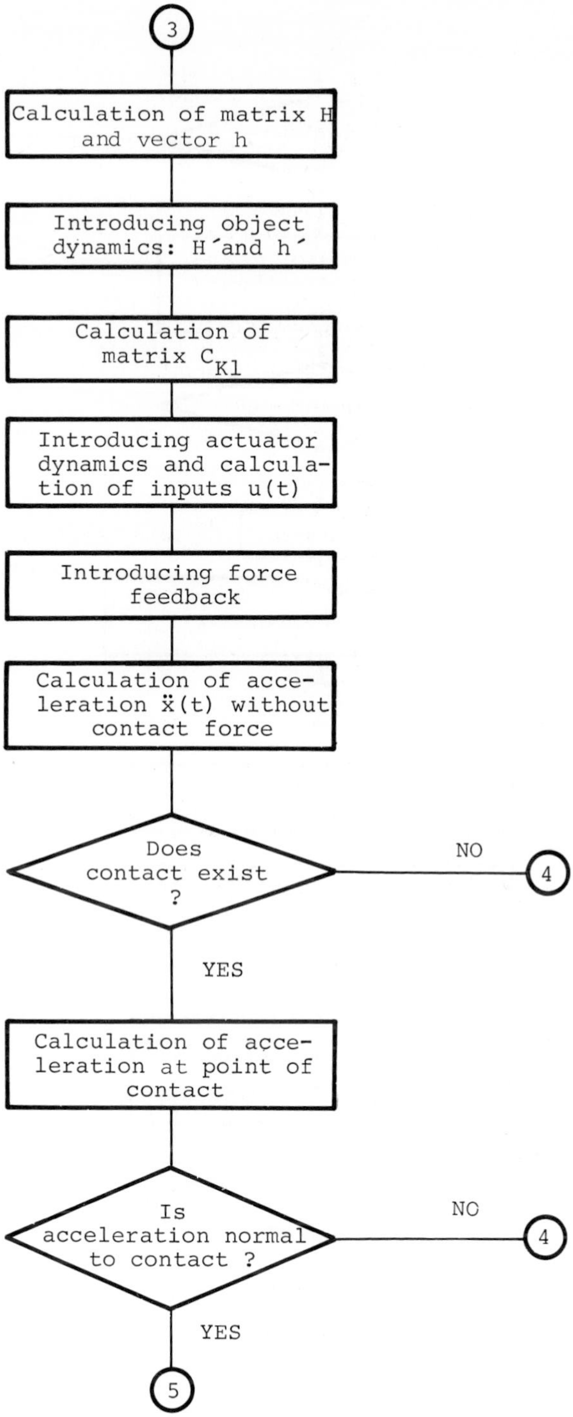

Fig. 3.A.1.2. (contd.)

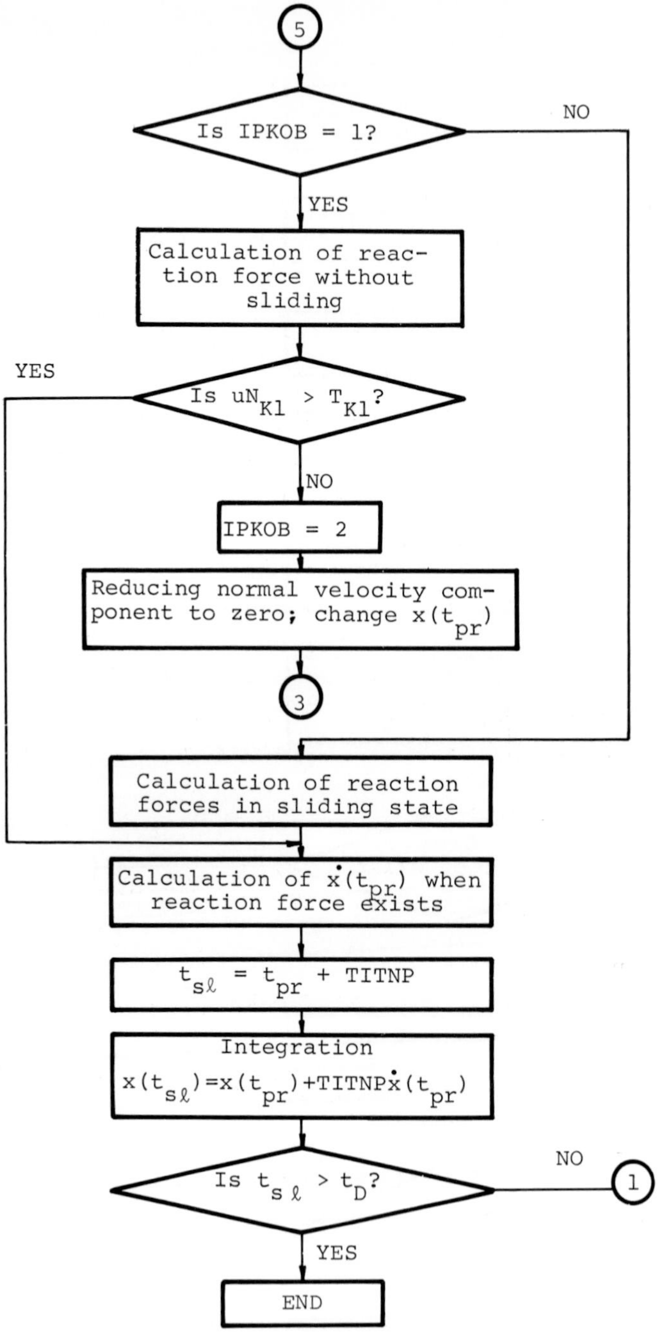

Fig. A.1.3. (contd.)

Subject Index

Scientific Fundamentals of Robotics 1

M. Vukobratović, V. Potkonjak

Dynamics of Manipulation Robots

Theory and Application

1982. 149 figures. XIII, 303 pages
(Communications and Control Engineering Series)
ISBN 3-540-11628-1

Contents: General Remarks about Robot and Manipulator Dynamics. – Computer-Aided Methods for Setting and Solving Mathematical Models of Active Mechanisms in Robotics. – Simulation of Manipulatory Dynamics and Adjusting to Functional Movements. – Dynamics of Manipulators with Elastic Segments. – Dynamical Method for the Evaluation and Choice of Industrial Manipulators. – Subject Index.

In this monograph, complete mathematical models of the dynamics of open active mechanisms as applied to robotics are treated and presented for the first time. It also presents the first parallel survey of computer-oriented methods for automatic construction of the dynamical equations of manipulation robots. These are based on a) general theorems of mechanics, b) second-order Lagrange's equations and c) Appel-Gibbs functions. Adjustment blocks are considered, thus enabling the dynamics of spatial mechanisms to be applied to concrete manipulation tasks. The dynamical nominal states (functional movements) of manipulation systems are synthesized and the elastic properties of the manipulation systems are considered.
The basic idea of this book is to describe dynamical models and functional requirements of manipulator dynamics, to describe procedures for computer-aided design of the mechanisms of manipulation robots, and, at the same time to introduce different criteria and corresponding constraints of practical relevance.

Springer-Verlag
Berlin
Heidelberg
New York